Welding
Skills, Processes and Practices for Entry-Level Welders

Book 1

- Occupational Orientation
- Safety and Health
- Drawing and Symbol Interpretation
- Thermal Cutting
- Weld Inspection Testing and Codes

First Edition

Larry Jeffus
Lawrence Bower

DELMAR
CENGAGE Learning™

Australia • Brazil • Japan • Korea • Mexico • Singapore • Spain • United Kingdom • United States

DELMAR
CENGAGE Learning

Welding: Skills, Processes and Practices for Entry-Level Welders: Book One
Larry Jeffus/Lawrence Bower

Vice President, Editorial: Dave Garza

Director of Learning Solutions:
Sandy Clark

Executive Editor: David Boelio

Managing Editor: Larry Main

Senior Product Manager: Sharon
Chambliss

Editorial Assistant: Lauren Stone

Vice President, Marketing: Jennifer
McAvey

Executive Marketing Manger:
Deborah S. Yarnell

Senior Marketing Manager:
Jimmy Stephens

Marketing Specialist: Mark Pierro

Production Director: Wendy Troeger

Production Manager: Mark Bernard

Content Project Manager: Cheri Plasse

Art Director: Benj Gleeksman

Technology Project Manager:
Christopher Catalina

Production Technology Analyst:
Thomas Stover

For product information and technology assistance, contact us at
**Professional & Career Group Customer Support,
1-800-648-7450**

For permission to use material from this text or product, submit a request online at **cengage.com/permissions**. Further permissions questions can be e-mailed to **permissionrequest@cengage.com**.

Library of Congress Control Number: 2008910433

ISBN-13: 978-1-4354-2788-4
ISBN-10: 1-4354-2788-2

Delmar
5 Maxwell Drive
Clifton Park, NY, 12065-2919
USA

Cengage Learning products are represented in Canada by Nelson Education, Ltd.

For your lifelong learning solutions, visit **delmar.cengage.com**
Visit our corporate website at **cengage.com**.

Notice to the Reader

Publisher does not warrant or guarantee any of the products described herein or perform any independent analysis in connection with any of the product information contained herein. Publisher does not assume, and expressly disclaims, any obligation to obtain and include information other than that provided to it by the manufacturer. The reader is expressly warned to consider and adopt all safety precautions that might be indicated by the activities described herein and to avoid all potential hazards. By following the instructions contained herein, the reader willingly assumes all risks in connection with such instructions. The publisher makes no representations or warranties of any kind, including but not limited to, the warranties of fitness for particular purpose or merchantability, nor are any such representations implied with respect to the material set forth herein, and the publisher takes no responsibility with respect to such material. The publisher shall not be liable for any special, consequential, or exemplary damages resulting, in whole or part, from the readers' use of, or reliance upon, this material.

Printed in the United States of America
1 2 3 4 5 XX 11 10 09

Brief Contents

Contents

Preface

ABOUT THE SERIES

Welding: Skills, Processes and Practices for Entry-Level Welders is an exciting new series that has been designed specifically to support the American Welding Society's (AWS) SENSE EG2.0 training guidelines. Offered in three volumes, these books are carefully crafted learning tools consisting of theory-based texts that are accompanied by companion lab manuals, and extensive instructor support materials. With a logical organization that closely follows the modular structure of the AWS guidelines, the series will guide readers through the process of acquiring and practicing welding knowledge and skills. For schools already in the SENSE program, for those planning to join, or for schools interested in obtaining certifiable outcomes based on nationally recognized industry standards order to comply with the latest Carl D. Perkins Career and Technical Education in requirements, *Welding: Skills, Processes and Practices for Entry-Level Welders* offers a turnkey solution of high quality teaching and learning aids.

Career and technical education instructors at the high school level are often called upon to be multi-disciplinary educators, teaching welding as only one of as many as five technical disciplines in any given semester. The *Welding: Skills, Processes and Practices for Entry-Level Welders* package provides these educators with a process-based, structured approach and the tools they need to be prepared to deliver high level training on processes and materials with which they may have limited familiarity or experience. Student learning, satisfaction and retention are the target of the logically planned practices, supplements and full color textbook illustrations. While the AWS standards for entry level welders are covered, students are also introduced to the latest in high technology welding equipment such as pulsed gas metal arc welding (GMAW-P). Career pathways and career clusters may be enhanced by the relevant mathematics applied to real world activities as well as oral and written communication skills linked to student interaction and reporting.

Book 1, the core volume, introduces students to the welding concepts covered in AWS SENSE Modules 1, 2, 3, 8 and 9 (Occupational Orientation, Safety and Health of Welders, Drawing and Welding Symbol Interpretation, Thermal Cutting, and Weld Inspection Testing and Codes). Book 1 contains all the material needed for a SENSE program that prepares students for qualification in Thermal Cutting processes. The optional Books 2 and 3 cover other important welding processes and are grouped in logical combinations. Book 2 corresponds to AWS SENSE Modules 5 and 6 (GMAW, FCAW), and Book 3 corresponds to AWS SENSE Modules 4 and 7 (SMAW, GTAW).

The texts feature hundreds of four-color figures, diagrams and tight shots of actual welds to speed beginners to an understanding of the most widely used welding processes.

FEATURES

- Produced in close collaboration with experienced instructors from established SENSE programs to maximize the alignment of the content with SENSE guidelines and to ensure 100% coverage of Level I-Entry Welder Key Indicators.
- Chapter introductions contain general performance objectives, key terms used, and the AWS SENSE EG2.0 Key Indicators addressed in the chapter.
- Coverage of Key Indicators is indicated in the margin by a torch symbol and a numerical reference.
- Contains scores of fully illustrated Practices, which are guided exercises designed to help students master processes and materials. Where applicable, the Practices reproduce and reference actual AWS technical drawings in order to help students create acceptable workmanship samples.
- Each section introduces students to the materials, equipment, setup procedures and critical safety information they need in order to weld successfully.
- Hundreds of four-color figures, diagrams and tight shots of actual welds to speed beginners to an understanding of the most widely used welding processes.
- End of chapter review questions develop critical thinking skills and help students to understand "why" as well as "how."

SUPPLEMENTS

Each book in the Welding Skills series is accompanied by a **Lab Manual** that has been designed to provide hands-on practice and reinforce the student's understanding of the concepts presented in the text. Each chapter contains practice exercises to reinforce the primary objectives of the lesson, including creation of workmanship samples (where applicable), and a quiz to test knowledge of the material. Artwork and safety precautions are included throughout the manuals.

Instructor Resources (on CD-ROM), designed to support Books 1-3 and the accompanying Lab Manuals, provide a wealth of time-saving tools, including:

- An Instructor's Guide with answers to end-of-chapter Review Questions in the texts and Lab Manual quizzes.
- Modifiable model Lesson Plans that aid in the design of a course of study that meets local or state standards and also maps to the SENSE guidelines.
- An extensive ExamView Computerized Test Bank that offers assessments in true/false, multiple choice, sentence completion and short answer formats. Test questions have been designed to expose students to the types of questions they'll encounter on the SENSE Level 1 Exams.
- PowerPoint Presentations with selected illustrations that provide a springboard for lectures and reinforce skills and processes covered in the texts. The PowerPoint Presentations can be modified or expanded as instructors desire, and can be augmented with additional illustrations from the Image Library.
- The Image Library contains nearly all (well over 1000!) photographs and line art from the texts, most in four-color.
- A SENSE Correlation Chart that shows the close alignment of the *Welding* series to the SENSE Entry Level 1 training guidelines. Each Key Indicator within each SENSE Module is mapped to the relevant text and lab manual page or pages.

TITLES IN THE SERIES

Welding: Skills, Processes and Practices for Entry-Level Welders: Book 1, Occupational Orientation, Safety and Health of Welders, Drawing and Welding Symbol Interpretation, Thermal Cutting, Weld Inspection Testing and Codes
(Order #: 1-4354-2788-2)
Lab Manual, Book One (Order #: 1-4354-2789-0)

Welding: Skills, Processes and Practices for Entry-Level Welders: Book 2, Gas Metal Arc Welding, Flux Cored Arc Welding (Order #:1-4354-2790-4)
Lab Manual, Book Two (Order #: 1-4354-2795-5)

Welding: Skills, Processes and Practices for Entry-Level Welders: Book 3, Shielded Metal Arc Welding, Gas Tungsten Arc Welding (Order #:1-4354-2796-3)
Lab Manual, Book Three (Order #: 1-4354-2797-1)

AWS Acknowledgment

The Authors and Publisher gratefully acknowledge the support provided by the American Welding Society in the development and publication of this textbook series. "American Welding Society," the AWS logo and the SENSE logo are the trade and service marks of the American Welding Society and are used with permission.

For more information on the American Welding Society and the SENSE program, visit **http://www.aws.org/education/sense/** or contact AWS at (800) 443-9353 ext. 455 or by email: **education@aws.org**.

Acknowledgments

The authors and publisher would like to thank the following individuals for their contributions to this series:

Garey Bish, *Gwinnett Technical College, Lawrenceville, GA*
Julius Blair, *Greenup County Area Technology Center, Greenup, KY*
Rick Brandon, *Pemiscot County Career & Technical Center, Hayti, MO*
Stephen Brandow, *University of Alaska Southeast, Ketchikan, Ketchikan, AK*
Francis X Brieden, *Career Technology Center of Lackawanna County, Scranton, PA*
John Cavenaugh, *Community College of Southern Nevada, Las Vegas, NV*
Clay Corey, *Washington-Saratoga BOCES, Fort Edward, NY*
Keith Cusey, *Institute for Construction Education, Decatur, IL*
Craig Donnell, *Whitmer Career Technology Center, Toledo, OH*
Steve Farnsworth, *Iowa Lakes Community College, Emmetsburg, IA*
Ed Harrell, *Traviss Career Center, Lakeland, FL*
Robert Hoting, *Northeast Iowa Community College, Sheldon, IA*
Steve Kistler, *Moberly Area Technical Center, Moberly, MO*
David Lynn, *Lebanon Technology & Career Center, Lebanon, MO*
Frank Miller, *Gadsden State Community College, Gadsden, AL*
Chris Overfelt, *Arnold R Burton Tech Center, Salem, VA*
Kenric Sorenson, *Western Technical College, LaCrosse, WI*
Pete Stracener, *South Plains College, Levelland, TX*
Bill Troutman, *Akron Public Schools, Akron, OH*
Norman Verbeck, *Columbia/Montour AVTS, Bloomsburg, PA*

About The Authors

Larry Jeffus is a dedicated teacher and author with over twenty years experience in the classroom and several Delmar Cengage Learning welding publications to his credit. He has been nominated by several colleges for the Innovator of the Year award for setting up nontraditional technical training programs. He was also selected as the Outstanding Post-Secondary Technical Educator in the State of Texas by the Texas Technical Society. Now retired from teaching, he remains very active in the welding community, especially in the field of education.

Lawrence Bower is a welding instructor at Blackhawk Technical College, an AWS SENSE School, in Janesville, Wisconsin. Mr. Bower is an AWS-certified Welding Inspector and Welding Educator. In helping to create *Welding: Skills, Processes and Practices for Entry-Level Welders*, he has brought to bear an excellent mix of training experience and manufacturing know-how from his work in industry, including fourteen years at United Airlines, and six years in the US Navy as an aerospace welder.

CHAPTER

1

Introduction to Welding

OBJECTIVES

After completing this chapter, the student should be able to

- describe how GMAW, FCAW, GTAW, and SMAW processes work
- list four factors that must be considered before a welding process is selected
- discuss three events in the history of welding
- describe the purpose of a welding procedure specification (WPS)
- define the terms *weld, forge welding, resistance welding, fusion welding, coalescence,* and *certification*
- list five welding or cutting processes an AWS SENSE entry-level welder must be able to demonstrate

KEY TERMS

American Welding Society (AWS)

automated operation

automatic operation

AWS Level I–Entry Welder

AWS Level II–Advanced Welder

AWS Level III–Expert Welder

AWS SENSE

certification

coalescence

flux cored arc welding (FCAW)

forge welding

fusion welding

gas metal arc welding (GMAW)

gas tungsten arc welding (GTAW)

layouts

machine operation

manual operation

oxyfuel (OF) gas

oxyfuel gas cutting (OFC)

oxyfuel gas welding (OFW)

qualification

resistance welding

semiautomatic operation

shielded metal arc welding (SMAW)

torch brazing (TB)

weld

welding

AWS SENSE EG2.0

Key Indicators Addressed in this Chapter:

Module 1: Occupational Orientation

Key Indicator 1: Prepares time or job cards, reports, and records

Key Indicator 3: Follows verbal instructions to complete work assignments

Key Indicator 4: Follows written details to complete work assignments

Module 3: Drawing and Weld Symbol Interpretation

Key Indicator 3: Fabricates parts from a drawing or sketch

INTRODUCTION

As methods of joining materials improved through the ages, so did the environment and mode of living for humans. Materials, tools, and machinery improved as civilization developed.

Fastening together the parts of work implements began when an individual attached a stick to a stone to make a spear or axe. Egyptians used stones to create temples and pyramids that were fastened together with a gypsum mortar. Some walls that still exist depict a space-oriented figure that was as appropriate then as now—an ibis-headed god named Thoth who protected the moon and was believed to cruise space in a vessel.

Other types of adhesives were used to join wood and stone in ancient times. However, it was a long time before the ancients discovered a method for joining metals. Workers in the Bronze and Iron Ages began to solve the problems of forming, casting, and alloying metals. Welding metal surfaces was a problem that long puzzled metalworkers of that period. Early metal-joining methods included forming a sand mold on top of a piece of metal and casting the desired shape directly on the base metal, so that the two parts fused together form a single piece of metal, Figure 1.1. Another early metal-joining method was to place two pieces of metal close together and

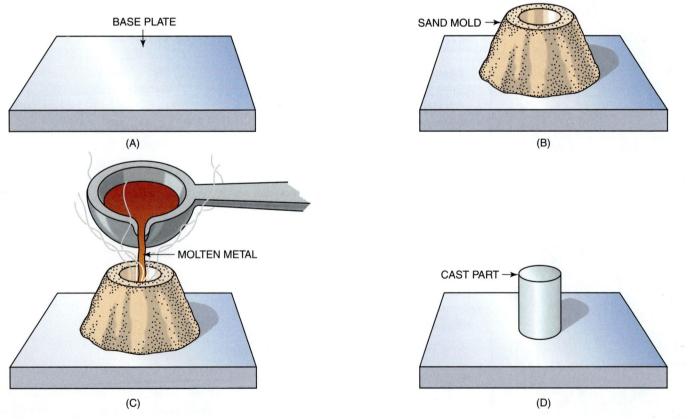

Figure 1.1 Direct casting
(A) Base plate to have part cast on it; (B) sand molded into shape desired; (C) pouring hot metal into mold; and (D) part cast is now part of the base plate

pour molten metal between them. When the edges of the base metal melted, the flow of metal was dammed using sand, and the molten metal was allowed to harden, Figure 1.2.

The Industrial Revolution, from 1750 to 1850, introduced a method of joining pieces of iron together known as **forge welding** or hammer welding. This process involved the use of a forge to heat the metal to a soft, plastic temperature. The ends of the iron pieces were then placed together and hammered until fusion took place.

Forge welding remained the primary welding method until Elihu Thomson, in 1886, developed the **resistance welding** process. This process produces a weld at the faying surfaces of a joint by the heat obtained from the resistance to the flow of welding current through the workpieces from electrodes that serve to concentrate the welding current and pressure in the weld area. This process provided a more reliable and faster way of joining metal than did previous methods.

As techniques were further developed, riveting was replaced in the United States and Europe by **fusion welding**, which melts together filler metal and base metal, or base metal only, to produce a weld. At that time the welding process was considered to be vital to military security: Welding repairs to ships damaged during World War I were carried out in great secrecy. Even today some aspects of welding are closely guarded secrets.

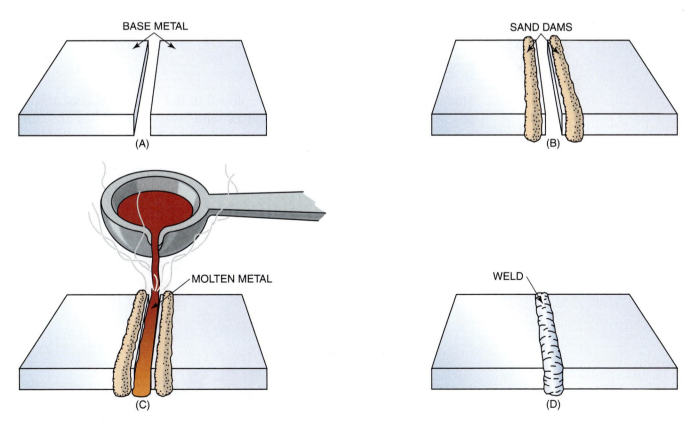

Figure 1.2 Flow welding
(A) Two pieces of metal plate; (B) sand dams to hold molten metal in place; (C) molten metal poured between metal plates; and (D) finished welded plate

Since the end of World War I, many welding methods have been developed for joining metals. These various methods are playing an important role in the expansion and production of the welding industry. Welding has become a dependable, efficient, and economical method for joining pieces of metal.

WELDING DEFINED

A **weld** is defined by the **American Welding Society (AWS)** as "a localized **coalescence** (the fusion or growing together of the grain structure of the materials being welded) of metals or nonmetals produced either by heating the materials to the required welding temperatures, with or without the application of pressure, or by the application of pressure alone, and with or without the use of filler materials." **Welding** is defined as "a joining process that produces coalescence of materials by heating them to the welding temperature, with or without the application of pressure or by the application of pressure alone, and with or without the use of filler metal." In less technical language, a weld is made when separate pieces of material are joined to form one piece by heating them to a temperature high enough to cause softening or melting and flow together. Pressure may or may not be used to force the pieces together. In some cases, pressure alone may be sufficient to force the separate pieces of material to combine and form one piece. Filler material is added when needed to form a completed weld in the joint. It is important to note that the word *material* is used because today welds can be made from a growing list of materials, including plastic, glass, and ceramics.

USES OF WELDING

Modern welding techniques are employed in the construction of numerous products. Ships, buildings, bridges, and recreational rides are fabricated using welding processes, Figures 1.3 and 1.4. Welding is often used to produce the machines that are used to manufacture new products.

Welding has made it possible for airplane manufacturers to meet the design demands of strength-to-weight ratios for both commercial and military aircraft.

Figure 1.3 Space shuttle being prepared for launch
Notice the large welded support structure
Source: Courtesy of NASA

Welded sculpture, Seattle, Washington

Welded joints are a critical component of structures

Roller coaster at Silver Dollar City, Branson, Missouri

Spiral staircase in Missouri City, Texas

Voyager of the Sea, Haiti

Figure 1.4 The uses of modern welding techniques
Source: Courtesy of Larry Jeffus

Voyager of the Sea dining room

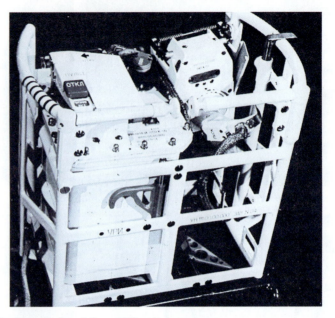

Figure 1.5 Machine designed for welding in space
Courtesy of E. O. Paton Electric Welding Institute, Commonwealth of Independent States, the former Soviet Union

The exploration of space would not be possible without modern welding techniques. From the earliest rockets to today's aerospace industry, welding has played an important role. The space shuttle's construction required the improvement of welding processes. Many of these improvements have helped improve our daily lives.

Many experiments aboard the space shuttle have involved welding and metal joining. The United States along with other nations are currently building a permanent space station. Someday welders will be required to build such a large structure in the vacuum of space. Figure 1.5 depicts a welding machine designed to be used in space. Figure 1.6 shows a

Figure 1.6 Cosmonaut making a weld outside a space ship
Courtesy of E. O. Paton Electric Welding Institute, Commonwealth of Independent States, the former Soviet Union

cosmonaut using the welder in open space. This specialized welder was developed at the E. O. Paton Electric Welding Institute in Kiev, Ukraine. As new welding techniques are developed for this major project, we will see them being used here on Earth to improve our world.

Welding is used extensively in the manufacture of automobiles, farm equipment, home appliances, computer components, mining equipment, and construction equipment. Railway equipment, furnaces, boilers, air-conditioning units, and hundreds of other products we use in our daily lives are also joined together by some type of welding process.

Items ranging from dental braces to telecommunications satellites are assembled by welding. Very little in our modern world is not produced using some form of this versatile process.

WELDING PROCESSES

The number of different welding processes has grown in recent years. These processes differ greatly in the manner in which heat, pressure, or both, are applied and in the type of equipment used. Table 1.1 lists over

Table 1.1 Master Chart of Welding and Allied Processes

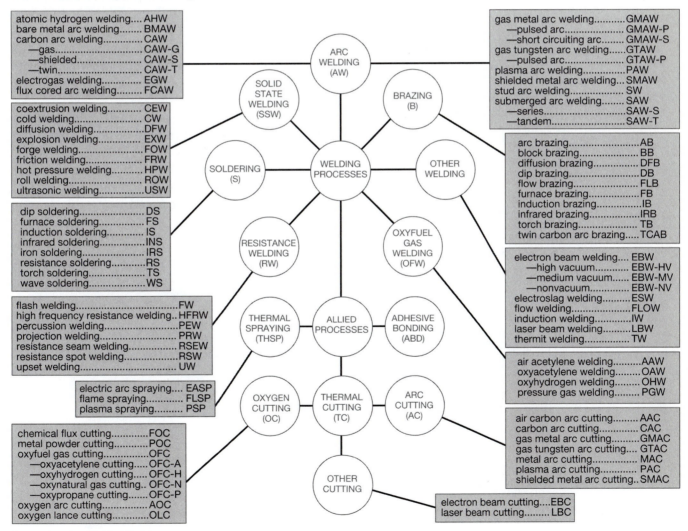

Source: American Welding Society

70 welding and allied processes. Some processes require hammering, pressing, or rolling to effect the coalescence in the weld joint. Other methods bring the metal to a fluid state so that the edges flow together.

The most popular welding processes are shielded metal arc welding (SMAW), gas tungsten arc welding (GTAW), gas metal arc welding (GMAW), flux cored arc welding (FCAW), torch or oxyfuel brazing (TB), and oxyacetylene welding (OAW).

The use of regional terms by skilled workers is a common practice in all trade areas, including welding. As an example, oxyacetylene welding is one part of the larger group of processes known as oxyfuel gas welding. The names used to refer to OAW include *gas welding* and *torch welding.* Shielded metal arc welding is often called *stick welding, rod welding,* or just *welding.* As you begin your career you will learn the various names used in your area, but you should keep in mind and use the more formal terms whenever possible. The AWS has published a book called *AWS A3.0 Standard Welding Terms and Definitions;* this book has the status of an American national standard. Welding terms used on contract documents and procedures often require this standard for their language.

Acetylene is the most commonly used fuel gas. It is widely used for **oxyfuel gas welding (OFW), oxyfuel gas cutting (OFC),** and **torch brazing (TB).** The **oxyfuel (OF) gas** processes are the most versatile of the welding processes. The equipment required is comparatively inexpensive, and the cost of operation is low, Figure 1.7.

Shielded metal arc welding (SMAW) has been a popular method of joining metal for years. High-quality welds can be made rapidly and with excellent uniformity. A variety of metal types and metal thicknesses can be joined with one machine, Figure 1.8.

Gas tungsten arc welding (GTAW) is easily performed on almost any metal. Its clean, high-quality welds often require little or no postweld finishing, Figure 1.9.

Gas metal arc welding (GMAW) is extremely fast and economical. This process is easily used for welding on thin-gauge metal as well as on heavy plate. The high welding rate and reduced postweld cleanup

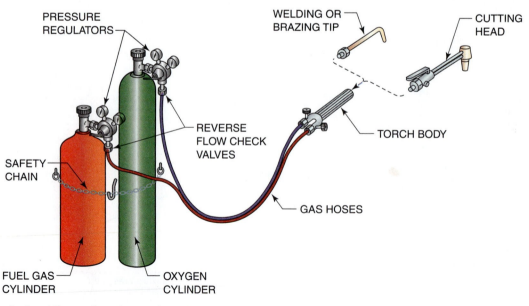

Figure 1.7 Oxyfuel welding and cutting equipment

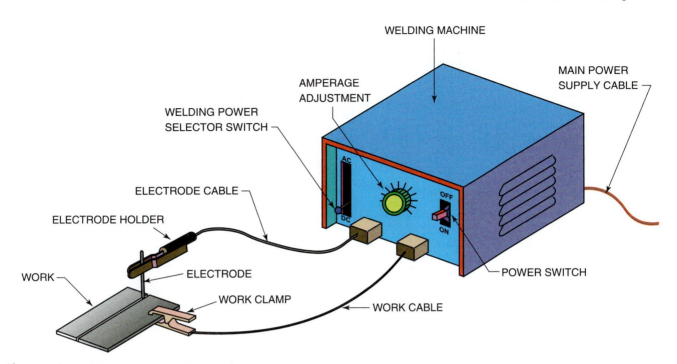

Figure 1.8 Shielded metal arc welding equipment

have made gas metal arc welding an outstanding welding process, Figure 1.10.

Flux cored arc welding (FCAW) uses the same type of equipment that is used for the gas metal arc welding process. A major advantage of this process is that with the addition of flux to the center of the filler wire it is often possible to make welds without the use of an external shielding gas. The introduction of smaller wire sizes and elimination of the shielding gas from some welds has resulted in an increase in the use of the FCAW process. Although slag must be cleaned from the welds after completion, the

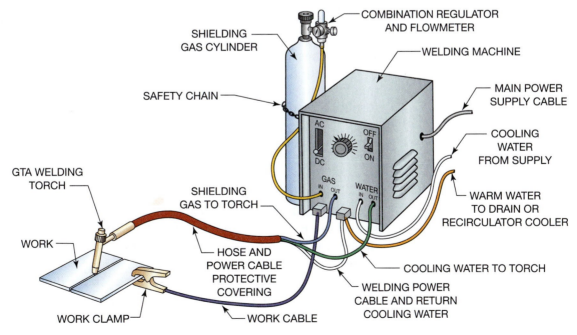

Figure 1.9 Gas tungsten arc welding equipment

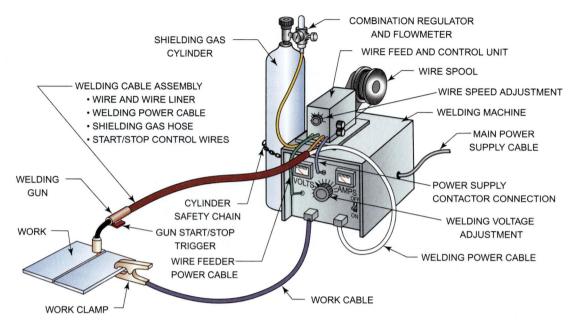

Figure 1.10 Gas metal arc welding equipment

process's advantages of high quality, versatility, out-of-position welding capability, and welding speed offset this requirement.

The introduction of new, low-cost equipment and improved availability of filler metals has resulted in GMAW and FCAW becoming the most commonly used welding processes. Many hardware stores, automotive shops, and farm supply stores carry this equipment and filler metal. Home hobby FCA welders can be purchased for less than $200.

The selection of the joining process for a particular job depends upon many factors. No single, specific rule controls the choice of welding process for a certain job. A few of the factors that must be considered when choosing a joining process are

- availability of equipment—What are the types, capacity, and condition of equipment available to make the welds?
- repetitiveness of the operation—How many welds will be required to complete the job, and are they all the same?
- quality requirements—Is this weld going to be used on a piece of furniture, to repair a piece of equipment, or to join a pipeline?
- location of work—Will the welding be done in a shop or on a remote job site?
- materials to be joined—Are the parts made out of a standard metal or some exotic alloy?
- appearance of the finished product—Will this be a weldment that is needed only to test an idea, or will it be a permanent structure?
- size of the parts to be joined—Are the parts small, large, or of different sizes, and can they be moved or must they be welded in place?
- time available for the work—Is this a rush job needing a fast repair, or is there time to allow for pre- and postweld cleanup?
- skill or experience of workers—Do the welders have the ability to do the job?

- cost of materials—Will the weldment be worth the expense of special equipment, materials, or finishing time?
- code or specification requirements—Often the selection of the process is dictated by a governing agency, codes, or standards.

The welding engineer and the welder must not only decide on the welding process but also select the method of applying it. The following methods are used to perform welding, cutting, or brazing operations:

- **manual operation**—The welder is required to manipulate the entire process.
- **semiautomatic operation**—The filler metal is added automatically, and all other manipulation is done manually by the welder.
- **machine operation**—Operations are done mechanically under the observation and correction of a welding operator.
- **automatic operation**—Operations are performed repetitively by a machine that has been programmed to do an entire operation without the intervention of the operator.
- **automated operation**—Operations are performed repetitively by a robot or other machine that is programmed flexibly to do a variety of processes.

OCCUPATIONAL OPPORTUNITIES IN WELDING

The American welding industry has contributed to the widespread growth of welding and allied processes. Without welding, many of the products we use on a daily basis could not be manufactured. The list of these products grows every day, thus increasing the number of jobs for people with welding skills. These well-paying jobs are not concentrated in major metropolitan areas but are found throughout the country and the world. Because of the diverse nature of the welding industry, the exact job duties of each skill area will vary. The following are general descriptions of the job classifications used in our profession; specific tasks may vary from one location to another.

- Welders perform manual or semiautomatic welding. They are the skilled craftspeople who, through their own labor, produce the welds on a variety of complex products, Figure 1.11.
- Tack welders perform manual or semiautomatic welding to produce tacks, which often help the welder by making small welds to hold parts in place. The tack weld must be correctly applied so that it is strong enough to hold the assembly but does not interfere with the finished welding.
- Welding operators run adaptive control, automatic, mechanized, or robotic welding equipment.
- Welders' helpers are employed in some welding shops to clean slag from the welds and help move and position weldments for the welder.
- Welder assemblers, or welder fitters, position all the parts in their proper places and make ready for the tack welders. These skilled workers must be able to interpret blueprints and welding

Figure 1.11 Welded structures
Amusement parks such as Silver Dollar City in Branson, Missouri, require many talented welders to produce such attractions as (A) an antique train engine to be used in a parade; (B) air-powered guns for launching toy balls; and (C) the *Branson Belle* paddle-wheel boat
Source: Courtesy of Larry Jeffus

procedures. They also must have knowledge of the effects of the contraction and expansion of various metals.

- Welding inspectors are often required to hold a special certification such as the one supervised by the American Welding Society known as Certified Welding Inspector (CWI). To become a CWI, candidates must pass a test covering the welding process, blueprint reading, weld symbols, metallurgy, codes and standards, and inspection techniques. Vision screening is also required on a regular basis once the technical skills have been demonstrated.

- Welding shop supervisors may or may not weld on a regular basis, depending on the size of the shop. In addition to their welding skills, they must demonstrate good management skills by effectively planning jobs and assigning workers.

- Welding salespeople may be employed by supply houses or equipment manufacturers. Sales jobs require a broad understanding of the welding process as well as good marketing skills. Good salespeople are able to provide technical information about their products to convince customers to make a purchase.

- Welding shop owners are often welders who have a high degree of skill and knowledge of small-business management and prefer to operate their own businesses. These individuals may specialize in one field, such as hardfacing, repair and maintenance, or specialty fabrications, or they may operate as subcontractors of manufactured items. A welding business can be as small as one person, one truck, and one portable welder or as large as a multimillion-dollar operation employing hundreds of workers.

- Welding engineers design, specify, and oversee the construction of complex weldments. The welding engineer may work with other engineers in areas such as mechanics, electronics, chemicals, or civil engineering in the process of bringing a new building, ship, aircraft, or product into existence. The welding engineer is required to know all of the welding process and metallurgy and to have good math, reading, communication, and design skills. This person usually has an advanced college degree and possesses a professional certification.

In many industries, the welder, welding operator, and tack welder must be able to pass a performance test to a specific code or standard.

The highest-paid welders are those who have the education and skills to read blueprints and do the required work to produce a weldment to strict specifications.

Large industrial concerns employ workers who serve as support for the welders. These engineers and technicians must have knowledge of chemistry, physics, metallurgy, electricity, and mathematics. Engineers are responsible for research, design, development, and fabrication of a project. Technicians work as part of the engineering staff. These individuals may oversee the actual work for the engineer by providing the engineer with progress reports as well as chemical, physical, and mechanical test results. Technicians may also require engineers to build prototypes for testing and evaluation.

Another group of workers employed by industry does **layouts** or makes templates. These individuals have drafting experience and a knowledge of such operations as punching, cutting, shearing, twisting, and forming. The layout is generally done directly on the material. A template is used for repetitive layouts and is made from sheet metal or other suitable materials.

The thermal-cutting processes are closely related to welding. Some operators use hand-held torches, and others are skilled operators of oxy-fuel, plasma arc or laser cutting machines. These machines range from simple mechanical devices to highly sophisticated, computer-controlled, multiple-head machines that are operated by specialists, Figure 1.12.

Employment of welders is expected to increase rapidly as a result of the wider use of various welding processes, including the use of robots. Many more skilled welders will be needed for maintenance and repair work in the expanding metalworking industries. The number of welders in production work is expected to increase in plants manufacturing sheet metal products, pressure vessels, boilers, railroad equipment, storage tanks, air-conditioning equipment, and ships and in the field of energy exploration and energy resources. The construction industry will need an ever-increasing number of good welders as the use of welded steel buildings grows.

Employment prospects for welding operators are expected to continue to be favorable because of the increased use of machine, automatic, and automated welding in the manufacture of aircraft, missiles, railroad cars, automobiles, and numerous other products.

Figure 1.12 Numerical-control oxygen cutting machine
Source: Courtesy of ESAB Welding & Cutting Products

TRAINING FOR WELDING OCCUPATIONS

Generally, several months of training are required to learn the basics of a welding process. To become a skilled welder, both welding school and on-the-job experience are required. Because of the diverse nature of the welding industry, no single list of skills can be given that meets every job's requirement. However, there are specific skills that are required of most entry-level welders. The American Welding Society, in partnership with the U.S. Department of Education, conducted a survey of the welding industry to define guidelines for knowledge and skills required in the welding workplace. In 1995, the AWS published a guideline curriculum for entry-level welders. The program is called SENSE, which stands for Students Excelling through National Standards in Education. This book covers those skill requirements.

In addition to welding skills, an entry-level welder must possess workplace skills. Workplace skills include proficiency in reading, writing, math, communication, and science as well as good work habits and an acceptance of close supervision. Some welding jobs also require a theoretical knowledge of welding, blueprint reading, welding symbols, metal properties, and electricity. A few of the jobs that require less skill can be learned after a few months of on-the-job training. However, the fabrication of certain alloys requires knowledge of metallurgical properties as well as the development of a greater skill in cutting and welding them.

Robotics and computer-aided manufacturing (CAM) both require more than a basic understanding of the welding process; they require that the student be computer literate.

A young person planning a career as a welder needs good eyesight, manual dexterity, hand and eye coordination, and an understanding of welding technology. For entry into manual welding jobs, employers prefer to hire young people who have high school or vocational training in welding processes. Courses in drafting, blueprint reading, mathematics, and physics are also valuable.

Beginners in welding who have no training often start in manual welding production jobs that require minimum skill. Some first work as helpers and are later moved into welding jobs. General helpers, if they show promise, may be given a chance to become welders by serving as helpers to experienced welders.

No formal apprenticeship is usually required for general welders. A number of large companies have welding apprenticeship programs. The military, at several of its installations, has programs in welding.

Skill and technical knowledge requirements are higher in some industries. In the fields of atomic energy, aerospace, and pressure vessel construction, high standards for welders must be met to ensure that weldments will withstand the critical forces that they will be subjected to in use.

After two years of training at a vocational school or technical institute, the skilled welder may qualify as a technician. Technicians are generally involved in the interpretation of engineers' plans and instructions.

Before being assigned a job where the service requirements of the weld are critical, welders usually must pass a certification test given by an employer. In addition, some localities require welders to obtain a license for certain types of outside construction.

After a welder, welding operator, or tack welder has received a **certification** or **qualification** by passing a standardized test, he or she is allowed to make only welds specifically covered by that test. The welding certification is very restrictive; it allows a welder to perform only code welds covered by that test. Certifications are usually good for a maximum of six months unless a welder is doing code-quality welds routinely. As a student, you should check into the acceptance of a welding qualification test before investing time and possibly money in the test.

QUALIFIED AND CERTIFIED WELDERS

Welder qualification and welder certification are often misunderstood. Being certified does not mean that a welder can weld anything, nor does it mean that every weld the welder makes is acceptable. It means that the welder has demonstrated the skills and knowledge necessary to make good welds in a particular process, on a specified alloy, in one or more positions. To ensure that a welder is consistently making welds that meet the standard, welds are inspected and tested. The more critical the welding, the more critical the inspection and the more extensive the testing of the welds.

All welding processes can be tested for qualification and certification. The testing can range from making spot welds with an electric resistance spot welder to making electron beam welds on aircraft. Being qualified or certified in one area of welding does not mean that a welder can make quality welds in other areas. Most qualifications and certifications are restricted to a single welding process, position, metal, and thickness range.

Individual codes control test requirements. Within these codes, changes in any of a number of essential variables can result in the need to recertify:

- Welders can be certified in each welding *process,* such as SMAW, GMAW, FCAW, GTAW, electron beam welding (EBW), and resistance spot welding (RSW). A separate test is required for each process.
- The type of *material* (e.g., steel, aluminum, stainless steel, titanium) being welded will require a change in certification. Even a change in the alloy within a base metal type can require a change in certification.
- Each certification is valid for a specific range of *thickness* of base metal. This range depends on the thickness of the metal used in the test. For example, if a 3/8-in. (9.5-mm) plain carbon steel plate is used in the test, then under some codes the welder would be qualified to make welds in a plate thickness range from 3/16 in. to 3/4 in. (4.7 mm to 19 mm).
- Changes in the classification and size of the *filler metal* can require recertification.
- If the process requires a *shielding gas,* then changes in gas type or mixture can affect the certification.
- In most cases, a weld test taken in a flat *position* limits certification to flat and possibly horizontal welding. A test taken in the vertical position, however, usually allows the welder to work in the flat, horizontal, and vertical positions, depending on the code requirements.
- Changes in *weld type,* such as groove or fillet welds, require a new certification. In addition, variations in *joint geometry,* such as groove type, groove angle, and number of passes, can also require retesting.
- In some cases, changing *welding current* from alternating current (AC) to direct current (DC) or changes to pulsed power and high frequency, for example, can affect the certification.

Any welder qualification or certification process must include the specific welding skill level to be demonstrated. The detailed information for a welding test is often given as part of a welding procedure specification (WPS) or similar set of welding specifications or schedules. Such standards inform everyone about which skills are required, enabling the welder to prepare for the welding test and to demonstrate welding skills to the company. Varying from the strict limitations in the WPS usually requires that a different test be taken.

Welder performance qualification testing is the demonstration of a welder's ability to produce welds meeting very specific prescribed standards. The form used to document this test is called the *Welder Performance Qualification Test Record.* The detailed written instructions to be followed by the welder are called the *Welder Qualification Procedure.* Welders who have passed this type of testing are often referred to as *qualified welders* or as *qualified.*

Welder certification is the written verification that a welder has produced welds meeting a prescribed standard of welder performance. A welder holding such a written verification is often referred to as *certified* or as a *certified welder*.

An AWS certified welder is one who has complied with all the provisions, requirements, and specifications of the AWS regarding certification. Very specific requirements must be met by any school or organization before it can offer this certification. Under the **AWS SENSE** program, the welder must pass a closed-book exam covering specific knowledge areas and a performance test. Written documentation, including the welder's name and social security number and test results, must be sent to the AWS, where such records are entered into the AWS National Registry for welders. The certification record expires after one year and is automatically deleted from the registry.

The AWS has developed three levels of qualification for welders through the SENSE program. The first level, **Level I–entry**, is for the beginning welder; **Level II–advanced** and **Level III–expert** are for more skilled welders. These qualifications are gaining widespread acceptance by industry. Welder performance qualification exams allow welders to prove their skills on a standard test. This textbook series will focus on the AWS entry-level welder qualification.

AWS ENTRY-LEVEL WELDER QUALIFICATION AND WELDER CERTIFICATION

The AWS entry-level welder qualification and certification program specifies a number of requirements not normally found in the traditional welder qualification and certification process. The additions to the AWS program have broadened the scope of the test. Areas such as practical knowledge have long been an assumed part of most certification programs but have not been a formal part of the testing process. Most companies have assumed that welders who could produce code-quality welds could understand enough of the technical aspects of welding.

Today, however, greater importance is placed on the technical knowledge of the process, code, and other aspects of the complete welding process. This change is due to the greater complexity of many welding processes and an increased responsibility of companies and their welders to ensure the quality and reliability of weldments. It is important not only that the weld be correctly performed but also that the welder know why it must be performed in such a specific manner. This is all intended to increase accuracy and reduce rejection of welds.

A written test must be passed with a minimum grade of 75% on all areas except safety. The safety questions must be answered with a minimum accuracy of 90%. The following subject areas, covered in this textbook series, are included in the test:

- welding and cutting theory
- welding and cutting inspection and testing
- welding and cutting terms and definitions (Glossary)
- base and filler metal identification

Figure 1.13 SkillsUSA welding assembly
Source: Courtesy of Larry Jeffus

- common welding process variables
- electrical fundamentals
- drawing and welding symbol interpretation
- fabrication principles and practices
- safe practices

Each year SkillsUSA sponsors a series of welding skill competitions for its student members. Students can begin by joining their local SkillsUSA chapter. They can then compete in local, regional, and state competitions. After each round, the students with the best welding skills and knowledge advance to the next level of competition. Contestants are challenged with a written test and must show their proficiency in welding and fabrication, Figure 1.13. A national SkillsUSA Olympics competition is held each year in Kansas City, Missouri. The winners at the national competition can then go on to the International Skill Olympics. The international competition is held in a different country each year. Like most professional organizations, SkillsUSA emphasizes community service and citizenship as key components of the philosophy of the organization.

EXPERIMENTS AND PRACTICES

This textbook series contains experiments and practices. These are intended to help you develop your welding knowledge and skills.

The experiments are designed to allow you to see what effect changes in the process settings, operation, or techniques have on the type of weld produced. When you do an experiment, you should observe and possibly take notes on how the change affected the weld. Often as you make a weld it will be necessary for you to make changes in your equipment settings or your technique to ensure that you are making an acceptable weld. By watching what happens when you make the changes in the welding shop, you will be better prepared to decide on changes required to make good welds on the job.

It is recommended that you work in a small group as you try the experiments. In a small group, one person can be welding, one adjusting the equipment, and the others recording the machine settings and weld effects. Group work also allows you to watch the weld change more closely if someone is welding as you look on. Then, as a group member, changing places will reinforce your learning.

The practices are designed to build your welding skills. Each practice tells you in detail what equipment, supplies, and tools you will need as you develop the specific skill. The practices will start off easy and become progressively more complex or difficult. Welding is a skill that requires that you develop in stages from the basic to the more complex.

Each practice gives the evaluation criteria or acceptable limits for the weld. All welds have some discontinuities, but if they are within the acceptable limits, they are called not *defects* but *flaws*. As you practice your welding, keep in mind the acceptable limits so that you can progress to the next level when you have mastered the process and weld you are working on.

Layout

The experiments and practices will require that you read and interpret simple drawings and sketches including welding symbols. You will fill out a bill of materials that will be required to fabricate the weldment. The material specifications must be given in both standard and SI units, requiring that you make the necessary conversions.

Once the bill of materials is complete, you must lay out on appropriate metal stock the individual parts that are to be cut out. The parts must be laid out to within a fractional tolerance of 1/16 in. (1.6 mm), with an angular tolerance of +10° to –5°. Be sure to leave an appropriate amount of space between parts for the kerf if the parts are not to be sheared.

Written Procedures

Each weldment drawing includes a written list of notes that must be followed. You must also follow a WPS for each weldment. Specific information relating to the cutting or welding process can be found in this text. Additional information is available in the AWS EG2.0 publication and in the references listed in Table 1.2. The material in this text follows the AWS guidelines but does not include all of the AWS material as provided in its publications.

Written Records

Learners will fill out time or job cards, reports, and other records as needed, Tables 1.3–1.5. Written records must be complete, neat, and legible. These records must be turned in with the completed weldment and will be considered in the overall evaluation of your skills. Similar records are required by most large welding companies to determine the productivity of welders and to ensure that each job is charged correctly for time and materials. These records help the company to stay profitable.

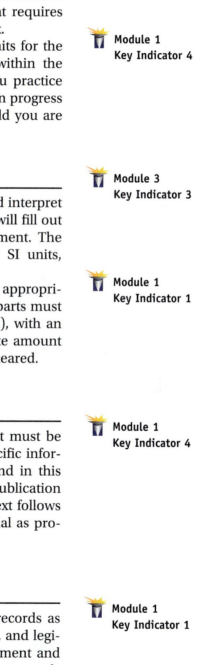

Module 1
Key Indicator 4

Module 3
Key Indicator 3

Module 1
Key Indicator 1

Module 1
Key Indicator 4

Module 1
Key Indicator 1

Table 1.2 Standard Welding Procedure Specifications

ANSI/AWS B2.1.001	Standard Welding Procedure Specification for Shielded Metal Arc Welding of Carbon Steel (M-1/P-1, Group 1 or 2), 3/16 through 3/4 inch in the As-Welded Condition, with Backing
ANSI/AWS B2.1.008	Standard Welding Procedure Specification for Gas Tungsten Arc Welding of Carbon Steel (M-1, Group 1), 10 through 18 Gauge, in the As-Welded Condition, with or without Backing
ANSI/AWS B2.1.009	Standard Welding Procedure Specification for Gas Tungsten Arc Welding of Austenitic Stainless Steel (M-8/P-8), 10 through 18 Gauge, in the As-Welded Condition, with or without Backing
ANSI/AWS B2.1.015	Standard Welding Procedure Specification for Gas Tungsten Arc Welding of Aluminum (M-22 or P-22), 10 through 18 Gauge, in the As-Welded Condition, with or without Backing
ANSI/AWS B2.1.019	Standard Welding Procedure Specification for CO_2 Shielded Flux Cored Arc Welding of Carbon Steel (M-1/P-1/S-1, Group 1 or 2), 1/8 through 1–1/2 inch Thick, E70T-1 and E71T-1, As-Welded Condition
ANSI/AWS B2.1.020	Standard Welding Procedure Specification for 75% Argon 25% CO_2 Shielded Flux Cored Arc Welding of Carbon Steel (M-1/P-1/S-1, Group 1 or 2), 1/8 through 1–1/2 inch Thick, E70T-1 and E71T-1, As-Welded Condition

Table 1.3 Useful Form for Keeping Track of Weldments

Time Card		
Name _____ Date _____ Job _____		
Starting Time	Ending Time	Total Time
Sign		Total

Verbal Instructions

Module 1
Key Indicator 3

In any working shop, verbal instructions must be given from time to time. These instructions are as important as written ones. Many welding shops assemble their welders at the beginning of a shift so that special instructions can be given verbally. Instructions may include things such as which stock to use, where scrap is to be deposited, and which welder to use. In some cases critical information such as safety concerns are given verbally. Your safety and the safety of others could depend on your ability to remember and follow verbal instructions.

Table 1.4 Useful Form for Keeping Track of Weldments

Inspection Report		Job: _____	
Inspection	Pass/fail	Inspector's Initials	Date
Layout			
Cutout			
Assembly			
Welding			
Tack			
Interpass			
Finish			
Overall Rating			
Accuracy			
Appearance			
Welder _____ Date _____			

Table 1.5 Useful Form for Keeping Track of Weldments

BILL OF MATERIALS		
Name _____ Date _____ Job _____		
Part ID	Size Determination	SI Determination
Material Specification _____		

WELDING VIDEO SERIES

Delmar Learning, in cooperation with the author, has produced a series of videotapes. Each of the four sets of tapes covers specific equipment setup and operation for welding, cutting, soldering, or brazing. When specific skills are covered both in this textbook and on a videotape, you will see a framed shot from the video, as shown in Figure 1.14. Reading the material, watching the video, and practicing should help you to develop your welding skills more rapidly.

Figure 1.14 Welding Principles and Practices on DVD series
This close-up of GMA welding can be found on DVD 2 of the Welding Principles and Practices on DVD series. It is an example of the type of close-ups shown in all of the Welding Series DVDs
Source: Courtesy of Larry Jeffus

METRIC UNITS

Both standard and metric (SI) units are given in this text. The SI units are in parentheses ( ) following the standard unit. When nonspecific values are used—for example, "set the gauge at 2 psig," where 2 is an approximate value—the SI units have been rounded off to the nearest whole number. Rounding-off is used in these cases to agree with the standard value and because whole numbers are easier to work with. SI units are not rounded off only when the standard unit is an exact measurement.

Often students have difficulty understanding metric units because exact conversions are used even when the standard measurement is an approximation. Rounding off the metric units makes understanding the metric system much easier, Table 1.6. By using this approximation method, you can make most standard-to-metric conversions in your head without needing to use a calculator.

Once you have learned to use approximations for metric units, you will find it easier to make exact conversions whenever necessary. Conversions

Table 1.6 Metric Conversions Approximations

1/4 inch = 6mm
1/2 inch = 13mm
3/4 inch = 18mm
1 inch = 25mm
2 inches = 50mm
1/2 gal = 2 L
1 gal = 4 L
1 lb = 1/2 K
2 lb = 1 K
1 psig = 7 kPa
1°F = 2°C

Approximate conversions of standard units to metric make it possible to quickly have an idea of how large or heavy an object is in the other units. For estimating purposes, exact conversions are not required.

must be exact in the shop when a part is dimensioned with one system's units and the other system must be used to fabricate the part. For that reason you must be able to make those conversions. Tables 1.7 and 1.8 are set up to be used with or without the aid of a calculator. Many calculators today have built-in standard-to-metric conversions. It is a good idea to know how to make these conversions with and without these aids, of course. Practice making such conversions whenever the opportunity arises.

Table 1.7 Conversions for U.S. Customary (Standard) Units and Metric Units (SI)

TEMPERATURE
Units

°F (each 1° change)	= 0.555°C (change)
°C (each 1° change)	= 1.8°F (change)
32°F (ice freezing)	= 0°Celsius
212°F (boiling water)	= 100°Celsius
–460°F (absolute zero)	= 0°Rankine
–273°C (absolute zero)	= 0°Kelvin

Conversions

°F to °C _____ °F – 32 = _____ × .555 = _____ °C
°C to °F _____ °C × 1.8 = _____ + 32 = _____ °F

LINEAR MEASUREMENT
Units

1 inch	= 25.4 millimeters
1 inch	= 2.54 centimeters
1 millimeter	= 0.0394 inch
1 centimeter	= 0.3937 inch
12 inches	= 1 foot
3 feet	= 1 yard
5280 feet	= 1 mile
10 millimeters	= 1 centimeter
10 centimeters	= 1 decimeter
10 decimeters	= 1 meter
1000 meters	= 1 kilometer

Conversions

in. to mm _____ in. × 25.4 = _____ mm
in. to cm _____ in. × 2.54 = _____ cm
ft to mm _____ ft × 304.8 = _____ mm
ft to m _____ ft × 0.3048 = _____ m
mm to in. _____ mm × 0.0394 = _____ in.
cm to in. _____ cm × 0.3937 = _____ in.
mm to ft _____ mm × 0.00328 = _____ ft
m to ft _____ m × 3.28 = _____ ft

AREA MEASUREMENT
Units

1 sq in.	= 0.0069 sq ft
1 sq ft	= 144 sq in.
1 sq ft	= 0.111 sq yd
1 sq yd	= 9 sq ft
1 sq in.	= 645.16 sq mm
1 sq mm	= 0.00155 sq in.
1 sq cm	= 100 sq mm
1 sq m	= 1000 sq cm

Conversions

sq in. to sq mm _____ sq in. × 645.16
= _____ sq mm
sq mm to sq in. _____ sq mm × 0.00155
= _____ sq in.

VOLUME MEASUREMENT
Units

1 cu in.	= 0.000578 cu ft
1 cu ft	= 1728 cu in.
1 cu ft	= 0.03704 cu yd
1 cu ft	= 28.32 L
1 cu ft	= 7.48 gal (U.S.)
1 gal (U.S.)	= 3.737 L
1 cu yd	= 27 cu ft
1 gal	= 0.1336 cu ft
1 cu in.	= 16.39 cu cm
1 L	= 1000 cu cm
1 L	= 61.02 cu in.
1 L	= 0.03531 cu ft
1 L	= 0.2642 gal (U.S.)
1 cu yd	= 0.769 cu m
1 cu m	= 1.3 cu yd

Conversions

cu in. to L _____ cu in. × 0.01638 = _____ L
L to cu in. _____ L × 61.02 = _____ cu in.
cu ft to L _____ cu ft × 28.32 = _____ L
L to cu ft _____ L × 0.03531 = _____ cu ft
L to gal _____ L × 0.2642 = _____ gal
gal to L _____ gal × 3.737 = _____ L

WEIGHT (MASS) MEASUREMENT
Units

1 oz	= 0.0625 lb
1 lb	= 16 oz
1 oz	= 28.35 g
1 g	= 0.03527 oz
1 lb	= 0.0005 ton
1 ton	= 2000 lb
1 oz	= 0.283 kg
1 lb	= 0.4535 kg
1 kg	= 35.27 oz
1 kg	= 2.205 lb
1 kg	= 1,000 g

(Continued)

Table 1.7 *Continued*

Conversions

lb to kg _____	lb × 0.4535	= _____ kg
kg to lb _____	kg × 2.205	= _____ lb
oz to g _____	oz × 0.03527	= _____ g
g to oz _____	g × 28.35	= _____ oz

PRESSURE AND FORCE MEASUREMENTS
Units

1 psig	= 6.8948 kPa
1 kPa	= 0.145 psig
1 psig	= 0.000703 kg/sq mm
1 kg/sq mm	= 6894 psig
1 lb (force)	= 4.448 N
1 N (force)	= 0.2248 lb

Conversions

psig to kPa _____	psig × 6.8948	= _____ kPa
kPa to psig _____	kPa × 0.145	= _____ psig
lb to N _____	lb × 4.448	= _____ N
N to lb _____	N × 0.2248	= _____ psig

VELOCITY MEASUREMENTS
Units

1 in./sec	= 0.0833 ft/sec
1 ft/sec	= 12 in./sec
1 ft/min	= 720 in./sec
1 in./sec	= 0.4233 mm/sec
1 mm/sec	= 2.362 in./sec
1 cfm	= 0.4719 L/min
1 L/min	= 2.119 cfm

Conversions

ft/min to in./sec _____	ft/min	× 720	= _____ in./sec
in./min to mm/sec _____	in./min	× .4233	= _____ mm/sec
mm/sec. to in./min _____	mm/sec	× 2.362	= _____ in./min
cfm to L/min _____	cfm	× 0.4719	= _____ L/min
L/min to cfm _____	L/min	× 2.119	= _____ cfm

Table 1.8 Abbreviations and Symbols

U.S. Customer (Standard) Units

°F	= degrees Fahrenheit
°R	= degrees Rankine
	= degrees absolute F
lb	= pound
psi	= pounds per square inch
	= lb per sq in.
psia	= pounds per square inch absolute
	= psi + atmospheric pressure
in.	= inches = in. = ′
ft	= foot or feet = ft = ″
sq in.	= square inch = in.
sq ft	= square foot = ft
cu in.	= cubic inch = in.
cu ft	= cubic foot = ft
ft-lb	= foot-pound
ton	= ton of refrigeration effect
qt	= quart

Metric Units (SI)

°C	= degrees Celsius
°K	= Kelvin
mm	= millimeter
cm	= centimeter
cm^2	= centimeter squared
cm^3	= centimeter cubed
dm	= decimeter
dm^2	= decimeter squared
dm^3	= decimeter cubed
m	= meter
m^2	= meter squared
m^3	= meter cubed
L	= liter
g	= gram
kg	= kilogram
J	= joule
kJ	= kilojoule
N	= newton
Pa	= pascal
kPa	= kilopascal
W	= watt
kW	= kilowatt
MW	= megawatt

Miscellaneous Abbreviations

P = pressure
sec = seconds
h = hours
r = radius of circle
D = diameter
π = 3.1416 (a constant used in determining the area of a circle)
A = area
V = volume
∞ = infinity

SUMMARY

Welding is a very diverse trade. Almost every manufactured product utilizes a welding or joining process in its production. Products that are produced by welding range from small objects, such as sunglasses and dental braces, to larger structures, such as buildings, ships, and space shuttles. Your knowledge and understanding of the various processes and their applications will provide you with employable skills that can result in a rich and rewarding career in the welding field. The art and science of joining metals has been around for centuries, and with minor changes and improvements in materials, equipment, and supplies, it will be with us through the remainder of the 21st century.

REVIEW

1. Explain how to forge weld.
2. What welding process is Elihu Thomson credited with developing in 1886?
3. During World War I, what process replaced riveting for ship repair?
4. What do we call the localized growing together of the grain structure during a weld?
5. List six items that use welding in their construction.
6. What three things differ greatly from one welding process to another?
7. Which gases are most commonly used for the OFW process?
8. What is the technically correct name for *gas welding*?
9. Which welding process is most commonly used to join metal?
10. What is the technically correct name for *stick welding*?
11. GTAW is the abbreviation for which process?
12. What is the ideal process for high welding rates on thin-gauge metal?
13. Flux inside the welding wire gives which process its name?
14. List six factors that may be considered in selecting a welding process.
15. Which method of welding application requires a welder to manipulate the entire process?
16. Which method of welding requires the welder to control everything except the adding of filler metal?
17. Which welding process is repeatedly performed by a machine that has been programmed?
18. A _____ works with engineers to produce prototypes for testing.
19. A _____ places parts together in their proper position for the tack welder.
20. A _____ is a piece of sheet metal cut to the shape of a part so that it may be repetitively laid out.
21. What are the approximate standard units for the following SI values?
    a. 13 mm
    b. 4 L
    c. 100°C
    d. 4 K

22. What are the exact standard units for the following SI values?
    a. 13 mm
    b. 4 L
    c. 100°C
    d. 4 K
23. What are some advantages of a working environment that involves groups?
24. How could good work habits create occupational opportunities for you?

Safety in Welding

OBJECTIVES

After completing this chapter, the student should be able to

- describe the three classifications of burns and the emergency steps that should be taken to treat each of them

- describe the dangers all three types of light pose to welding and how to protect yourself and others from these dangers

- explain how to use personal protective equipment (PPE) to avoid eye and ear injuries

- avoid dangerous fumes and gases by providing ventilation to the welding area

- explain the purpose of material safety data sheets (MSDSs) and where they can be found

- describe what type of general work clothing should be worn in a welding shop

- describe how to handle, secure, and store compressed gas cylinders

- explain electrical safety practices and list rules for extension cords and portable power tools

- describe four ways to safely lift heavy welded assemblies

KEY TERMS

acetone

acetylene

carbon dioxide extinguisher

confined spaces

dry chemical extinguisher

earmuffs

earplugs

electric shock

electrical ground

electrical resistance

exhaust pickups

flash burn

flash glasses

foam extinguisher

forced ventilation

full face shield

goggles

ground-fault circuit interrupter (GFCI or GFI)

hot work permit

infrared light

material safety data sheet (MSDS)

natural ventilation

safety glasses

type A fire extinguisher

type B fire extinguisher

type C fire extinguisher

type D fire extinguisher

ultraviolet

valve protection cap

ventilation

visible

warning label

welding helmet

INTRODUCTION

Accident prevention is the main intent of this chapter. The safety information included in this text is intended as a guide. There is no substitute for caution and common sense. A safe job is no accident; it takes work to make the job safe. Each person must take personal responsibility for his or her own safety and the safety of others on the job.

Welding, like other heavy industrial jobs, has a number of potential safety hazards. These hazards need not result in personal injury. Learning to work safely to avoid these hazards is as important as learning to be a skilled welder.

Most large welding manufacturers have mandatory safety classes that you must successfully complete before beginning work. These classes may cover company-specific regulations, local and state regulations, as well as federal Occupational Safety and Health Administration (OSHA) regulations. From time to time, additional safety classes may be required. Companies provide all of this safety training to protect you, others, and the business from injury and losses in production resulting from accidents. Violation of company safety policies and practices may result in your being suspended or terminated. There are some dangers that are not covered in this text; you can get specific safety information from your supervisor, the shop safety officer, or other workers.

BURN CLASSIFICATION

Burns are among the most common and painful injuries that occur in the welding shop. Burns can be caused by ultraviolet light rays as well as by contact with hot welding material. The chance of infection is high with burns because of the resulting dead tissue. It is important that all burns receive proper medical treatment to reduce the chance of infection.

Burns are divided into three classifications, depending upon the degree of severity. The three classifications are first-degree, second-degree, and third-degree burns. Whether burns are caused by hot material or by light, they can be avoided if proper clothing and personal protective equipment (PPE) are worn.

First-Degree Burns

First-degree burns occur when the surface of the skin is reddish in color, tender, and painful, but the burn does not involve any broken skin. The first step in treating a first-degree burn is to immediately put the burned area under cold water (not ice water) or apply a cold-water compress (a clean lint-free towel, washcloth, or handkerchief soaked in cold water) until the pain decreases. Then cover the area with sterile bandages or a clean cloth. Do not apply butter or grease. Do not apply any other home remedies or medications without a doctor's recommendation. See Figure 2.1.

Second-Degree Burns

Second-degree burns occur when the surface of the skin is severely damaged, resulting in the formation of blisters and possible breaks in the skin. Again, the most important first step in treating a second-degree burn is to put the area under cold water (not ice water) or apply a cold-water compress until the pain decreases. Gently pat the area dry with a clean lint-free towel, and cover the area with a sterile bandage or clean cloth to prevent infection. Seek medical attention. If the burns are

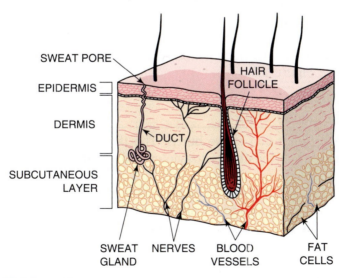

Figure 2.1 First-degree burn
Only the skin surface (epidermis) is affected

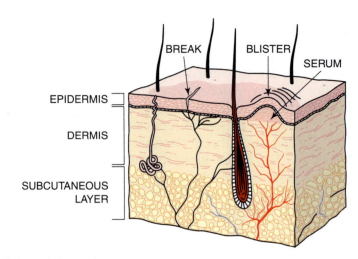

Figure 2.2 Second-degree burn
The epidermal layer is damaged, forming blisters or shallow breaks

around the mouth or nose, or involve singed nasal hair, breathing problems may develop. Do not apply ointments, sprays, antiseptics, or home remedies. Note: in an emergency, any cold liquid you drink—for example, water, cold tea, soft drinks, or milk shake—can be poured on a burn. The purpose is to reduce the skin temperature as quickly as possible to reduce tissue damage. See Figure 2.2.

Third-Degree Burns

Third-degree burns occur when the surface of the skin and possibly the tissue below the skin appear white or charred. Initially, there may be little pain because nerve endings have been destroyed. Do not remove any clothes that are stuck to the burn. Do not put ice water or ice on the burns; this could intensify the shock reaction. Do not apply ointments, sprays, antiseptics, or home remedies to the burns. If the victim is on fire, smother the flames with a blanket, rug, or jacket. Breathing difficulties are common with burns around the face, neck, and mouth; be sure that the victim is breathing. Place a cold cloth or cool (not ice) water on burns of the face, hands, or feet to cool the burned areas. Cover the burned area with thick, sterile, nonfluffy dressings. Call for an ambulance immediately; people with even small third-degree burns need to consult a doctor. See Figure 2.3.

Burns Caused by Light

Some types of light can cause burns. The three types of light are **ultraviolet**, **infrared**, and **visible**. Ultraviolet and infrared light are not visible to the unaided human eye, and they are types of light that can cause burns. During welding, one or more of the three types of light may be present. Arc welding produces all three types of light, but gas welding produces visible and infrared light only.

The light from the welding process can be reflected from walls, ceilings, floors, or any other large surface. This reflected light is as dangerous as the direct welding light. To reduce the danger from reflected light, the

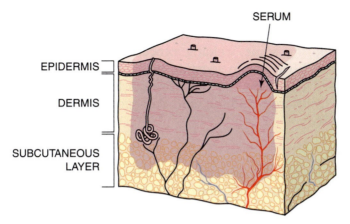

Figure 2.3 Third-degree burn
The epidermis, dermis, and subcutaneous layers of tissue are destroyed

welding area, if possible, should be painted with a flat, dark-colored or black paint. Flat black will reduce the reflected light by absorbing more of it than any other color. When the welding is to be done on a job site, in a large shop, or other area that cannot be painted, weld curtains can be placed to absorb the welding light, Figure 2.4. These special portable welding curtains may be either transparent or opaque. Transparent welding curtains are made of a special high-temperature, flame-resistant plastic that will prevent the harmful light from passing through.

Ultraviolet Light
Ultraviolet light waves are the most dangerous. They can cause first-degree and second-degree burns to a welder's eyes or to any exposed skin. Because a welder cannot see or feel ultraviolet light while being exposed to it, the welder must stay protected when in the area of any of

> ## Caution
> Welding curtains must always be used to protect other workers in the area who might be exposed to the welding light.

Figure 2.4 Portable welding curtains

the arc welding processes. The closer a welder is to the arc and the higher the current, the quicker a burn can occur. The ultraviolet light is so intense during some welding processes that a welder's eyes can receive a **flash burn** within seconds, and the skin can be burned within minutes. Ultraviolet light can pass through loosely woven clothing, thin clothing, light-colored clothing, and damaged or poorly maintained arc welding helmets.

Infrared Light

Infrared light is the light wave that is felt as heat. Although infrared light can cause burns, a person will immediately feel this type of light. Therefore, burns can easily be avoided. When you are welding and you feel infrared light, you are probably being exposed to ultraviolet light at the same time; therefore, protective action should be taken to cover yourself.

Visible Light

Visible light is the light that we see. It is produced in varying quantities and colors during welding. Too much visible light may cause temporary night blindness (poor eyesight under low light levels). Too little visible light may cause eye strain, but visible light is not hazardous.

FACE, EYE, AND EAR PROTECTION

Module 2
Key Indicator 1

Face and Eye Protection

Eye protection must be worn in the shop at all times. Eye protection can be **safety glasses**, with side shields, Figure 2.5, **goggles**, or a **full face shield**. To give better protection when working in brightly lit areas or outdoors, some welders wear **flash glasses**, which are special, lightly tinted, safety glasses. These safety glasses provide protection from both flying debris and reflected light.

Suitable eye protection is important because eye damage caused by excessive exposure to arc light is not noticed. Welding light damage occurs often without warning, just as a sunburn's effect is felt the following day. Therefore, welders must take appropriate precautions in selecting filters or goggles that are suitable for the process being used, Table 2.1. Selecting the correct shade of lens is also important, because both extremes of too light or too dark can cause eye strain. New welders

SIDE SHIELDS

Figure 2.5 Safety glasses with side shields

Table 2.1 Huntsman® Selector Chart

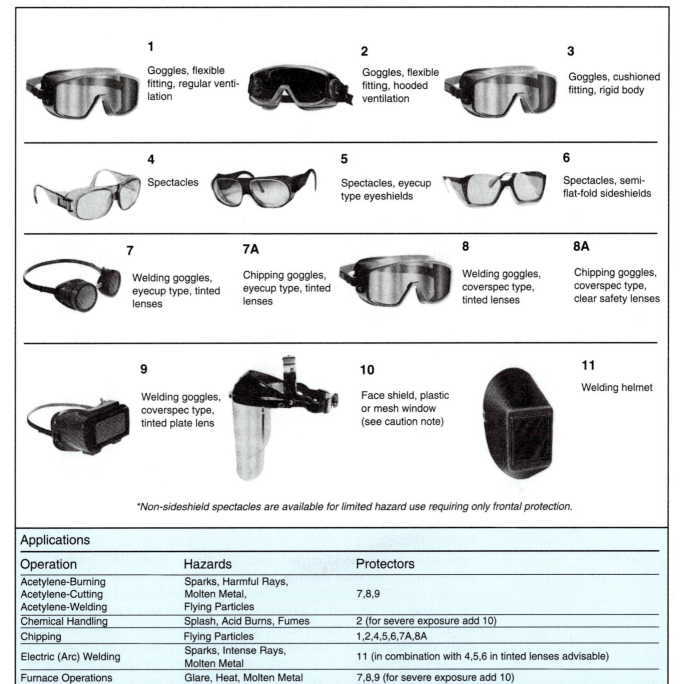

1. Goggles, flexible fitting, regular ventilation
2. Goggles, flexible fitting, hooded ventilation
3. Goggles, cushioned fitting, rigid body
4. Spectacles
5. Spectacles, eyecup type eyeshields
6. Spectacles, semi-flat-fold sideshields
7. Welding goggles, eyecup type, tinted lenses
7A. Chipping goggles, eyecup type, tinted lenses
8. Welding goggles, coverspec type, tinted lenses
8A. Chipping goggles, coverspec type, clear safety lenses
9. Welding goggles, coverspec type, tinted plate lens
10. Face shield, plastic or mesh window (see caution note)
11. Welding helmet

Non-sideshield spectacles are available for limited hazard use requiring only frontal protection.

Applications

Operation	Hazards	Protectors
Acetylene-Burning Acetylene-Cutting Acetylene-Welding	Sparks, Harmful Rays, Molten Metal, Flying Particles	7,8,9
Chemical Handling	Splash, Acid Burns, Fumes	2 (for severe exposure add 10)
Chipping	Flying Particles	1,2,4,5,6,7A,8A
Electric (Arc) Welding	Sparks, Intense Rays, Molten Metal	11 (in combination with 4,5,6 in tinted lenses advisable)
Furnace Operations	Glare, Heat, Molten Metal	7,8,9 (for severe exposure add 10)
Grinding-Light	Flying Particles	1,3,5,6 (for severe exposure add 10)
Grinding-Heavy	Flying Particles	1,3,7A,8A (for severe exposure add 10)
Laboratory	Chemical Splash, Glass Breakage	2 (10 when in combination with 5,6)
Machining	Flying Particles	1,3,5,6 (for severe exposure add 10)
Molten Metals	Heat, Glare, Sparks, Splash	7,8 (10 in combination with 5,6 in tinted lenses)
Spot Welding	Flying Particles, Sparks	1,3,4,5,6 (tinted lenses advisable, for severe exposure add 10)

CAUTION:
Face shields alone do not provide adequate protection. Plastic lenses are advised for protection against molten metal splash.
Contact lenses, of themselves, do not provide eye protection in the industrial sense and shall not be worn in a hazardous environment without appropriate covering safety eyewear.

*Source: Courtesy of Kedman Co., Huntsman Product Division

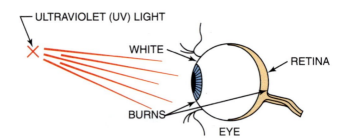

Figure 2.6 Eye damage from ultraviolet light
Ultraviolet light can burn the eye on the white or on the retina

often select too dark a lens, assuming it will give them better protection, but this results in eye strain in the same manner as if they were trying to read in a poorly lit room. Any approved arc welding lenses will filter out the harmful ultraviolet light. Select a lens that lets you see comfortably. At the very least, the welder's eyes must not be strained by excessive glare from the arc.

Ultraviolet light can burn the eye in two ways. This light can injure either the white of the eye or the retina, which is the back of the eye. Burns on the retina are not painful but may cause some loss of eyesight. The whites of the eyes are very sensitive, and burns are very painful, Figure 2.6. The eyes are easily infected because, as with any burn, many cells are killed. These dead cells in the moist environment of the eyes will promote the growth of bacteria that cause infection. When the eye is burned, it feels as though there is something in the eye. Without a professional examination, however, it is impossible to tell whether there is something in the eye. Because there may actually be something in the eye and because of the high risk of infection, home remedies or unprescribed medicines should never be used for eye burns. Any time you receive an eye injury, you should see a doctor.

Welding Helmets

Even with quality **welding helmets**, like those shown in Figure 2.7, the welder must check for potential problems that may occur from accidents or daily use. Small, undetectable leaks of ultraviolet light in an arc welding helmet can cause a welder's eyes to itch or feel sore after a day of welding. To prevent these leaks, make sure the lens gasket is installed correctly, Figure 2.8. The outer and inner clear lenses must be plastic. As shown in Figure 2.9, the lens can be checked for cracks by twisting it between your fingers. Worn or cracked spots on a helmet must be repaired. Tape can be used as a temporary repair until the helmet can be replaced or permanently repaired. Approved safety glasses with side shields should always be worn under your welding hood, even for small jobs or tack welds.

Safety Glasses

Safety glasses with side shields are adequate for general use, but if heavy grinding, chipping, or overhead work is being done, goggles or a full face shield should be worn in addition to safety glasses, Figure 2.10. Safety glasses are best for general protection. They must always be worn under an arc welding helmet and at all times in the shop or on the work site.

Figure 2.7 Typical arc welding helmets
Helmets are used to provide eye and face protection during welding
Source: Courtesy of Hornell, Inc.

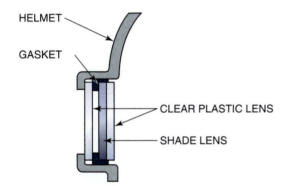

Figure 2.8 Gasket placement
The correct placement of the gasket around the shade lens
is important because it can stop ultraviolet light from
bouncing around the lens assembly

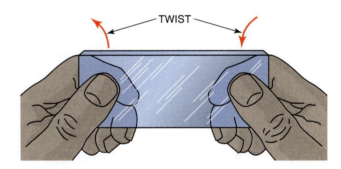

Figure 2.9 Shade lens
To check the shade lens for possible cracks, gently twist it

Ear Protection

The welding environment can be very noisy. The sound level is at times high enough to cause pain and some loss of hearing if the welder's ears are unprotected. Hot sparks can also drop into an open ear, causing severe burns.

Ear protection is available in several forms. One form of protection is **earmuffs** that cover the outer ear completely, Figure 2.11. Another form of protection is **earplugs** that fit into the ear canal, Figure 2.12. Both protect a person's hearing, but only the earmuffs protect the outer ear from burns.

RESPIRATORY PROTECTION

All welding and cutting processes produce undesirable by-products, such as harmful dusts, fogs, fumes, mists, gases, smokes, sprays, or vapors. For your own safety and the safety of others, your primary objective will be to

Caution

Damage to your hearing caused by high sound levels may not be detected until later in life, and the resulting loss in hearing is permanent. Your hearing will not improve with time, and each exposure to high levels of sound will further damage your hearing.

**Module 2
Key Indicator 1**

Figure 2.10 Full face shield

Figure 2.11 Earmuffs
Earmuffs provide complete ear protection and can be worn under a welding helmet
Source: Courtesy of Mine Safety Appliances Company

prevent these contaminants from forming and collecting in the shop atmosphere. This will be accomplished as much as possible by engineering and design control measures such as water tables for cutting, general and local ventilation, thorough cleaning of surface contaminants before starting work, and confinement of the operation to outdoor or open spaces.

Production of welding by-products cannot be avoided. They are created when the temperature of metals and fluxes is raised above the temperatures at which they vaporize or decompose. Most of the by-products are recondensed in the weld. However, some do escape into the atmosphere, producing the haze that occurs in improperly ventilated welding

Figure 2.12 Earplugs
Earplugs are used as protection from noise only
Source: Courtesy of Mine Safety Appliances Company

shops. Some fluxes used in welding electrodes produce fumes that can irritate the welder's nose, throat, and lungs.

When welders must work in an area where effective general controls to remove air-borne welding by-products are not feasible, respirators should be provided by employers when they are necessary to protect the welders' health. The respirators supplied by the welding shop must be suitable for the purpose intended. Where respirators are necessary to protect welders' health or whenever respirators are required by the welding shop, the shop should establish and implement a written respiratory protection program with worksite-specific procedures. Welders are responsible for following the welding shop's established written respiratory protection program. Guidelines for the respiratory protection program are available from the OSHA office in Washington, DC.

Training

Training must be a part of the welding shop's respiratory protection program. This training should include instruction on

- procedures for proper use of respirators, including techniques for putting them on and removing them
- schedules for cleaning, disinfecting, storing, inspecting, repairing, discarding, and performing other types of maintenance of the respiratory protection equipment
- selection of the proper respirators for use in the workplace and any respiratory equipment limitations
- procedures for testing the proper fitting of respirators
- proper use of respirators in both routine and reasonably foreseeable emergency situations
- regular evaluation of the effectiveness of the program

Equipment

All respiratory protection equipment used in a welding shop should be certified by the National Institution for Occupational Safety and Health (NIOSH). The following list indicates some of the types of respiratory protection equipment that may be used:

- Air-purifying respirators have an air-purifying filter, cartridge, or canister that removes specific air contaminants by passing ambient air through the air-purifying element.
- Atmosphere-supplying respirators supply breathing air from a source independent of the ambient atmosphere; this type of respirator includes both supplied-air respirators (SARs) and self-contained breathing apparatus (SCBA) units.
- SCBA units are atmosphere-supplying respirators for which the breathing-air source is designed to be carried by the user.
- SARs, or airline respirators, are atmosphere-supplying respirators for which the source of breathing air is not designed to be carried by the user.
- Demand respirators are atmosphere-supplying respirators that admit breathing air to the facepiece only when a negative pressure is created inside the facepiece by inhalation.

FRESH AIR

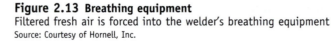

Figure 2.13 Breathing equipment
Filtered fresh air is forced into the welder's breathing equipment
Source: Courtesy of Hornell, Inc.

- Positive-pressure respirators are respirators in which the pressure inside the respiratory inlet covering exceeds the ambient air pressure outside the respirator.
- Powered air-purifying respirators (PAPRs) are air-purifying respirators that use a blower to force the ambient air through air-purifying elements to the inlet covering, Figure 2.13.

Respiratory protection equipment used in many welding applications is of the filtering facepiece (dust mask) type, Figure 2.14. These masks use the negative pressure as you inhale to draw air through a filter that is an integral part of the facepiece. In areas of severe contamination, you can use a hood-type respirator that covers your head and neck and may even cover portions of your shoulders and torso.

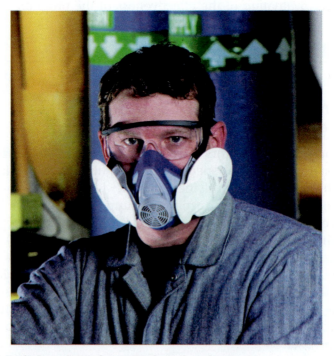

Figure 2.14 Typical respirator for contaminated environments
The filters can be selected for specific types of contaminants
Source: Courtesy of Mine Safety Appliances Company

Fume Sources

Some materials used as paints, coating, or plating on metals to prevent rust or corrosion can cause respiratory problems. Other potentially hazardous materials might be used as alloys in metals to give them special properties.

Before it is welded or cut, any metal that has been painted or has any grease, oil, or chemicals on its surface must be thoroughly cleaned. This cleaning can be done by grinding, sand blasting, or applying an approved solvent. It may not be possible to clean metals that are plated or alloyed before welding or cutting begins.

Most paints containing lead have been removed from the market. But some industries still use these lead-based paints, as in marine or shipping applications. Often old machinery and farm equipment surfaces still have lead-based paint coatings. Solder often contains lead alloys. The welding and cutting of lead-bearing alloys or metals whose surfaces have been painted with lead-based paint can generate lead oxide fumes. Inhalation and ingestion of lead oxide fumes and other lead compounds will cause lead poisoning. Symptoms include a metallic taste in the mouth, loss of appetite, nausea, abdominal cramps, and insomnia. In time, anemia and a general weakness, chiefly in the muscles of the wrists, develop.

Cadmium and zinc are plating materials used to prevent iron or steel from rusting. Cadmium is often used on bolts, nuts, hinges, and other hardware, and it gives the surface a yellowish-gold appearance. Acute exposure to high concentrations of cadmium fumes can produce severe lung irritation. Long-term exposure to low levels of cadmium in air can result in emphysema (a disease affecting the lungs' ability to absorb oxygen) and can damage the kidneys.

Zinc, often in the form of galvanizing, may be found on pipes, sheet metal, bolts, nuts, and many other types of hardware. Zinc plating that is thin may appear as a shiny, metallic patchwork or crystal pattern; thicker, hot-dipped zinc appears rough and may look dull. Zinc is used in large quantities in the manufacture of brass and is found in brazing rods. Inhalation of zinc oxide fumes can occur when welding or cutting on these materials. Exposure to these fumes is known to cause metal fume fever, whose symptoms are very similar to those of common influenza.

Some concern has been expressed about the possibility of lung cancer being caused by some of the chromium compounds that are produced in the welding of stainless steels. OSHA released its final Hexavalent Chromium Standard in February 2006.

Chromium hexavalent (CrVI) compounds, often called hexavalent chromium, Hex Chrome or Chrome 6, exist in several forms. Industrial uses of hexavalent chromium compounds include chromate pigments in dyes, paints, inks, and plastics; chromates added as anticorrosive agents to paints, primers, and other surface coatings; and chromic acid electroplated onto metal parts to provide a decorative or protective coating. Hex chrome can also be formed when performing "hot work" such as welding on stainless steel or melting chromium metal. In these situations, the chromium is not originally hexavalent, but the extreme temperatures involved in the process result in oxidation that can convert the chromium to a hexavalent state. Employers working with this material must

Caution

Extreme care must be taken to avoid the fumes produced when welding is done on dirty or used metal. Any chemicals that are on the metal will become mixed with the welding fumes, a combination that can be extremely hazardous. All metal must be cleaned before welding to avoid this potential problem.

establish a safety program that measures Hex Chrome levels and provides adequate respiratory protection for workers. Further respirator information and guidance can be found in OSHA 1910.134.

Rather than take chances, welders should recognize that fumes of any type, regardless of their source, should not be inhaled. The best way to avoid problems is to provide adequate **ventilation**. If this is not possible, breathing protection should be used. Protective devices for use in poorly ventilated or confined areas were shown in Figures 2.13 and 2.14.

Vapor Sources

Potentially dangerous gases also can be present in a welding shop. Proper ventilation or respirators are necessary when welding in **confined spaces**, regardless of the welding process being used. Ozone is a gas that is produced by the ultraviolet radiation in the air in the vicinity of arc welding and cutting operations. Ozone is irritating to all mucous membranes, with excessive exposure producing pulmonary edema, or fluid on the lung, making it difficult to breathe. Severe cases of pulmonary edema may require immediate care. Other effects of exposure to ozone include headache, chest pain, and dryness in the respiratory tract.

Phosgene is formed when ultraviolet radiation decomposes chlorinated hydrocarbons. Chlorinated hydrocarbon fumes can come from solvents such as those used for degreasing metals and from refrigerants from air-conditioning systems. They decompose in the arc to produce a potentially dangerous chlorine acid compound. This compound reacts with the moisture in the lungs to produce hydrogen chloride, which in turn destroys lung tissue. For this reason, any use of chlorinated solvents should be well away from welding operations in which ultraviolet radiation or intense heat is generated. Any welding or cutting on refrigeration or air-conditioning piping must be done only after the refrigerant has been completely removed in accordance with Environmental Protection Agency (EPA) regulations.

Care must be taken to avoid the infiltration of any fumes or gases, including argon or carbon dioxide, into a confined working space, such as when welding in tanks. The collection of some fumes and gases in a work area can go unnoticed by the welders. Concentrated fumes or gases can cause a fire or explosion if they are flammable, asphyxiation if they replace the oxygen in the air, or death if they are toxic.

Despite these fumes and other potential hazards in welding shops, welders have been found to be as healthy as workers employed in other industrial occupations.

Module 2
Key Indicator 3

VENTILATION

The welding area should be well ventilated. Excessive fumes, ozone, or smoke may collect in the welding area; ventilation should be provided for their removal. **Natural ventilation** is best, but forced ventilation may be required. Areas that have 10,000 cu ft (283 m^3) or more per welder or that have ceilings 16 ft (4.9 m) high or higher, Figure 2.15, may not require forced ventilation unless fumes or smoke begin to collect.

Small shops or shops with large numbers of welders require forced ventilation. **Forced ventilation** can be general or localized using fixed

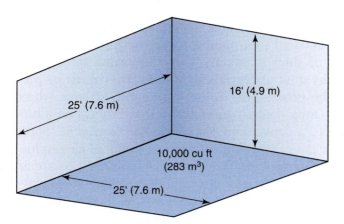

Figure 2.15 Forced ventilation
A room with a ceiling 16 ft (4.9 m) high may not require forced ventilation for one welder

Figure 2.16 An exhaust pickup
Source: Courtesy of Larry Jeffus

or flexible **exhaust pickups**, Figure 2.16. General room ventilation must be at a rate of 2000 cu ft (56 m^3) or more per person welding. Localized exhaust pickups must have a draft strong enough to provide 100 linear feet (30.5 m) per minute air velocity pulling welding fumes away from the welder. Local, state, or federal regulations may require that welding fumes be treated to remove hazardous components before they are released into the atmosphere.

Any system of ventilation should draw the fumes or smoke away before they rise past the level of the welder's face.

Forced ventilation is always required when welding on metals that contain zinc, lead, beryllium, cadmium, mercury, copper, austenitic manganese, or other materials that give off dangerous fumes.

CONFINED SPACES

Module 2
Key Indicator 5

Work in confined spaces requires special precautions. Owners, contractors, and workers all need to be familiar with written confined-space working procedures. Often work is supervised by a specially trained person. Asphyxiation (lack of breathing air) causes unconsciousness and even death without warning. Confined-space atmospheres that are oxygen enriched will greatly intensify combustion, which can cause rapid, severe, and often fatal burns.

Confined spaces must also be tested for toxic or flammable gases, dusts, and vapors, and for excessive or inadequate oxygen levels before entering and during operations. These same precautions apply to areas such as tank bottoms, pits, low areas, and spaces near floors when heavier-than-air gases and vapors are present, and to high areas such as tank tops and near ceilings when lighter-than-air gases are present. Gases such as carbon dioxide, argon, and propane are heavier than air; gases such as natural gas and helium are lighter than air. If possible, a continuous monitoring system with audible alarms should be used for confined-space work. Further precautions can be found in ANSI Z117.1 and OSHA 29 CFR 1910.146.

Caution

If you feel you have been injured while using a product, you should, if possible, take the material's MSDS with you when you seek medical treatment.

MATERIAL SAFETY DATA SHEETS (MSDSs)

All manufacturers of potentially hazardous materials must provide to the users of their products detailed information regarding possible hazards resulting from the use of their products. These **material safety data sheets** are often known by their abbreviation, **MSDSs**. They must be provided to anyone using the product or anyone working in an area where the products are in use. Often companies will post these sheets on a bulletin board or put them in a convenient place near the work area. Some states have right-to-know laws that require specific training of all employees who handle or work in areas with hazardous materials. MSDSs are an important part of an overall hazard communication program. Other types of hazard communications are product labeling, special danger symbols (see Figure 2.41), flashing lights, and audible signals.

WASTE MATERIAL DISPOSAL

Welding shops generate a lot of waste materials. Much of the waste is scrap metal. All scrap metal, including electrode stubs, can easily be recycled. Recycling metal is good for the environment and can be a source of revenue for the welding shop.

Other forms of waste, such as burned flux, cleaning solvents, and dust collected in shop air-filtration systems, may be considered hazardous materials. Check with the material manufacturer or an environmental consultant to determine whether any waste material is considered hazardous. Throwing hazardous waste material into the trash, pouring it on the ground, or dumping it down the drain is illegal. Before you dispose of any welding shop waste that is considered hazardous, you must first consult local, state, and/or federal regulations. Protecting our environment from pollution is everyone's responsibility.

GENERAL WORK CLOTHING

Special protective clothing cannot be worn at all times. It is, therefore, important to choose general work clothing that will minimize the possibility of getting burned because of the high temperature and amount of hot sparks, metal, and slag produced during welding, cutting, or brazing.

Work clothing must also stop ultraviolet light from passing through it. This is accomplished if the material chosen is dark, thick, and tightly woven. The best choice is 100% wool, but it is difficult to find. Another good choice is 100% cotton clothing, the most popular fabric used.

You must avoid wearing synthetic materials, including nylon, rayon, and polyester. They can easily melt or catch fire. Some synthetics produce a hot sticky residue that can make burns more severe. Others may produce poisonous gases.

The following are some guidelines for selecting work clothing:

- Shirts must be long-sleeved to protect the arms, have a high-buttoned collar to protect the neck, Figure 2.17, be long enough to tuck into the pants to protect the waist, and have flaps on the pockets to keep sparks out (or have no pockets).

Figure 2.17 Clothing
The top button of a welder's shirt should always be buttoned to avoid severe burns to the neck
Source: Courtesy of Larry Jeffus

STEEL

Figure 2.18 Footwear
Safety boots with steel toes are required by many welding shops

- Pants must have legs long enough to cover the tops of the boots and must be without cuffs that would catch sparks.
- Boots must have high tops to keep out sparks, have steel toes to prevent crushed toes, Figure 2.18, and have smooth tops to prevent sparks from being trapped in seams.
- Caps should be thick enough to prevent sparks from burning the top of a welder's head.

All clothing must be free of frayed edges and holes. The clothing must be relatively tight-fitting to prevent excessive folds or wrinkles that might trap sparks.

Some welding clothes have pockets on the inside to prevent the pockets from collecting sparks. However, it is not safe to carry butane lighters or matches in these or any pockets while welding. Lighters and matches can easily catch on fire or explode if they are subjected to the heat and sparks of welding.

SPECIAL PROTECTIVE CLOTHING

General work clothing is worn by each person in the shop. In addition to this clothing, extra protection is needed for each person who is in direct contact with hot materials. Leather is often the best material to use, as it is lightweight and flexible, resists burning, and is readily available. Synthetic insulating materials are also available. Ready-to-wear leather protection includes capes, jackets, aprons, sleeves, gloves, caps, pants, knee pads, and spats.

Hand Protection

All-leather, gauntlet-type gloves should be worn when doing any welding, Figure 2.19. Gauntlet gloves that have a cloth liner for insulation are best for hot work. Noninsulated gloves will give greater flexibility for fine work. Some leather gloves are available with a canvas gauntlet top; they should be used for light work only.

> **Caution**
>
> There is no safe place to carry butane lighters or matches while welding or cutting. They can catch fire or explode if subjected to welding heat or sparks. Butane lighters may explode with the force of one-fourth of a stick of dynamite. Matches can erupt into a ball of fire. Both butane lighters and matches must always be removed from the welder's pockets and placed a safe distance away before any work is started.

Module 2
Key Indicator 4

Figure 2.19 All-leather, gauntlet-type welding gloves
Source: Courtesy of Larry Jeffus

Figure 2.20 Soft leather gloves
For welding that requires a great deal of manual dexterity, soft leather gloves can be worn
Source: Courtesy of Larry Jeffus

When a great deal of manual dexterity is required for gas tungsten arc welding, brazing, soldering, oxyfuel gas welding, and other delicate processes, soft leather gloves may be used, Figure 2.20. All-cotton gloves are sometimes used when doing very light welding.

Body Protection

Full-leather jackets and capes will protect a welder's shoulders, arms, and chest, Figure 2.21. A jacket, unlike a cape, protects a welder's back and complete chest. A cape is open and much cooler but offers less protection. The cape can be used with a bib apron to provide some additional protection while leaving the back cooler. Either the full jacket or the cape with a bib apron should be worn for any out-of-position work.

Figure 2.21 Full-leather jacket
Source: Courtesy of Larry Jeffus

Waist and Lap Protection

Bib aprons or full aprons will protect a welder's lap. Welders will especially need to protect their laps if they squat or sit while working and when they bend over or lean against a table.

Arm Protection

For some vertical welding, a full or half sleeve can protect a person's arm, Figure 2.22. The sleeves work best if the work level is not above the welder's chest. Work levels higher than this usually require a jacket or cape to keep sparks off the welder's shoulders.

Leg and Foot Protection

When heavy cutting or welding is being done and a large number of sparks are falling, leather pants and spats should be used to protect the welder's legs and feet. If the weather is hot and full-leather pants are uncomfortable, leather aprons with leggings are available. Leggings can be strapped to the legs, leaving the back open. Spats will prevent sparks from burning through the front of lace-up boots.

Figure 2.22 Full-leather sleeve

Module 2
Key Indicator 7

HANDLING AND STORING CYLINDERS

Oxygen and fuel-gas cylinders or other flammable materials must be stored separately. The two storage areas must be separated by 20 ft (6.1 m) or by a wall 5 ft (1.5 m) high with at least a half-hour burn rating, Figure 2.23. The purpose of the distance or wall is to keep the heat of a small fire from causing the compressed gas cylinder safety valve to release. If the safety valve were to release the oxygen, a small fire would become a raging inferno.

Inert-gas cylinders may be stored separately or with oxygen cylinders.

Empty cylinders must be stored separately or with full cylinders of the same gas in the same room or area. All cylinders must be stored vertically and have the protective caps screwed on firmly.

Securing Gas Cylinders

Cylinders must be secured with a chain or other device so that they cannot be knocked over accidentally. Cylinders attached to a manifold or stored in a special room used only for cylinder storage should be chained.

Storage Areas

Cylinder storage areas must be located away from halls, stairwells, and exits so that in case of an emergency they will not block an escape route. Storage areas should also be located away from heat, radiators, furnaces, and welding sparks. The location of storage areas should be

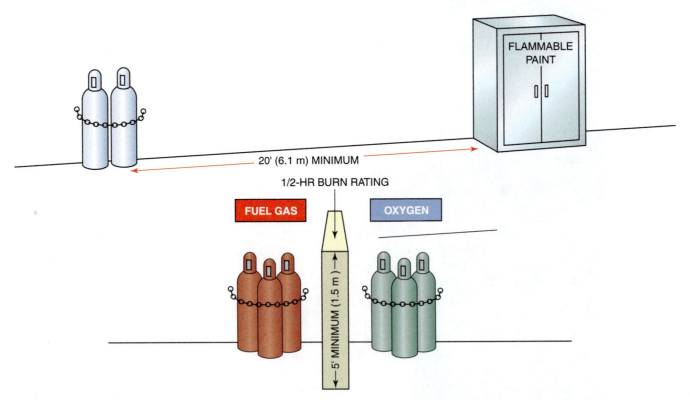

Figure 2.23 Fuel-gas cylinder storage
Stored fuel-gas cylinders should be separated from any flammable material by a minimum distance of 20 ft (6.1 m) or a wall 5 ft (1.5 m) high

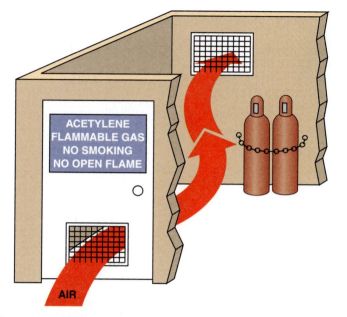

Figure 2.24 Acetylene storage
A separate room used to store acetylene must have good ventilation and should have a warning sign posted on the door

such that unauthorized people cannot tamper with the cylinders. A warning sign that reads "Danger—No Smoking, Matches, or Open Lights," or with similar wording, must be posted in the storage area, Figure 2.24.

Cylinders with Valve Protection Caps

Cylinders equipped with a **valve protection cap** must have the cap in place unless the cylinder is in use. The protection cap prevents the valve from being broken off if the cylinder is knocked over. If the valve of a full high-pressure cylinder (argon, oxygen, carbon dioxide, mixed gases) is broken off, the cylinder can fly around the shop like a missile if it has not been secured properly.

Never lift a cylinder by the safety cap or the valve. The valve can easily break off or be damaged.

When cylinders are moved, the valve protection cap must be on, especially if the cylinders are mounted on a truck or trailer for out-of-shop work. The cylinders must never be dropped or handled roughly.

General Precautions

Use warm (not boiling) water to loosen cylinders that are frozen to the ground. Any cylinder that leaks, has a bad valve, or has damaged threads must be identified and reported to the supplier. A piece of soapstone can be used to write the problem on the cylinder. If the leak cannot be stopped by closing the cylinder valve, the cylinder should be moved to a vacant lot or an open area. The pressure should then be slowly released after posting a warning sign, Figure 2.25.

Acetylene cylinders that have been lying on their sides must stand upright for four hours or more before they are used. The acetylene is absorbed in **acetone**, and the acetone is absorbed in a filler. The filler

Figure 2.25 Fuel-gas cylinder leaks
Move a leaking fuel-gas cylinder out of the building or any work area. The pressure should slowly be released after posting a warning of the danger

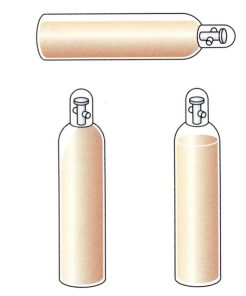

Figure 2.26 Acetone settling
The acetone in an acetylene cylinder must have time to settle before the cylinder can be used safely

does not allow the liquid to settle back away from the valve very quickly, Figure 2.26. If the cylinder has been in a horizontal position, using it too soon after it is placed in a vertical position may draw acetone out of the cylinder. Acetone lowers the flame temperature and can damage regulator or torch valve settings.

Module 2
Key Indicator 4

FIRE PROTECTION

Fire is a constant danger to the welder. Welding is considered to be "hot work" by the National Association of Fire Prevention. When performing welding outside of a shop, the welder may be required to obtain a **hot work permit** from the local fire marshal. Most hot work permits provide a checklist of items that must be on the job site or be inspected before hot work begins. Floors must be swept and floor drains must be checked with an explosimeter tool, which detects and measures concentrations of combustible gases or vapors in the air. Walls, floor openings, and any ductwork in the area must also be inspected.

Hotwork permits (Figure 2.27) usually require at least two signatures, one from the area supervisor and one from the fire watcher. In particularly hazardous situations, the local fire marshal should be notified before and after hot work begins. Fire watchers must remain on duty at least 30 minutes after hot work operations are finished.

During times when burn bans are in effect, performing hot work without a permit can be a violation of state or local laws. Even with a permit, the welder can be held liable for any damage resulting from a fire caused by his or her welding. Highly combustible materials should be 35 ft (10.7 m) or more away from any welding. When it is necessary to weld within 35 ft (10.7 m) of combustible materials, when sparks can reach materials farther than 35 ft (10.7 m) away, or when anything more than a minor fire might start, a fire watch is required.

HOT WORK PERMIT
(Sample)

Instructions
1. Evaluate if the hot work can be avoided or completed in a safer way.
2. Follow precautions listed to the right.
3. Complete permit and display in area where work is being done. [Type text]

Hot work done by:

Name of employee————————————

Name of Contractor:————————————

Job Date _____ Job No. _____

Location (Building & Floor)

————————————————————

Type of Job:

————————————————————

Name of Person Performing Work:

————————————————————

I verify that the above location has been examined, that precautions on the checklist have been take to prevent fire, and that permission is authorized to perform the work.

Signed by:
Safety Supervisor _____
              (signature)
Area Supervisor:_____
             (signature)
Fire Watch:_____
           (signature)
Permit Expires_____ _____
     (Date)   (Time)

Recommended Precautions Checklist

☐ Available sprinklers, hose streams and extinguishers are in service and good repair.
☐ Hot work Equipment is in good repair

Requirements within 35 FT. (11M.) of Work
☐ Combustible floors wet down, covered with damp sand or other shields
☐ Floors swept clean of combustibles.
☐ Explosive atmosphere in area eliminated
☐ Flammable liquids, dust, lint, and oily deposits removed.
☐ All wall and floor openings covered to prevent sparks from passing thru.
☐ Ducts and conveyor systems that might carry sparks to distant combustibles are protected or shut down.
☐ Fire-resistant covers suspended beneath work.
☐ Other combustible materials removed or covered with fire-resistant covers.

Work on Walls and Ceilings
☐ Combustibles on the other side of the wall moved away.
☐ Construction is noncombustible and without combustible coverings.

Work on Enclosed Equipment
☐ Equipment cleaned of all combustibles
☐ Equipment purged of all flammable vapors

Fire Watch and Work Area Monitoring
☐ Fire watch should be provided during, and for at least 30 minutes after work is completed.
☐ Fire watch trained on facility alarms and equipped with fire extinguishers
☐ Fire watch may be required above, below and in adjacent areas
☐ Other precautions taken_____

Figure 2.27 Hot work permit

Fire Watch

A fire watch can be provided by any person who knows how to sound the alarm and use a fire extinguisher. The fire extinguisher must be the type required to put out a fire on the type of combustible materials near the welding. Combustible materials that cannot be removed from the welding area should be soaked with water or covered with sand or noncombustible insulating blankets, whichever is available.

Fire Extinguishers

The four types of fire extinguishers are type A, type B, type C, and type D. Each type is designed to put out fires on certain types of materials. Some fire extinguishers can be used on more than one type of fire. However, using the wrong type of fire extinguisher can be dangerous, either causing the fire to spread, causing electric shock, or causing an explosion.

Type A Extinguishers

Type A fire extinguishers are used for combustible solids (articles that burn), such as paper, wood, and cloth. The symbol for a type A extinguisher is a green triangle with the letter *A* in the center, Figure 2.28.

Type B Extinguishers

Type B fire extinguishers are used for combustible liquids, such as oil, gas, and paint thinner. The symbol for a type B extinguisher is a red square with the letter *B* in the center, Figure 2.29.

Type C Extinguishers

Type C fire extinguishers are used for electrical fires. For example, they are used on fires involving motors, fuse boxes, and welding machines. The symbol for a type C extinguisher is a blue circle with the letter *C* in the center, Figure 2.30.

Type D Extinguishers

Type D fire extinguishers are used on fires involving combustible metals, such as zinc, magnesium, and titanium. The symbol for a type D extinguisher is a yellow star with the letter *D* in the center, Figure 2.31.

Location of Fire Extinguishers

Fire extinguishers should be suitable for use on the types of combustible materials located nearby, Figure 2.32. The extinguishers should be placed so that they can be easily removed without reaching over combustible material. They should also be placed at a level low enough to be easily lifted off the mounting, Figure 2.33. The location of fire extinguishers should be marked with red paint and signs, high enough so that their location can be seen from a distance over people and equipment. The location should also be marked near the floor so that they can be found even if a room is full of smoke, Figure 2.34.

Figure 2.28 Type A fire extinguisher symbol

Figure 2.29 Type B fire extinguisher symbol

Figure 2.30 Type C fire extinguisher symbol

Figure 2.31 Type D fire extinguisher symbol

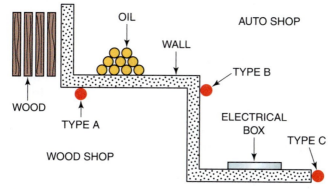

Figure 2.32 Choice of fire extinguisher
The type of fire extinguisher provided should be appropriate for the materials being used in the surrounding area

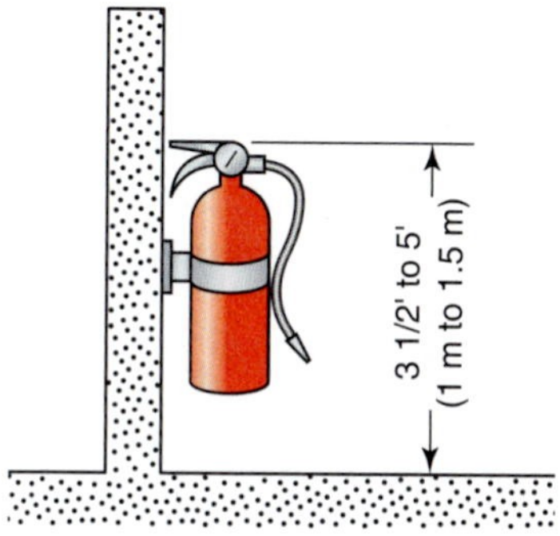

Figure 2.33 Mounting fire extinguishers
Mount the fire extinguisher so that it can be lifted easily in an emergency

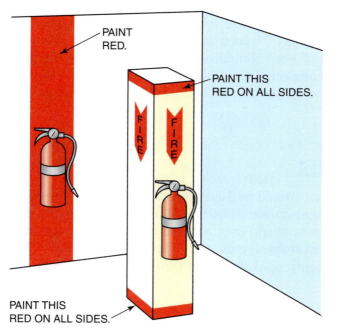

Figure 2.34 Locating fire extinguishers
The locations of fire extinguishers should be marked so they can be found easily in an emergency

Use of Fire Extinguishers

A fire extinguisher works by breaking the "fire triangle" of heat, fuel, and oxygen. Most extinguishers both cool the fire and remove the oxygen. They use a variety of materials to extinguish the fire. The majority of fire extinguishers found in welding shops uses foam, carbon dioxide, a pump tank, or dry chemicals.

When using a **foam extinguisher**, do not spray the stream directly into the burning liquid. Allow the foam to fall lightly on the base of the fire.

When using a **carbon dioxide extinguisher**, direct the discharge as close to the fire as possible, first at the edge of the flames and gradually to the center.

When using a **dry chemical extinguisher**, direct the extinguisher at the base of the flames. In the case of type A fires, follow up by directing the dry chemicals at the remaining material still burning. The extinguisher must be directed at the base of the fire, where the fuel is located, Figure 2.35.

Figure 2.35 Using fire extinguishers
Point the extinguisher at the material burning, not at the flames

EQUIPMENT MAINTENANCE

A routine schedule for planned maintenance of equipment will aid in detecting potential problems such as leaking coolant, loose wires, poor grounds, frayed insulation, or split hoses. Fixing small problems in time can prevent the loss of valuable time due to equipment breakdown or injury.

Any maintenance beyond routine external maintenance should be referred to a trained service technician. In most areas, it is against the law for anyone but a licensed electrician to work on arc welders and for anyone but a factory-trained repair technician to work on regulators. Electric shock and exploding regulators can cause serious injury or death.

Hoses

Hoses must be used only for the gas or liquid for which they were designed. Green hoses are to be used only for oxygen, and red hoses are to be used only for acetylene or other fuel gases. Using unnecessarily long lengths of hose should be avoided. Never use oil, grease, or other pipe-fitting compounds on any joints. Hoses should also be kept out of the direct line of sparks. Any leaking or bad joints in gas hoses must be repaired.

Module 1
Key Indicator 2

WORK AREA

The work area should be kept picked up and swept clean. Collections of steel, welding electrode stubs, wire, hoses, and cables are difficult to work around and easy to trip over. An electrode caddy can be used to hold the electrodes and stubs, Figure 2.36. Hooks can be made to hold hoses and cables, and scrap steel should be thrown into scrap bins.

Figure 2.36 Electrode caddy
An easy-to-build electrode caddy can be used to hold both electrodes and stubs
Source: Courtesy of Larry Jeffus

Arc welding areas should be painted with a flat dark-colored or black finish to absorb as much of the ultraviolet light as possible. Portable screens should be used whenever arc welding is to be done outside of a welding booth.

If a piece of hot metal is going to be left unattended, write the word *hot* on it before leaving. This procedure can also be used to warn people of hot tables, vises, firebricks, and tools.

HAND TOOLS

Hand tools are used by the welder to assemble and disassemble parts for welding as well as to perform routine equipment maintenance.

The adjustable wrench is the most popular tool used by the welder. The wrench should be adjusted tightly on the nut and pushed so that most of the force is on the fixed jaw, Figure 2.37. When a wrench is being used on a tight bolt or nut, the wrench should be pushed with the palm of an open hand or pulled to prevent injuring the hand. If a nut or bolt is too tight to be loosened with a wrench, obtain a longer wrench. A cheater bar should not be used.

The fewer points a box end wrench or socket has, the stronger it is and the less likely it is to slip or damage the nut or bolt, Figure 2.38.

Striking a hammer directly against a hard surface such as another hammer face or anvil may result in chips flying off and causing injury.

The mushroomed heads of chisels, punches, and the faces of hammers should be ground off, Figure 2.39. Chisels and punches that are going to be hit harder than a slight tap should be held in a chisel holder or with pliers to eliminate the danger of injuring your hand.

PUSH IN THIS DIRECTION ONLY.

Figure 2.37 Adjustable wrench
The adjustable wrench is stronger when used in the direction indicated

Figure 2.38 Wrench slippage
The fewer the points, the less likely the wrench is to slip
Source: Courtesy of Larry Jeffus

CORRECTLY GROUND

MUSHROOMED

Figure 2.39 Punches and chisels
Any mushroomed heads on punches or chisels must be ground off
Source: Courtesy of Larry Jeffus

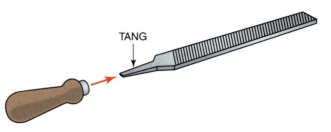

TANG

Figure 2.40 Files
To protect yourself from the sharp tang of a file, always use a handle
on the file

A handle should be placed on the tang of a file to avoid injuring your hand, Figure 2.40. A file can be kept free of chips by rubbing a piece of soapstone on it before it is used.

It is important to remember to use the correct tool for the job. Do not try to force a tool to do a job it was not designed to do.

ELECTRICAL SAFETY

Module 2
Key Indicator 2

Injuries and even death can be caused by **electric shock** unless proper precautions are taken. Most welding and cutting operations involve electrical equipment in addition to the arc welding power supply. Grinders, electric motors on automatic cutting machines, and drills are examples. Most electrical equipment in a welding shop is powered by AC sources with input voltages ranging from 115 volts to 460 volts. However, fatalities have occurred when people were working with equipment operating at less than 80 volts. Most electric shocks in the welding industry are a result of accidental contact with bare or poorly insulated conductors. **Electrical resistance** is lowered in the presence of water or moisture, so welders must take special precautions when working under damp or wet conditions, including perspiration. Figure 2.41 shows a typical **warning label** attached to welding equipment.

The workpiece being welded and the frame or chassis of all electrically powered machines must be connected to a good **electrical ground**. The work lead from the welding power supply is not an electrical ground and is not sufficient. A separate lead is required to ground the workpiece and power source.

Electrical connections must be tight. Terminals for welding leads and power cables must be shielded from accidental contact by personnel or by metal objects. Cables must be used within their current-carrying and duty-cycle capacities; otherwise, they will overheat and break down the insulation rapidly. Cable connectors for lengthening leads must be insulated.

Cables must be checked periodically to be sure that they have not become frayed; if they have, they must be replaced immediately.

Welders should not allow the metal parts of electrodes or electrode holders to touch their skin or wet coverings on their bodies. Dry gloves in good condition must always be worn. Rubber-soled shoes are advisable. Precautions must be taken against accidental contact with bare conducting surfaces when the welder is required to work in cramped

Caution

Welding cables must never be spliced within 10 ft (3 m) of the electrode holder.

Welding Safety Checklist

Hazard	Factors to Consider	Precaution Summary
Electric shock can kill 	• **Wetness** • **Welder in or on workpiece** • **Confined space** • **Electrode holder and cable insulation**	• Insulate welder from workpiece and ground using *dry* insulation. Rubber mat or dry wood. • Wear *dry, hole-free* gloves. (Change as necessary to keep dry.) • Do not touch electrically "hot" parts or electrode with bare skin or wet clothing. • If wet area and welder cannot be insulated from workpiece with dry insulation, use a semiautomatic, constant-voltage welder or stick welder with voltage reducing device. • Keep electrode holder and cable insulation in good condition. Do not use if insulation damaged or missing.
Fumes and gases can be dangerous 	• **Confined area** • **Positioning of welder's head** • **Lack of general ventilation** • **Electrode types, i.e., manganese, chromium, etc. See MSDS** • **Base metal coatings, galvanize, paint**	• Use ventilation or exhaust to keep air breathing zone clear, comfortable. • Use helmet and positioning of head to minimize fume in breathing zone. • Read warnings on electrode container and material safety data sheet (MSDS) for electrode. • Provide additional ventilation/exhaust where special ventilation requirements exist. • Use special care when welding in a confined area. • Do not weld unless ventilation is adequate.
Welding sparks can cause fire or explosion 	• **Containers which have held combustibles** • **Flammable materials**	• Do not weld on containers which have held combustible materials (unless strict AWS F4.1 procedures are followed). Check before welding. • Remove flammable materials from welding area or shield from sparks, heat. • Keep a fire watch in area during and after welding. • Keep a fire extinguisher in the welding area. • Wear fire retardant clothing and hat. Use earplugs when welding overhead.
Arc rays can burn eyes and skin 	• **Process: gas-shielded arc most severe**	• Select a filter lens which is comfortable for you while welding. • Always use helmet when welding. • Provide non-flammable shielding to protect others. • Wear clothing which protects skin while welding.
Confined space 	• **Metal enclosure** • **Wetness** • **Restricted entry** • **Heavier than air gas** • **Welder inside or on workpiece**	• Carefully evaluate adequacy of ventilation especially where electrode requires special ventilation or where gas may displace breathing air. • If basic electric shock precautions cannot be followed to insulate welder from work and electrode, use semiautomatic, constant-voltage equipment with cold electrode or stick welder with voltage reducing device. • Provide welder helper and method of welder retrieval from outside enclosure.
General work area hazards	• **Cluttered area**	• Keep cables, materials, tools neatly organized.
	• **Indirect work (welding ground) connection**	• Connect work cable as close as possible to area where welding is being performed. Do *not* allow alternate circuits through scaffold cables, hoist chains, ground leads.
	• **Electrical equipment**	• Use only double insulated or properly grounded equipment. • Always disconnect power to equipment before servicing.
	• **Engine-driven equipment**	• Use in only open, well ventilated areas. • Keep enclosure complete and guards in place. • See Lincoln service shop if guards are missing. • Refuel with engine off. • If using auxiliary power, OSHA may require GFI protection or assured grounding program (or isolated windings if less than 5KW).
	• **Gas cylinders**	• Never touch cylinder with the electrode. • Never lift a machine with cylinder attached. • Keep cylinder upright and chained to support.

Figure 2.41 Warning labels
Note the warning information contained on this typical label, which may be attached to welding equipment or printed in the equipment owner's manual

Source: Courtesy of the Lincoln Electric Company

kneeling, sitting, or lying positions. Insulated mats or dry wooden boards are desirable protection from being grounded.

Welding circuits must be turned off when the work station is left unattended. When working on the welder, welding leads, electrode holder, torches, the wire feeder, guns, or other parts, the main power supply must be turned off and locked or tagged to prevent electrocution. Since the electrode holder is energized when changing coated electrodes, the welder must wear dry gloves.

ELECTRICAL SAFETY SYSTEMS

For protection from electric shock, standard portable power tools are built with either of two safety systems: external grounding or double insulation.

A tool with external grounding has a wire that runs from the housing through the power cord to a third prong on the power plug. When this third prong is connected to a grounded, three-hole electrical outlet, the grounding wire will carry any current that leaks past the electrical insulation of the tool away from the user and into the ground. In most electrical systems, the three-prong plug fits into a three-prong, grounded receptacle. If the tool is operated at less than 150 volts, it has a plug like that shown in Figure 2.42A. If it is used at 150 to 250 volts, it has a plug like that shown in Figure 2.42B. In either type, the green (or green and yellow) conductor in the tool cord is the grounding wire. Never connect the grounding wire to a power terminal.

A double-insulated tool has an extra layer of electrical insulation that eliminates the need for a three-prong plug and grounded outlet. Double-insulated tools do not require grounding and, therefore, have a two-prong plug. In addition, double-insulated tools are always labeled as such on their nameplate or case, Figure 2.43.

VOLTAGE WARNINGS

Before connecting a tool to a power supply, be sure the voltage supplied is the same as that specified on the nameplate of the tool. A power source with a voltage greater than that specified for the tool can lead to serious injury to the user as well as damage to the tool. Using a power source with a voltage lower than the rating on the nameplate is harmful to the motor.

Tool nameplates also bear a number with the abbreviation *amps* (for amperes, a measure of electric current). This refers to the current-drawing requirement of the tool. The higher the input current, the more powerful the motor.

EXTENSION CORDS

Module 2
Key Indicator 7

If the power source is some distance from the work area or if the portable tool is equipped with a stub power cord, an extension cord must be used. When extension cords are used on portable power tools, the conductors must be large enough to prevent an excessive drop in voltage. The voltage drop is the difference between the voltage at the power tool and the voltage at the supply. This drop in voltage occurs because of resistance to electrical flow in the wire. A voltage drop causes loss of power, overheating, and possible motor damage. Table 2.2 shows the

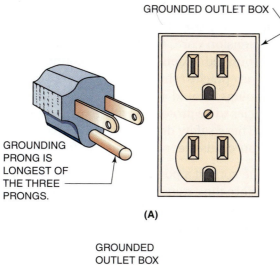

GROUNDED OUTLET BOX

GROUNDING PRONG IS LONGEST OF THE THREE PRONGS.

(A)

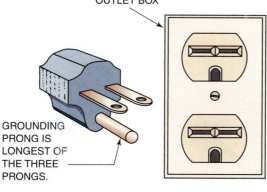

GROUNDED OUTLET BOX

GROUNDING PRONG IS LONGEST OF THE THREE PRONGS.

(B)

Figure 2.42 Grounding plugs
(A) A three-prong grounding plug for use with up to 150-volt tools and (B) a grounding plug for use with 150- to 250-volt tools

Figure 2.43 Typical portable power tool nameplate

Table 2.2 Recommended Extension Cord Sizes for Use with Portable Electrical Tools

Nameplate Amperes	Cord Length in Feet							
	25	50	75	100	125	150	175	200
1	16	16	16	16	16	16	16	16
2	16	16	16	16	16	16	16	16
3	16	16	16	16	16	16	14	14
4	16	16	16	16	16	14	14	12
5	16	16	16	16	14	14	12	12
6	16	16	16	14	14	12	12	12
7	16	16	14	14	12	12	12	10
8	14	14	14	14	12	12	10	10
9	14	14	14	12	12	10	10	10
10	14	14	14	12	12	10	10	10
11	12	12	12	12	10	10	10	8
12	12	12	12	12	10	10	8	8

Note: Wire sizes shown are American Wire Gauge (AWG)

correct size of extension cord to use based on cord length and nameplate amperage rating. If in doubt, use the next larger size. The smaller the gauge number of an extension cord, the larger the cord.

Only three-wire, grounded extension cords connected to properly grounded, three-wire receptacles should be used. Two-wire extension cords with two-prong plugs should not be used. Current specifications require outdoor receptacles to be protected with **ground-fault circuit interrupter (GFCI)** devices. These safety devices are often referred to as GFIs.

When using extension cords, keep in mind the following safety tips:

- Always connect the cord of a portable electric power tool into the extension cord before the extension cord is connected to the outlet. Always unplug the extension cord from the receptacle before unplugging the cord of the portable power tool from the extension cord.
- Extension cords should be long enough to make connections without being pulled taut, which creates unnecessary strain or wear, but they should not be excessively long.

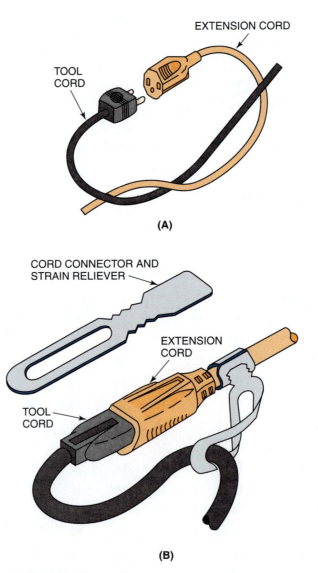

Figure 2.44 Connecting extension cords
(A) A knot or (B) a cord connector will prevent the extension cord from accidentally pulling apart from the tool cord during operation

- Be sure that the extension cord does not come in contact with sharp objects or hot surfaces. The cord should not be allowed to kink, nor should it be dipped in or splattered with oil, grease, or chemicals.
- Before using a cord, inspect it for loose or exposed wires and damaged insulation. If a cord is damaged, replace it. This also applies to the tool's power cord.
- Extension cords should be checked frequently while in use to detect unusual heating. Any cable that feels more than slightly warm to a bare hand placed outside the insulation should be checked immediately for overloading.
- See that the extension cord is positioned so that no one trips or stumbles over it.
- To prevent the accidental separation of a tool cord from an extension cord during operation, make a knot as shown in Figure 2.44A or use a cord connector as shown in Figure 2.44B.
- Extension cords that go through dirt and mud must be cleaned before storing.

SAFETY RULES FOR PORTABLE ELECTRIC TOOLS

**Module 2
Key Indicator 2**

In all tool operation, safety is simply the removal of any element of chance. The following are a few safety precautions that should be observed. These are general rules that apply to all power tools. They should be strictly obeyed to avoid injury to the operator and damage to the power tool.

- Know the tool. Learn the tool's applications and limitations as well as its specific potential hazards by reading the manufacturer's literature.
- Ground the portable power tool unless it is double-insulated. If the tool is equipped with a three-prong plug, it must be plugged into a three-hole electric receptacle. Never remove the third prong.
- Do not expose the power tool to water or rain. Do not use a power tool in wet locations.
- Keep the work area well lighted. Avoid chemical or corrosive environments.
- Because electric tools spark, portable electric tools should never be started or operated in the presence of propane, natural gas, gasoline, paint thinner, acetylene, or other flammable vapors that could cause a fire or explosion.
- Do not force a cutting tool to cut faster. It will do the job better and more safely if operated at the cutting rate for which it was designed.
- Use the right tool for the job. Never use a tool for any purpose other than that for which it was designed.
- Wear eye protectors. Safety glasses or goggles will protect the eyes while you operate power tools.
- Wear a face or dust mask if the operation creates dust.
- Take care of the power cord. Never carry a tool by its cord or yank it to disconnect it from the receptacle.
- Secure your work with clamps. It is safer than using your hands, and it frees both hands to operate the tool.
- Do not overreach when operating a power tool. Keep proper footing and balance at all times.

- Maintain power tools. Follow the manufacturer's instructions for lubricating and changing accessories. Replace all worn, broken, or lost parts immediately.
- Disconnect the tools from the power source when they are not in use.
- Form the habit of checking to see that any keys or wrenches are removed from the tool before turning it on.
- Avoid accidental starting. Do not carry a plugged-in tool with your finger on the switch. Be sure the switch is off when plugging in the tool.
- Be sure accessories and cutting bits are attached securely to the tool.
- Do not use tools with cracked or damaged housings.
- When operating a portable power tool, give it your full and undivided attention; avoid dangerous distractions.
- Never use a power tool with its safeties or guards removed or inoperable.

Grinders

Grinding using a pedestal grinder or a portable grinder is required to do many welding jobs correctly. Often it is necessary to grind a groove, remove rust, or smooth a weld. Grinding stones have their maximum revolutions per minute (RPM) listed on the paper blotter, Figure 2.45. They must never be used on a machine with a higher rated RPM. If grinding stones are turned too fast, they can explode.

Grinding Stone

Before a grinding stone is put on the machine, it should be tested for cracks. This is done by tapping the stone in four places and listening for a sharp ring, which indicates that it is good, Figure 2.46. A dull sound indicates that the grinding stone is cracked and should not be used. Once a stone has been installed and has been used, it may need to be trued and balanced using a special tool designed for that purpose, Figure 2.47. Truing keeps the stone face flat and sharp for better results.

Types of Grinding Stones

Each grinding stone is made for grinding specific types of metal. Most stones are for ferrous metals, meaning iron, cast iron, steel, and stainless steel, among others. Some stones are made for nonferrous metals such as aluminum, copper, and brass. If a ferrous stone is used to grind nonferrous metal, the stone will become glazed (the surface clogs with metal) and may explode as a result of frictional heat building up on the surface. If a nonferrous stone is used to grind ferrous metal, the stone will be quickly worn away.

When the stone wears down, keep the tool rest adjusted to within 1/16 in. (2 mm), Figure 2.48, so that the metal being ground cannot be pulled between the tool rest and the stone surface. Stones should not be used when they are worn down to the size of the paper blotter. If small parts become hot from grinding, pliers can be used to hold them. Gloves should never be worn when grinding. If a glove gets caught in a stone, the whole hand may be drawn in.

The sparks from grinding should be directed down and away from other people or equipment.

Figure 2.47 Balancing a grinding stone
Use a grinding stone redressing tool as needed to keep the stone in balance
Source: Courtesy of Larry Jeffus

Figure 2.45 Grinding stone compatibility
Always check that the grinding stone and the grinder are compatible before installing a stone
Source: Courtesy of Larry Jeffus

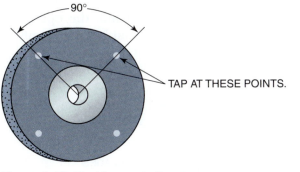

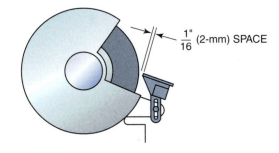

Figure 2.48 Tool rests
Keep the tool rest adjusted

Figure 2.46 Checking a grinding stone
Grinding stones should be checked for cracks before they are installed

Drills

Holes should be center punched before they are drilled to help stop the drill bit from wandering. If the bit gets caught, stop the motor before trying to remove the bit. All metal being drilled on a drill press should be securely clamped to the table.

The sharp metal shavings should be avoided as they come out of the hole. If they start to become long, stop the downward pressure until the shaving breaks. Then the hole can be continued.

METAL-CUTTING MACHINES

Many types of mechanical metal-cutting machines are used in the welding shop—for example, shears, punches, cut-off machines, and band saws. Their advantages over thermal cutting include little or no requirement for postcutting cleanup, the wide variety of metals that can be cut, and the fact that the metal is not heated.

Shears and Punches

Welders frequently use shears and punches in the fabrication of metal for welding. These machines can be operated either by hand or by powerful motors. Hand-operated equipment is usually limited to thin sheet stock or small bar stock. Powered equipment can be used on material an inch or more in thickness and several feet wide, depending on its rating. Their power is a potential danger if these machines are not used correctly. Both shears and punches are rated by the thickness, width, and type of metal that they can be safely used to work. Failure to follow these limitations can result in damage to the equipment, damage to the metal being worked, and injury to the operator.

Shears work like powerful scissors. The correct placement of the metal being cut is as close to the pivot pin as possible, Figure 2.49. The metal being sheared must be securely held in place by the clamp on the shear before it is cut. If you are cutting a long piece of metal that is not being supported by the shear table, then portable supports must be used. As the metal is being cut it may suddenly move or bounce around; if you are holding on to it, this can cause a serious injury.

Power punches are usually either hydraulic or flywheel operated. Both types move quickly, but usually only the hydraulic type can be stopped midstroke. Once the flywheel-type punch has been engaged, in contrast, it will make a complete cycle before it stops. Because punches move quickly or may not be stopped, it is very important that the operator's two hands be clear of the machine and that the metal be held firmly in place by the machine clamps before the punching operation is started.

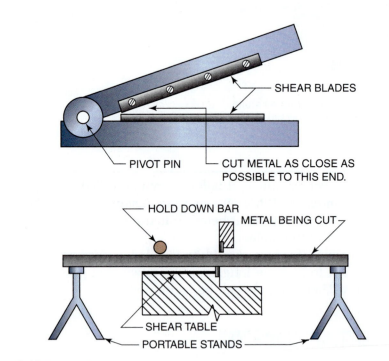

SHEAR BLADES

PIVOT PIN

CUT METAL AS CLOSE AS POSSIBLE TO THIS END.

HOLD DOWN BAR

METAL BEING CUT

SHEAR TABLE

PORTABLE STANDS

Figure 2.49 Power shear

Cut-off Machines

Cut-off machines may use abrasive wheels or special saw blades to make their cuts. Most abrasive cut-off wheels spin at high speeds (high RPMs) and are used dry (without coolant). Most saws operate much more slowly and with a liquid coolant. Both types of machines produce quality cuts in a variety of bar- or structural-shaped metals. The cuts require little or no postcut cleanup. Always wear eye protection when operating these machines. Before a cut is started, the metal must be clamped securely in the machine vise. Even the slightest movement of the metal can bind or break the wheel or blade. If the machine has a manual feed, the cutting force must be applied at a smooth and steady rate. Apply only enough force to make the cut without dogging down the motor. Use only reinforced abrasive cut-off wheels that have an RPM rating equal to or higher than the machine-rated speed.

Band Saws

Band saws can be purchased as vertical or horizontal, and some can be used in either position. Some band saws can be operated with a cooling liquid and are called *wet saws;* most small saws operate dry. The blade guides must be adjusted as closely as possible to the metal being cut. The cutting speed and cutting pressure must be low enough to prevent the blade from overheating. When using a vertical band saw with a manual feed, you must keep your hands away from the front of the blade so that, if your hand slips, it will not strike the moving blade. If the blade breaks, sticks, or comes off the track, turn off the power, lock it off, and wait for the band saw drive wheels to come to a complete stop before touching the blade. Be careful of hot flying chips.

MATERIAL HANDLING

Proper lifting, moving, and handling of large, heavy welded assemblies are important to the safety of workers and the weldment. Improper work habits can cause serious personal injury and damage to equipment and materials.

Lifting

When you are lifting a heavy object, the weight of the object should be distributed evenly between both hands, and your legs should be used to lift, not your back, Figure 2.50. Do not try to lift a large or bulky object without help if the object is heavier than you can lift with one hand.

Hoists or Cranes

The capacity of hoists or cranes should be checked before trying to lift a load. They can be accidentally overloaded with welded assemblies. Keep any load as close to the ground as possible while it is being moved. Pushing a load on a crane is better than pulling a load. It is advisable to stand to one side of ropes, chains, and cables that are being used to move or lift a load, Figure 2.51: If they break and snap back, they will miss you. If it is necessary to pull a load, use a rope, Figure 2.52.

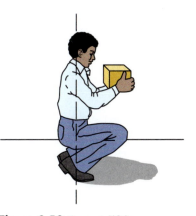

Figure 2.50 Correct lifting
Lift with your legs, not your back

Figure 2.51 Moving a load
Never stand in line with a rope, chain, or cable that is being used to move or lift a load

Figure 2.52 Moving an overhead load
When moving a load overhead, stay out of the way of the load in case it falls

LADDER SAFETY

Improper use of ladders is often a factor in falls. Always keep this in mind when erecting a ladder; even short step stools can pose a potential fall hazard. Never approach a climb assuming that because it is not high, it cannot be that dangerous. All ladder use poses a danger to your safety. Some welders think that if a ladder starts to fall they will just "jump clear." You cannot jump clear if the ladder under you has given way, because there is nothing solid under your feet for you to jump from. When a ladder falls, you fall. Keep the area around the base of the ladder clear so that if you do fall, it will not be into debris or equipment.

Types of Ladders

Both step ladders and straight ladders are used extensively in welding fabrication. Straight ladders may be single-section or extension-type

Table 2.3 Major Advantages and Disadvantages of Typical Ladder Materials

Material	Advantages	Disadvantages
Wood	Electrically non-conductive	Long-term exposure to weather will cause rotting
Aluminum	Light weight Weather resistant	Electrically conductive Shakier than wood or fiberglass
Fiberglass	Electrically non-conductive Weather resistant	Heavier than aluminum and wood Fiberglass splinters

ladders. Most ladders are made from wood, aluminum, or fiberglass, and each type has its advantages and disadvantages, Table 2.3. All ladders used in welding should be listed with the American National Standards Institute (ANSI) and Underwriters Laboratories (UL) to ensure that they are constructed to a standard of safety.

Ladder Inspection

Over time, ladders can become worn or damaged and should be inspected each time they are used. Look for loose or damaged steps, rungs, rails, braces, and safety feet. Check to see that all hardware is tight, including hinges, locks, nuts, bolts, screws, and rivets. Wooden ladders must be checked for cracks, rot, or wood decay. Never use a defective ladder. Make any necessary repairs before it is used or, if it cannot be repaired, replace it.

Rules for Ladder Use

Read the entire ladder manufacturer's list of safety rules before using the ladder for the first time. Step ladders must be locked in the full opened position with the spreaders. Straight or extension ladders must be used at the proper angle; either too steep or too flat is dangerous, Figure 2.53.

The following are general safety and usage rules for ladders:

- Follow all recommended practices for safe use and storage.
- Do not exceed the manufacturer's recommended maximum weight limit for the ladder.
- Before setting up a ladder, make certain that it will be erected on a level, solid surface.
- Never use a ladder in a wet or muddy area where water or mud will be tracked up the ladder's steps or rungs. Only climb or descend ladders when you are wearing clean, dry shoes.
- Wear well-fitted shoes or boots.
- Tie the ladder securely in place.
- Climb and descend the ladder cautiously.
- Do not carry tools and supplies in your hand as you climb or descend a ladder. Use a rope to raise or lower the items once you are safely in place.
- Never use ladders around live electrical wires.
- Never use a ladder that is too short for the job so you have to reach or stand on the top step.

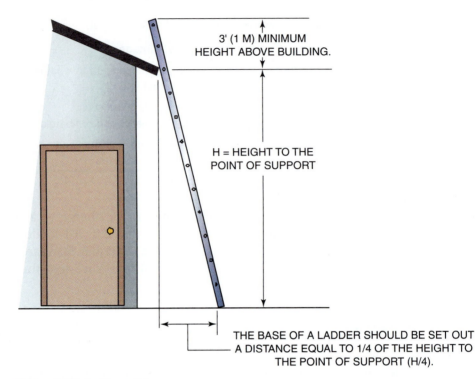

3' (1 M) MINIMUM
HEIGHT ABOVE BUILDING.

H = HEIGHT TO THE
POINT OF SUPPORT

THE BASE OF A LADDER SHOULD BE SET OUT
A DISTANCE EQUAL TO 1/4 OF THE HEIGHT TO
THE POINT OF SUPPORT (H/4).

Figure 2.53 Ladder safety
Make sure the ladder is leaning at the proper angle

SUMMARY

The safety of the welder working in industry is of utmost importance to the industry. A sizable amount of money is spent for the protection of welders. Usually manufacturers have a safety department with one individual in charge of plant safety. The safety officer's job is to make sure that all welders comply with safety rules during production. The proper clothing, shoes, and eye protection to be worn are emphasized in these plants. Any worker who does not follow established safety rules is subject to dismissal.

If an accident does occur, it is important that appropriate and immediate first aid steps be taken. All welding shops should have established plans for actions to take in case of accidents. You should take time to learn the proper procedure for accident response and reporting before you need to respond in an emergency. After the situation has been properly taken care of, you should fill out an accident report.

Equipment should be periodically checked to be sure that it is safe and in proper working condition. Maintenance workers are employed to see that the equipment is in proper working condition at all times.

Further safety information is available in *Safety for Welders*, by Larry F. Jeffus, published by Delmar Learning, and from the American Welding Society or the U.S. Department of Labor (OSHA) Regulations.

REVIEW

1. What is the key to preventing accidents in a welding shop?
2. Who is ultimately responsible for the welder's safety?

3. Describe the three classifications of burns.
4. What emergency steps should be taken to treat burns?
5. List the three types of light that may be present during welding.
6. Which type of light is the most likely to cause burns? Why?
7. What can be done on the job site to reduce the danger of reflected light?
8. In what two ways can ultraviolet light burn the eyes?
9. What is the name of the eye burn that can occur in a fraction of a second?
10. Why is it important to seek medical treatment for eye burns?
11. Why must eye protection be worn at all times in the welding shop?
12. According to Table 2.1, what eye and/or face protection should be used for each of the following?
    a. acetylene welding
    b. chipping
    c. electric arc welding
    d. spot welding
13. What types of injuries can occur to the ears during welding?
14. What types of protection are available to protect the ears during welding?
15. What types of information should be covered in a respirator training program? *Use, cleaning schedule, application, testing procedures, Emergency application, Program evaluation.*
16. Name two types of respirators and describe how they work. *Air purifying, atmosphere supplying.*
17. Name a common metal for which a welder would encounter *Stainless steel and chromium metal.* hexavalent chromium fumes while welding or grinding that metal.
18. List the materials that can give off dangerous fumes during welding and require forced ventilation.
19. Name two gases that are lighter than air and two gases that are heavier than air. *Ozone Phosgene; Argon and carbon monoxide*
20. Why must metal that has been used before be cleaned prior to welding?
21. Under what conditions can natural ventilation be used? *10,000 cubic feet, Cielings 16 ft or higher.*
22. Who must be provided with material safety data sheets (MSDSs)? *Everyone in Shop.*
23. Name two advantages of recycling scrap metal. *Money, Environement.*
24. What fabrics are the best choice for general work clothing in a welding shop? *Leather, Synthetic materials.*
25. Describe the ideal work shirt, pants, boots, and caps that should be worn in a welding shop.
26. Why is it unsafe to carry butane lighters or matches in your pockets while welding? *Explosions.*
27. What special protective items can be worn to provide extra protection for a welder's hands, arms, body, waist, legs, and feet?
28. Describe an acceptable storage area for a cylinder of fuel gas. *away from Stairwells, halls and exits. radiators furnaces or Sparks.*
29. How must high-pressure gas cylinders be stored so they cannot accidentally be knocked over? *With chain or other device.*
30. What should be done with a leaking cylinder if the leak cannot be stopped? *Taken outside w/ sign.*
31. Why is it important that acetylene cylinders not be stored horizontally? *To keep acetone and filler out of valves*
32. How far away should highly combustible materials be from any welding or cutting? *35 ft (10.7m)*

A: Flammables
B: Cumbustibles
C: Electrical
D: Metal Chemical

33. List the four types of fire extinguishers and what type of burning material they are used to extinguish.
34. What is hot work? *Any work that is at risk to fires.*
35. When is a fire watch needed? *Less than 35ft from spark tacks.*
36. Why is it important to have a planned maintenance program for tools and equipment? *maintainance.*
37. Why is it important to keep a welding area clean? *minimize hazards*
38. What should you do if you have to leave a piece of hot metal unattended? *Write hot on it.*
39. Why must a mushroomed chisel or hammer be reground?
40. What causes most electric shock in the welding industry? *Poor insulation.*
41. What can happen if too much power is being carried by a cable? *Fire.*
42. Why must equipment be turned off and unplugged before working on the electrical terminals? *To prevent Shock.*
43. According to the welding safety checklist in Figure 2.41, what are the factors necessary for a confined-space hazard?
44. According to Table 2.2, what gauge wire size would be needed for a power tool that has a nameplate amperage of 9 and a cord length of 100 ft?
45. What is a GFCI?
46. Why is it important to not weld when everything is wet?
47. List five safety tips for safe extension cord use.
48. List 10 safety rules for the safe use of portable electric tools.
49. List two types of grinders used by welders.
50. How close to the grinding stone face should the tool rest be adjusted?
51. Name metal-cutting machines used in the welding shop and what their advantages are.
52. Describe how a person should safely lift a heavy object.
53. List the things that should be inspected on a ladder.
54. List and explain five ladder use safety rules.

CHAPTER 3

Drawing and Welding Symbol Interpretation

OBJECTIVES

After completing this chapter, the student should be able to

- list five basic factors related to joint design
- identify the major parts of a welding symbol
- explain groove preparation
- describe how nondestructive test symbols are used
- list the five major types of joints
- identify groove welds on pipe and plate in the flat, horizontal, vertical, and overhead orientations
- identify fillet welds on pipe and plate in the flat, horizontal, vertical, and overhead positions

KEY TERMS

combination symbol	joint dimensions	weld location
edge preparation	joint type	weld types
fillet (F)	projection drawings	welding position
grooves (G)	weld joint	welding symbols

AWS SENSE EG2.0

Key Indicators Addressed in this Chapter:

Module 3: Drawing & Weld Symbol Interpretation

Key Indicator 1: Interprets basic elements of a drawing or sketch
Key Indicator 2: Interprets welding symbol information

INTRODUCTION

Joint design affects the quality and cost of the completed weld. Selecting the most appropriate joint design for a welding job requires special attention and skill. The eventual design selection can be influenced by a number of factors, including (but not limited to) the welding process to be used, whether the joint is to be welded in the field or in a shop, and whether the joint is a one-time weld or is to be mass produced.

Every weld joint selection for a job requires some compromises. For example, the compromises may be between strength and cost, between equipment available and welder skill, or among a number of variables. Because there are so many factors, a good design requires experience. Even with experience, trial welds are necessary before selecting the final joint configuration and welding parameters.

This chapter will familiarize welders with the most important factors and give some appreciation of joint design. Experience in the welding field will help a welder become a better joint designer and fabricator.

Welding symbols are the language used to let the welder know exactly what welding is needed. A welding symbol is used as a shorthand and can provide the welder with all of the required information to make the correct weld. This chapter emphasizes the use and interpretation of welding symbols so the welder will develop a welder's "vocabulary."

WELD JOINT DESIGN

The selection of the best joint design for a specific weldment requires careful consideration of a variety of factors. If individual factors are considered in isolation, the result may be a part that cannot be fabricated.

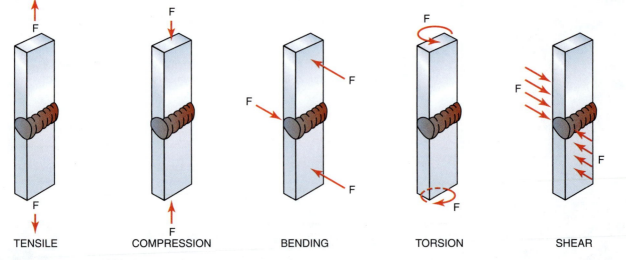

| TENSILE | COMPRESSION | BENDING | TORSION | SHEAR |

Figure 3.1 Forces on a weld

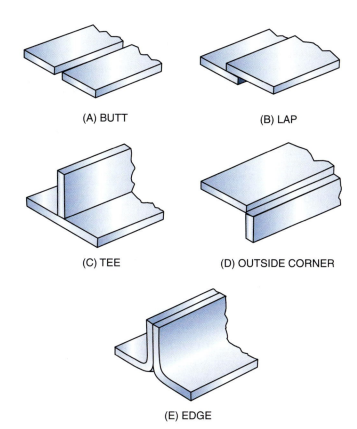

(A) BUTT

(B) LAP

(C) TEE

(D) OUTSIDE CORNER

(E) EDGE

Figure 3.2 Types of joints

For example, a narrower joint angle requires less filler metal, and that results in lower welding cost. But if the angle is too small for the welding process being used, the weld cannot be made.

The purpose of a **weld joint** is to join parts together so that the stresses are distributed. The forces causing stresses in welded joints are tensile, compression, bending, torsion, and shear, Figure 3.1. The ability of a welded joint to withstand these forces depends upon both the joint design and the weld integrity. Some joints can withstand certain types of forces better than others.

The basic parts of a weld joint design that can be changed include

- **joint type**—The type of joint is chosen by analyzing the way that the joint members come together, Figure 3.2.
- **edge preparation**—The faying surfaces (the surfaces of materials in contact with each other) of the mating members that form the joint are shaped for that specific joint. This preparation may be the same on both members of the joint, or each side may be shaped differently, Figure 3.3.
- **joint dimensions**—The depth and/or angle of the preparation and the joint spacing can be changed to make the weld, Figure 3.4.

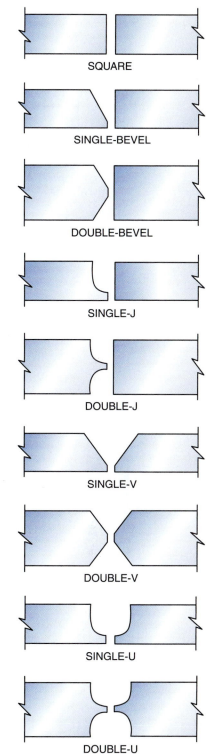

SQUARE

SINGLE-BEVEL

DOUBLE-BEVEL

SINGLE-J

DOUBLE-J

SINGLE-V

DOUBLE-V

SINGLE-U

DOUBLE-U

Figure 3.3 Edge preparation

Welding Process

The welding process to be used has a major effect on the selection of the joint design. Each welding process has characteristics that affect its performance. Some processes are easily used in any position; others may be

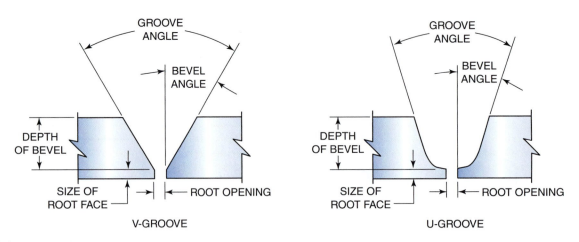

Figure 3.4 Groove terms

restricted to particular positions. The rate of travel, penetration, deposition rate, and heat input also affect the welds used on some joint designs. For example, a square butt joint can be made in very thick plates using either electroslag or electrogas welding, but not many other processes can be used on such a joint design.

Base Metal

Because some metals present specific problems in terms of, for example, thermal expansion, crack sensitivity, or distortion, the joint selected must control these problems. Thus, magnesium is very susceptible to postweld stresses, and the U-groove works best for thick sections.

Plate Welding Positions

The ideal **welding position** for most joints is the flat position, because it allows for larger molten weld pools to be controlled. Usually, the larger the weld pool, the faster the joint can be completed. It is not always possible to position a part so that all the welds can be made in the flat position. Special joint designs may be used for certain types of out-of-position welding. For example, the single bevel joint is often the best choice for horizontal welding, Figure 3.5.

The American Welding Society has divided plate welding into four basic positions for **grooves (G)** and **fillet (F)** welds as follows:

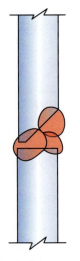

Figure 3.5 Bead positions for a horizontal weld

- flat 1G or 1F—Welding is performed from the upper side of the joint, and the face of the weld is approximately horizontal, Figure 3.6.
- horizontal 2G or 2F—The axis of the weld is approximately horizontal, but the type of weld dictates the complete definition. For a fillet weld, welding is performed on the upper side of an approximately vertical surface. For a groove weld, the face of the weld lies in an approximately vertical plane, Figure 3.7.
- vertical 3G or 3F—The axis of the weld is approximately vertical, Figure 3.8.
- overhead 4G or 4F—Welding is performed from the underside of the joint, Figure 3.9.

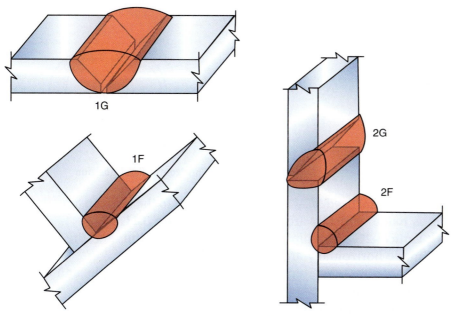

Figure 3.6 Plate flat position   **Figure 3.7** Plate horizontal position

Pipe Welding Positions

The American Welding Society lists five basic positions for pipe welding:

- horizontal rolled 1G—The pipe is rolled either continuously or intermittently so that the weld can be performed within 0° to 15° of the top of the pipe, Figure 3.10.

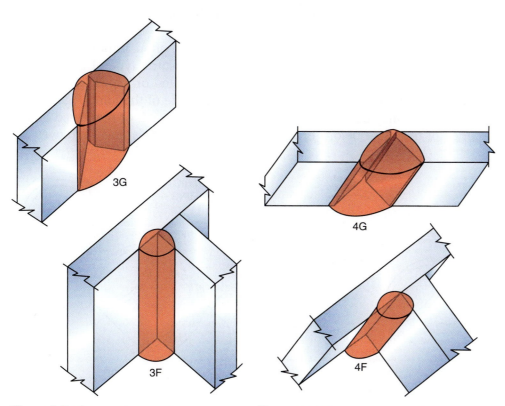

Figure 3.8 Plate vertical position   **Figure 3.9** Plate overhead position

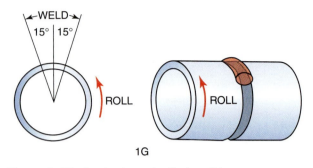

Figure 3.10 Pipe horizontal rolled position

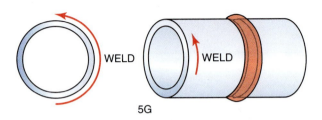

Figure 3.11 Pipe horizontal fixed position

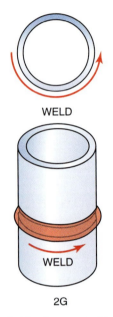

Figure 3.12 Pipe vertical position

- horizontal fixed 5G—The pipe is parallel to the horizon, and the weld is made vertically around the pipe, Figure 3.11.
- vertical 2G—The pipe is vertical to the horizon, and the weld is made horizontally around the pipe, Figure 3.12.
- inclined 6G—The pipe is fixed at a 45° inclined angle, and the weld is made around the pipe, Figure 3.13.
- inclined with a restriction ring 6GR—The pipe is fixed at a 45° inclined angle, and a restricting ring is placed around the pipe below the weld groove, Figure 3.14.

Metal Thickness

As metal thickness increases, the joint design must change. On thin sections, it is often possible to make full-penetration welds using a square butt joint. But with thicker plates or pipe the edge must be prepared with a groove on one or both sides. The edge may be shaped with a bevel, V-groove, J-groove, or U-groove. The choice of shape depends on the type of metal, its thickness, and whether it is made before or after assembly.

When welding on thick plate or pipe, it is often impossible for the welder to get 100% penetration without using some type of groove. The groove may be cut into either one of the plates or pipes or both. On some plates it can be cut both inside and outside of the joint, Figure 3.15. The groove may be ground, flame-cut, gouged, sawed, or machined on the edge of the plate before or after the assembly. Bevels and V-grooves are

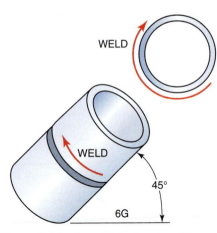

Figure 3.13 Pipe 45° inclined position

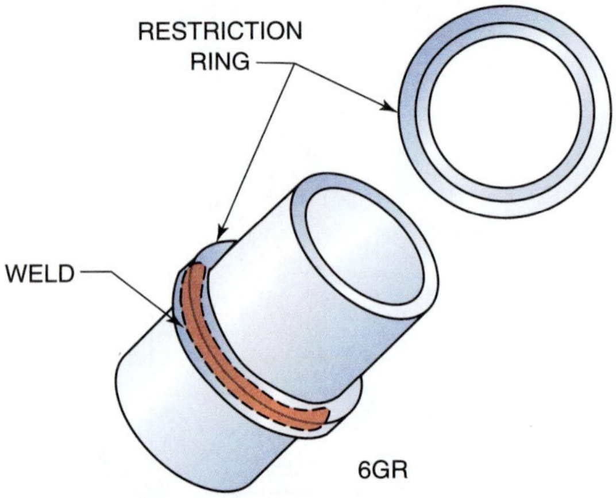

Figure 3.14 Pipe 45° inclined position with a restricting ring

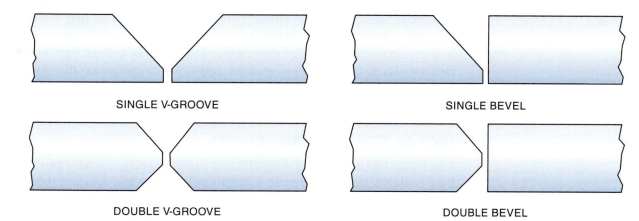

Figure 3.15 V-groove and bevel joint types

best if they are cut before the parts are assembled; J-grooves and U-grooves can be cut either before or after assembly, Figure 3.16. A lap joint is seldom prepared with a groove, because little or no strength can be gained by grooving this joint.

For most welding processes, plates that are thicker than 3/8 in. (10 mm) may be grooved on both the inside and outside of the joint. Whether to groove one or both sides is most often determined by joint design, position, code, and application. Plates in the flat position are usually grooved on only one side unless they can be repositioned or must be welded on both sides. Tee joints in thick plates are easier to weld and exhibit less distortion if they are grooved on both sides.

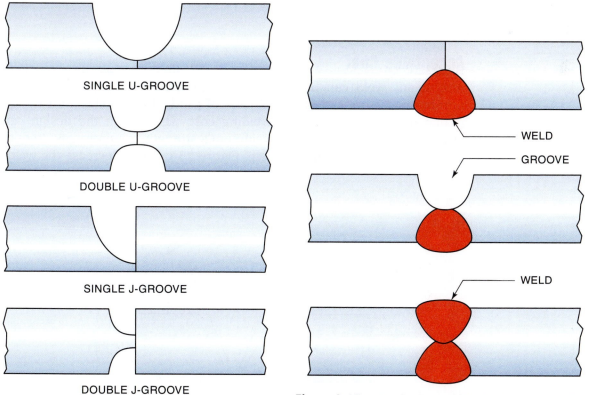

Figure 3.16 U-groove and J-groove joint types

Figure 3.17 Back-gouging a weld joint to ensure 100% joint penetration

Sometimes plates are either grooved and welded or just welded on one side and then back-gouged and welded, Figure 3.17. Back-gouging is a process of cutting a groove in the back side of a joint that has been welded. Back-gouging can ensure 100% joint fusion at the root and remove discontinuities of the root pass.

Code or Standards Requirements

The type, depth, angle, and location of the groove are usually determined by a code or standard that has been qualified for the specific job. Organizations such as the American Welding Society, the American Society of Mechanical Engineers, and the American Bureau of Ships are among the agencies that issue such codes and specifications. The most common codes or standards are the AWS D1.1 and the ASME Boiler and Pressure Vessel (BPV), Section IX.

The joint design for a particular set of specifications often must be *prequalified*. Such joints have been tested and found to be reliable for the weldments for specific applications. The joint design can be modified, but the cost to have the new design accepted under the standard being used is often prohibitive.

Welder Skill

Often the skills or abilities of the welder are a limiting factor in joint design. A joint must be designed in such a way that the welders can reliably reproduce it. Some joints have been designed without adequate room for the welder to see the molten weld pool or to get the electrode or torch into the joint.

Acceptable Cost

Almost any weld can be made in any material in any position, but a number of factors can affect the cost of producing a weld. Joint design is one major way to control welding cost. Changes in the design can reduce cost yet still meet the weldment's strength requirements. Reducing the groove angle can also help, Figure 3.18. It will decrease the weld-

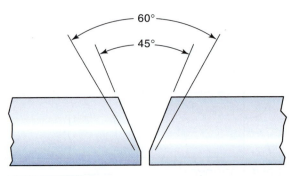

Figure 3.18 Groove angle
A smaller groove angle reduces both weld time and weld metal

ing filler metal required to complete the weld as well as decrease the time required to fill the groove opening. Joint design must be a consideration for any project to be competitive and cost-effective.

MECHANICAL DRAWINGS

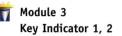 **Module 3**
Key Indicator 1, 2

Mechanical drawings have been around for centuries. Leonardo da Vinci (1452–1519) used mechanical drawings extensively in his inventive works. Many of his drawings still exist today and are as easily understood now as when they were drawn. For that reason, mechanical drawings have been called the universal language: They are produced in a similar format worldwide. Despite the few differences in how the views may be laid out, Figure 3.19, the drawings can still be understood. Notwithstanding different languages and measuring systems, the basic shape of an object and the location of components can be determined from any good drawing.

A group of drawings, known as a *set of drawings,* should contain enough information to enable a welder to produce the weldment. The set of drawings may contain various pages showing different aspects of the project to aid in its fabrication. The pages may include the following: title page, pictorial, assembly drawing, detailed drawing, and exploded view, Figure 3.20.

In addition to the shape as described by the various lines, a set of drawings may contain information such as the title box and bill of materials. The *title box,* which appears in one corner of the drawing, should contain the name of the part, the company name, the scale of the drawing, the date of the drawing, the name of the person who made the drawing, the drawing number, the number of drawings in the set, and tolerances.

A *bill of materials* can also be included in the set of drawings. This is a list of the various items that will be needed to build the weldment, Table 3.1.

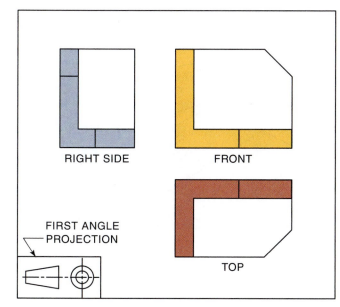

Figure 3.19 Two different methods used to rotate drawing views

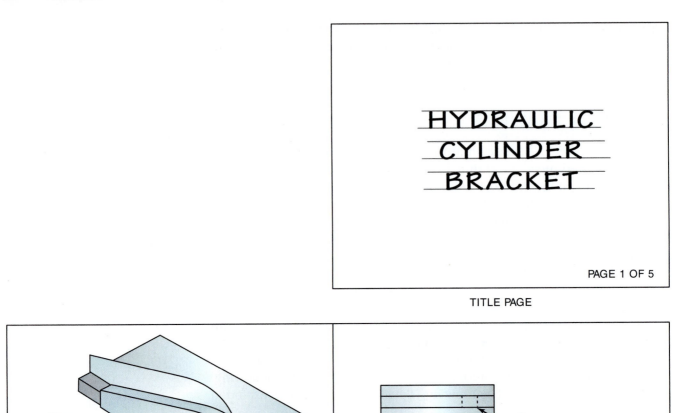

HYDRAULIC
CYLINDER
BRACKET

PAGE 1 OF 5

TITLE PAGE

PAGE 2 OF 5

PICTORIAL

PAGE 3 OF 5

ASSEMBLY

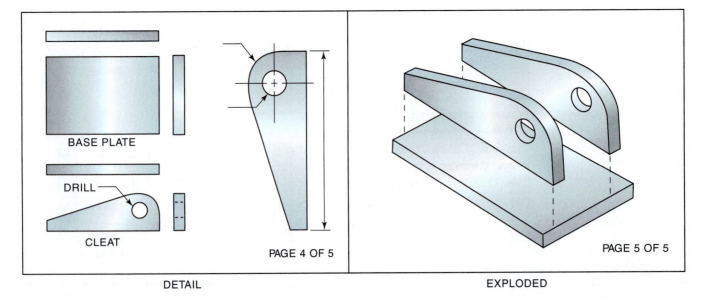

BASE PLATE

DRILL

CLEAT

PAGE 4 OF 5

DETAIL

PAGE 5 OF 5

EXPLODED

Figure 3.20 Drawings that can make up a set of drawings

Table 3.1 Bill of Materials

Part	Number Required	Type of Material	Size (Standard Units)	(SI Units)
Base	1	Hot roll steel	1/2″ × 5″ × 8″	12.7 mm × 127 mm × 203.2 mm
Cleat	2	Hot roll steel	1/2″ × 4″ × 8″	12.7 mm × 101.6 mm × 203.2 mm

Lines

To understand drawings, you need to know what the different types of lines represent. The language of drawing uses lines for its alphabet and the various parts of the object being illustrated. The various line types are collectively known as the *alphabet of lines,* Table 3.2 and Figure 3.21.

Table 3.2 Alphabet of Lines

Line Type	Description	Purpose
OBJECT LINE	Solid bold line	To show the intersection of surfaces or the extent of a curved surface
HIDDEN LINE	Broken medium line	To show the intersection of surfaces or the extent of a curved surface that occurs below the surface and hidden from view
CENTER LINE	Fine broken line made up of longer line sections on both sides of a short, dashed line	To show the center of a hole, curve, or symmetrical object
EXTENSION / DIMENSION LINE / LINE	Extension lines (fine line) extending from near the surface of the object	Extension lines extend from an object line or a hidden line to locate dimension points.
	Dimension lines (medium line) extending between extension lines or object lines	Dimension lines touch the extension lines and/or object lines that represent the points being dimensioned.
CUTTING PLANE LINE	Bold broken lines with arrowheads pointing in the direction of the cut surface	These lines extend all the way across the surface that is being imaginarily cut. The arrowhead ends point in the direction in which the cut surface willl be shown in the sectional drawing.
SECTION LINES / STEEL / CAST IRON	Series of fine drawn at an angle to the object lines. The line angle usually changes from one part to another. The cast iron section lines are used universally for most sections.	Used to indicate a surface that has been imaginarily cut or broken. The spacing and pattern can be used to indicate the type of material that is being viewed.
LEADER OR ARROW LINE	Medium line with an arrowhead at one end	Leader and arrow lines are used to locate points on the drawing to which a specific note, dimension, or welding symbol refers.
LONG BREAK LINE	Bold straight line with intermittent zigzag	To indicate that a portion of the part has not been included in the drawing either to conserve space or because the omitted portion was not significant to this specific drawing
SHORT BREAK LINE	Bold freehand irregular line	Used for the same purposes as the long break above except on parts not wide enough to allow the long break lines with their zigzags to be used clearly

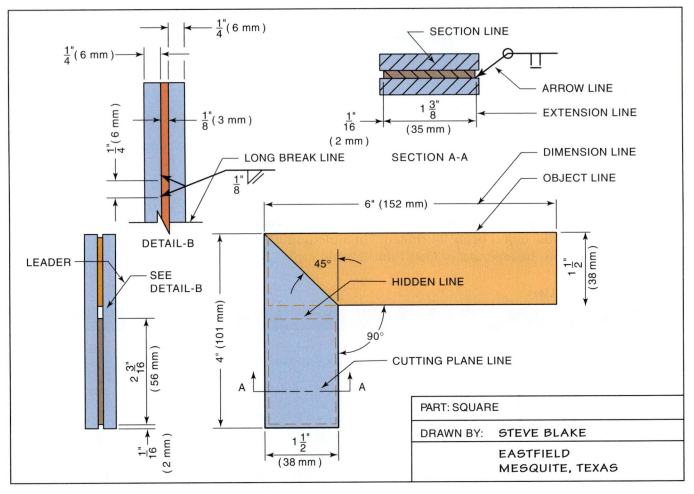

Figure 3.21 Alphabet of lines

Types of Drawings

The drawings used for most welding projects can be divided into two categories: orthographic projections and pictorial. **Projection drawings** are made as though one were looking through the sides of a glass box at the object and tracing its shape on the glass, Figure 3.22. If all the sides of the object were traced and the box unfolded and laid out flat, six basic views would be shown, Figure 3.23.

Pictorial drawings present the object in a more realistic or understandable form and usually appear as one of two types: isometric or cavalier, Figure 3.24. The more realistic, perspective drawing form is seldom used for welding projects.

Projection Drawings

Seldom are all of the six projections or views required to build a weldment. Only those needed are normally provided, usually only the front, right side, and top views. Sometimes only one or two of these views are required.

The front view is not necessarily the front of the object. A view is selected as the front view because an object's overall shape is best described when it is viewed from this direction. As an example, the front view of a car or truck would probably not be the view in a drawing because viewing the vehicle from its front may not show enough detail to let you know whether it is a car, light truck, station wagon, or van. From the front most vehicles look similar.

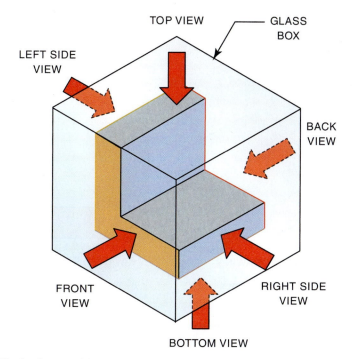

Figure 3.22 **Viewing an object as if it were inside a glass box**

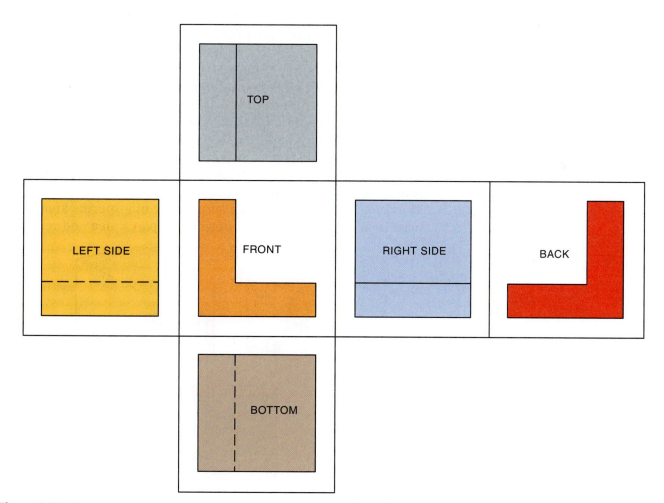

Figure 3.23 **The arrangement of views for an object if the glass box were unfolded**

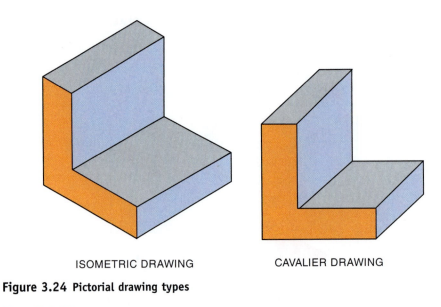

ISOMETRIC DRAWING   CAVALIER DRAWING

Figure 3.24 Pictorial drawing types

Special Views

Special views may be included on a drawing to help describe the object so it can be made accurately. Special views on some drawings may include:

- The *section view* is drawn as if part of the object were sawn away to reveal internal details, Figure 3.25. This view is useful when the internal details would not be as clear if they were shown as hidden lines. Sections can be either fully across the object or just partially across it. The imaginary cut surface is set off from other noncut surfaces by section lines drawn at an angle on the cut surfaces. The location of this imaginary cut is shown using a cutting plane line, Figure 3.26. Some drawings use specific types of section lines to illustrate the type of material the part was made with.
- The *cut-away view* is used to show detail within a part that would be obscured by the part's surface. Often a free-hand break line is used to outline the area that has been imaginarily removed to reveal the inner workings.
- The *detail view* is usually an external view of a specific area of a part. Detail views show small details of a part's area and remove

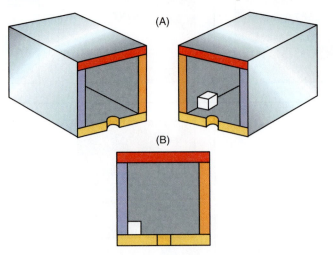

Figure 3.25 Section drawing
A section drawing is made as if the part were cut in two (A), so that you can see inside of it (B)

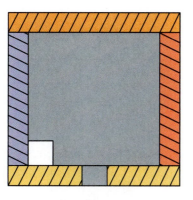
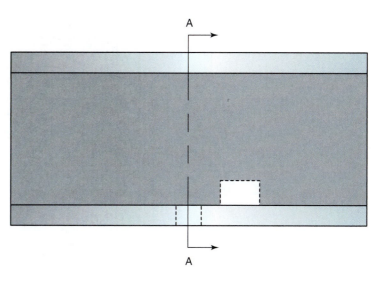

SECTION A-A

Figure 3.26 Cutting plane line and section

the need to draw an enlargement of the entire part. If only a small portion of a view has significance, this area can be shown in a detail view, either at the same scale or larger if needed. By showing only what is needed within the detail, the part drawn can be clearer and does not require such a large page.

- A *rotated view* can be used to show a surface of the part that would not normally be drawn square to any of the six normal view planes. If a surface is not square to the viewing angle, then lines may be distorted. For example, when viewed at an angle, a circle looks like an ellipse, Figure 3.27.

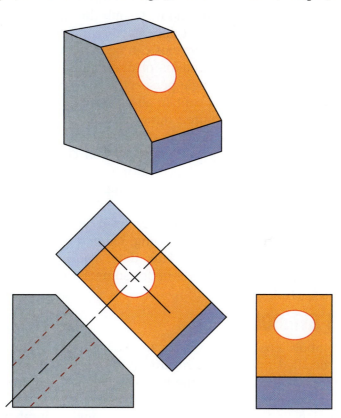

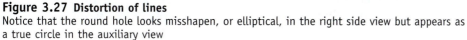

Figure 3.27 Distortion of lines
Notice that the round hole looks misshapen, or elliptical, in the right side view but appears as a true circle in the auxiliary view

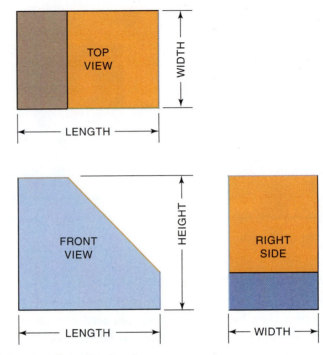

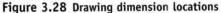

Figure 3.28 Drawing dimension locations

Dimensioning

Often it is necessary to look at other views to locate all of the dimensions required to build the object. If a welder knows how the views are arranged, it becomes easier to locate dimensions. Length dimensions can be found on the front and top views. Height dimensions can be found on the front and right side views. Width dimensions can be found on the top and right side views, Figure 3.28. The locations of dimensions on these views are consistent with both the first angle perspective and third angle layouts. A properly executed drawing will contain all necessary dimensions.

If you cannot find the required dimensions on the drawings, do not try to obtain them by measuring the drawing itself. Even if the original drawing was made accurately, the paper it is on changes with changes in humidity. Copies of the original drawing are never the exact same size. The most acceptable way of determining missing dimensions is to contact the person who made the drawing.

Keep the drawing clean and well away from any welding. Avoid writing or doing calculations on the drawing. Often a drawing will be filed following the project for use at a later date. The better care you take with the drawings, the easier it will be for someone else to use them.

Module 3
Key Indicator 1, 2

WELDING SYMBOLS

Welding symbols enable a designer to indicate clearly to the welder important, detailed information regarding the weld. The information in the welding symbol can include the following details for the weld: length, depth of penetration, height of reinforcement, groove type, groove dimensions, location, process, filler metal, strength, number of welds, weld shape, and surface finishing. All this information would normally be included on the welding assembly drawings.

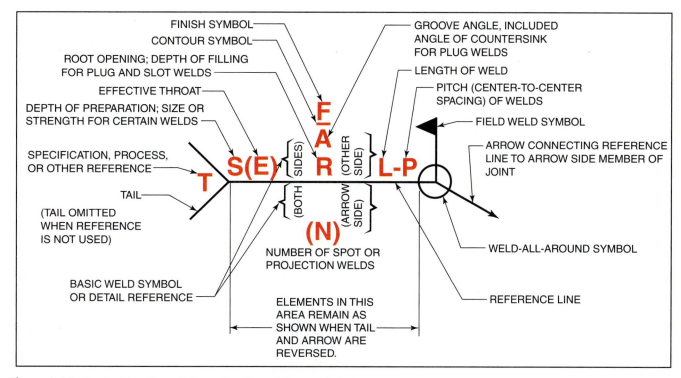

Figure 3.29 Standard location of elements of a welding symbol
Source: Courtesy of the American Welding Society

Welding symbols are a shorthand language for the welder. They save time and money and serve to ensure understanding and accuracy. Welding symbols have been standardized by the American Welding Society. Some of the more common symbols for welding are reproduced in this chapter. More information about symbols and how they apply to all forms of manual and automatic machine welding can be found in the complete material, *Standard Symbols for Welding, Brazing and Nondestructive Examination,* ANSI/AWS A2.4, published as an American National Standard by the American Welding Society.

Figure 3.29 shows the basic components of welding symbols. The symbols are based on a reference line with an arrow at one end. Other information relating to features of the weld is shown by symbols, abbreviations, and figures located around the reference line. A tail is added to the basic symbol as necessary for the placement of specific information.

Indicating Types of Welds

Weld types are classified as follows: fillets, grooves, flange, plug or slot, spot or projection, seam, back or backing, and surfacing. Each type of weld is indicated on drawings by a specific symbol. A fillet weld, for example, is designated by a right triangle. A plug weld is indicated by a rectangle. All of the basic symbols are shown in Figure 3.30.

Weld Location

Welding symbols are applied to the joint as the basic reference. All joints have an arrow side (near side) and another side (far side). Accordingly, the terms *arrow side, other side,* and *both sides* are used to indicate the **weld location** with respect to the joint. The reference line is always drawn

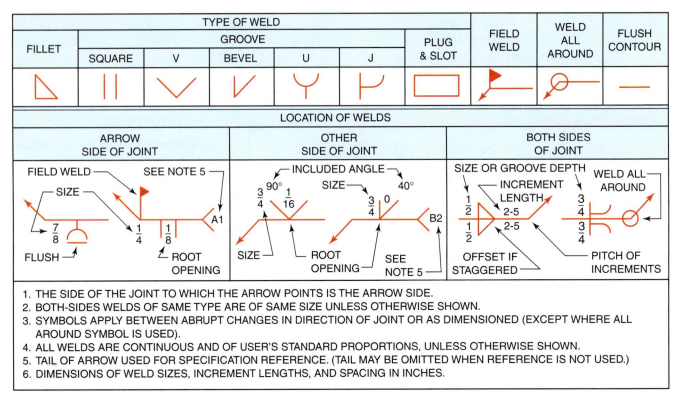

Figure 3.30 Welding symbols for different types of welds

horizontally. An arrow line is drawn from one end or both ends of a reference line to the location of the weld. The arrow line can point to either side of the joint and extend either upward or downward.

If the weld is to be deposited on the arrow side of the joint (near side), the proper weld symbol is placed below the reference line, Figure 3.31A. If the weld is to be deposited on the other side of the joint (far side), the weld symbol is placed above the reference line, Figure 3.31B. When welds are to be deposited on both sides of the same joint, the same weld symbol appears above and below the reference line, Figure 3.31C and D, along with detailed information.

A tail is added to the basic welding symbol to give welding specifications, procedures, or other supplementary information required to make the weld, Figure 3.32. The notation placed in the tail of the symbol may indicate the welding process to be used, the type of filler metal needed, whether or not peening or root chipping is required, and other information pertaining to the weld. If notations are not used, the tail of the symbol is omitted.

For joints that are to have more than one weld, a symbol is shown for each weld.

Significance of Arrow Location

In the case of fillet and groove welding symbols, the arrow connects the welding symbol reference line to one side of the joint. The surface of the joint the arrow point touches is considered to be the arrow side of the joint. The side opposite the arrow side of the joint is considered to be the other (far) side of the joint.

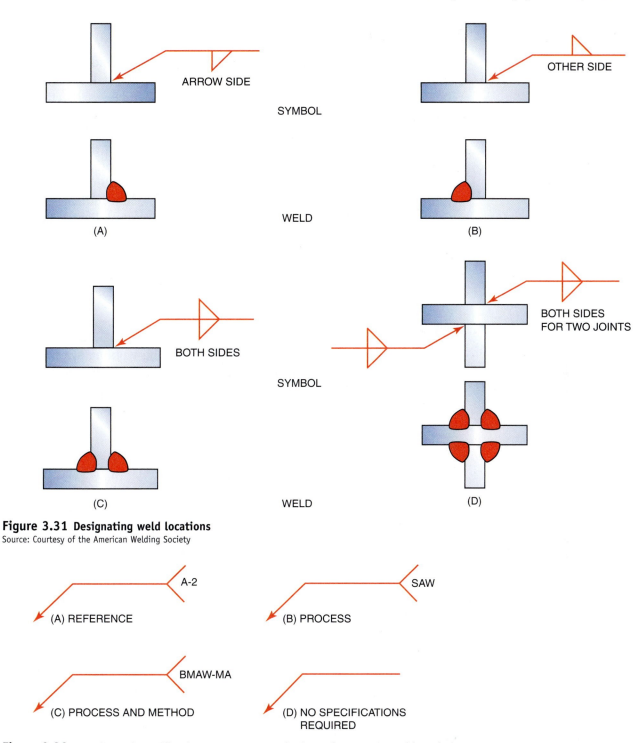

Figure 3.31 Designating weld locations
Source: Courtesy of the American Welding Society

Figure 3.32 Locations of specifications, processes, and other references on weld symbols

On a drawing, when a joint is illustrated by a single line and the arrow of a welding symbol is directed to the line, the arrow side of the joint is considered to be the near side of the joint.

For welds designated by the plug, slot, spot, seam, resistance, flash, upset, or projection welding symbols, the arrow connects the welding symbol reference line to the outer surface of one of the members of the joint at the center line of the desired weld. The member to which the arrow points is considered to be the arrow side member. The remaining member of the joint is considered to be the other side member.

FILLET WELDS

The dimensions of fillet welds are shown on the same side of the reference line as the weld symbol and to the left of the symbol, Figure 3.33A. When both sides of a joint have the same-size fillet welds, they are dimensioned as shown in Figure 3.33B. When the two sides of a joint have different-size fillet welds, both are dimensioned, Figure 3.33C. When the dimensions of one or both welds differ from the dimensions given in the general notes, both welds are dimensioned. The size of a fillet weld with unequal legs is shown in parentheses to the left of the weld symbol, Figure 3.33D. The length of a fillet weld, when indicated on the welding symbol, is shown to the right of the weld symbol, Figure 3.33E. In intermittent fillet welds, the length and pitch increments are placed to the right of the weld symbol, Figure 3.34. The first number represents the length of the weld, and the second number represents the pitch, or the distance between the centers of two welds.

Plug Welds

Holes in the arrow side member of a joint for plug welding are indicated by placing the weld symbol below the reference line. Holes in the other side member of a joint for plug welding are indicated by placing the weld

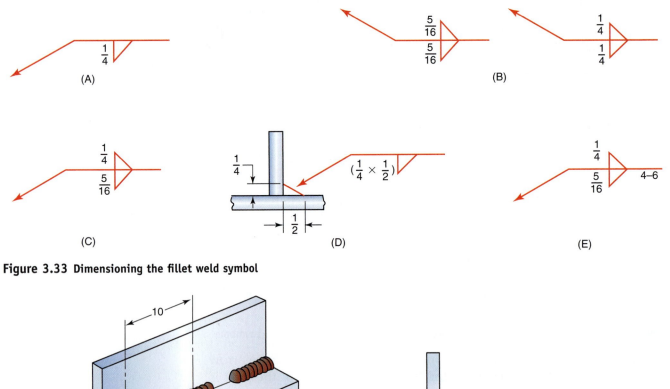

Figure 3.33 Dimensioning the fillet weld symbol

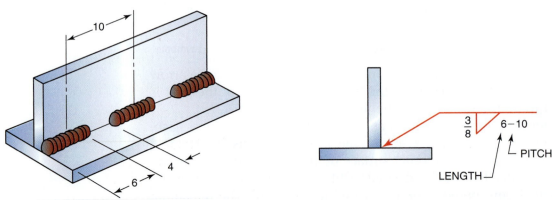

Figure 3.34 Dimensioning intermittent fillet welds

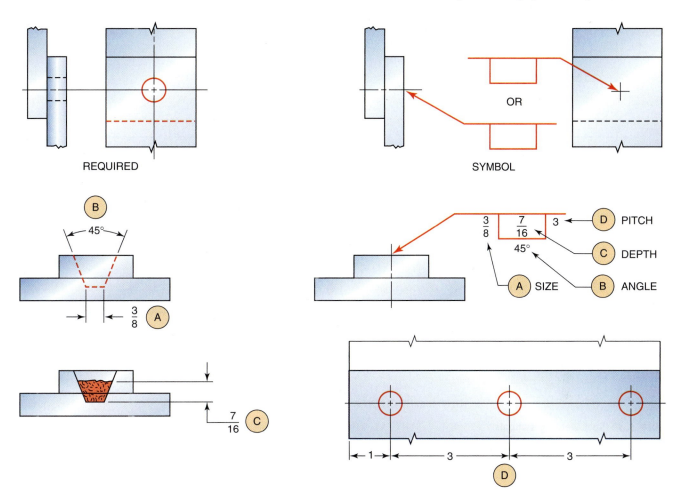

Figure 3.35 Applying dimensions to plug welds

symbol above the reference line, Figure 3.35. Refer to Figure 3.35 for the location of the dimensions used on plug welds. The diameter or size is located to the left of the symbol (A). The angle of the sides of the hole, if not square, is given above the symbol (B). The depth of buildup, if not completely flush with the surface, is given in the symbol (C). The center-to-center dimensioning, or pitch, is located on the right of the symbol (D).

Spot Welds

The dimensions of resistance spot welds are indicated on the same side of the reference line as the weld symbol, Figure 3.36. Such welds are dimensioned either by size or by strength. The size is designated as the diameter of the weld expressed in fractions or in decimal hundredths of an inch. The size is shown with or without inch marks to the left of the weld symbol. The center-to-center spacing (pitch) is shown to the right of the symbol.

The strength of spot welds is shown as the minimum shear strength in pounds (newtons) per spot and is shown to the left of the symbol, Figure 3.37A. When a specific number of spot welds is desired in a certain joint, the quantity is placed above or below the weld symbol in parentheses, Figure 3.37B.

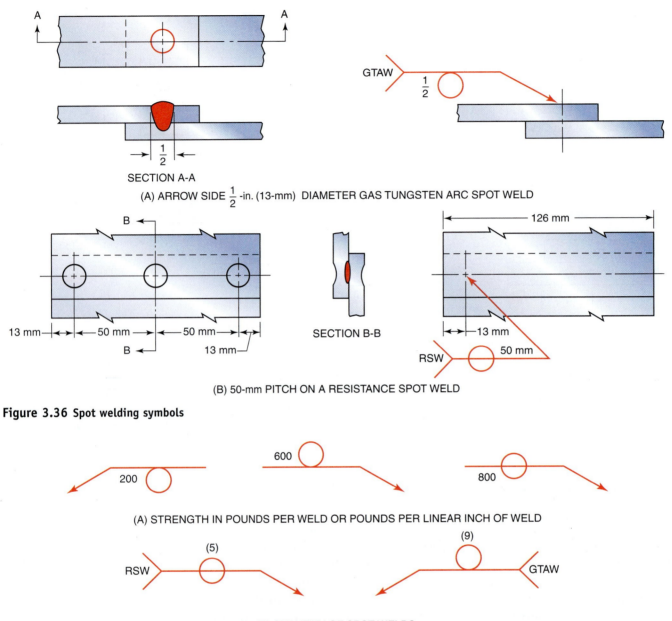

SECTION A-A

(A) ARROW SIDE $\frac{1}{2}$-in. (13-mm) DIAMETER GAS TUNGSTEN ARC SPOT WELD

SECTION B-B

(B) 50-mm PITCH ON A RESISTANCE SPOT WELD

Figure 3.36 Spot welding symbols

(A) STRENGTH IN POUNDS PER WELD OR POUNDS PER LINEAR INCH OF WELD

(B) QUANTITY OF SPOT WELDS

Figure 3.37 Designating strength and number of spot welds

Seam Welds

The dimensions of seam welds are shown on the same side of the reference line as the weld symbol. Dimensions relate to either size or strength. The size of seam welds is designated as the width of the weld expressed in fractions or decimal hundredths of an inch. The size is shown with or without the inch marks to the left of the weld symbol, Figure 3.38A. When the length of a seam weld is indicated on the symbol, it is shown to the right of the symbol, Figure 3.38B. When seam welding extends for the full distance between abrupt changes in the direction of welding, no length dimension is required on the welding symbol.

The strength of seam welds is designated as the minimum acceptable shear strength in pounds per linear inch. The strength value is placed to the left of the weld symbol, Figure 3.39.

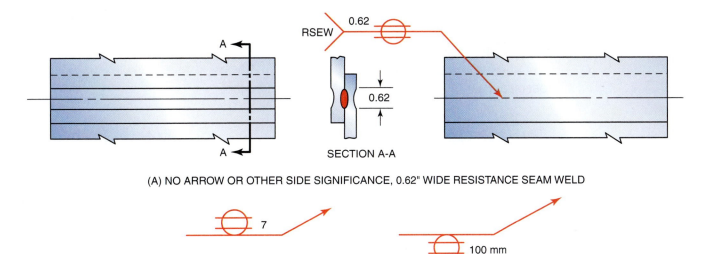

(A) NO ARROW OR OTHER SIDE SIGNIFICANCE, 0.62" WIDE RESISTANCE SEAM WELD

(B) LENGTH OF SEAM WELD

Figure 3.38 Designating the size of a seam weld

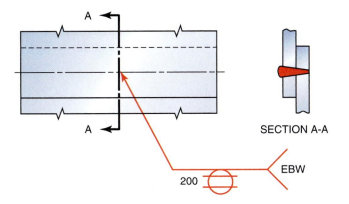

Figure 3.39 Strength of a seam weld made with an electron beam

Groove Welds

Joint strengths can be improved by making some type of groove preparation before the joint is welded. There are seven types of grooves. The groove can be made in one or both plates or on one or both sides. When a groove is cut in the plate, the weld can penetrate deeper into the joint. This helps to increase the joint strength without restricting weldment flexibility.

Grooves can be cut in base metal in a number of different ways using an oxyfuel cutting torch, air carbon arc cutting, plasma arc cutting, machining, or sawing.

The types of groove welds are classified as follows:

- single-groove and symmetrical double-groove welds that extend completely through the members being joined. No size is included on the weld symbol, Figure 3.40.
- groove welds that extend only part way through the parts being joined. The size as measured from the top of the surface to the bottom (not including reinforcement) is included to the left of the welding symbol, Figure 3.41.

The size of groove welds with a specified effective throat is indicated by showing the depth of groove preparation with the effective throat appearing

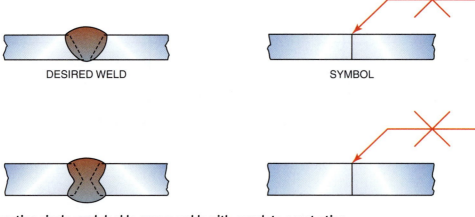

Figure 3.40 Designating single- and double-groove welds with complete penetration
Source: Courtesy of the American Welding Society

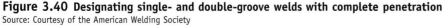

Figure 3.41 Designating the size of grooved welds with partial penetration
Source: Courtesy of the American Welding Society

in parentheses and placed to the left of the weld symbol, Figure 3.42. The size of square groove welds is indicated by showing the root penetration. The depth of chamfering and the root penetration are read in that order from left to right along the reference line.

The main purpose of the root face is to minimize the burn-through that can occur with a feather edge. The size of the root face is important to ensure good root fusion, Figure 3.43.

The size of flare groove welds is considered to extend only to the tangent points of the members, Figure 3.44.

The root opening of groove welds is the user's standard unless otherwise indicated. The root opening of groove welds, when not the user's standard, is shown inside the weld symbol, Figure 3.45.

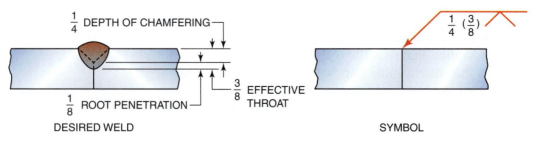

Figure 3.42 Showing size and root penetration of grooved welds
Source: Courtesy of the American Welding Society

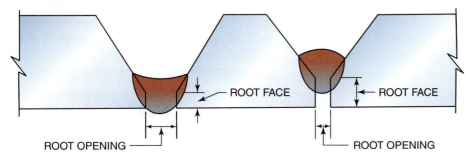

Figure 3.43 Effect of root dimensioning on groove penetration

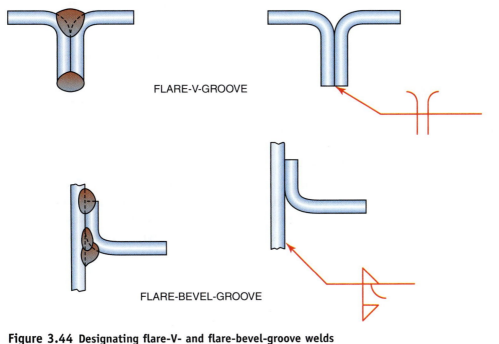

FLARE-V-GROOVE

FLARE-BEVEL-GROOVE

Figure 3.44 Designating flare-V- and flare-bevel-groove welds
Source: Courtesy of the American Welding Society

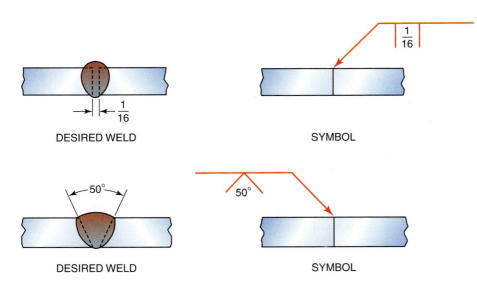

DESIRED WELD

SYMBOL

DESIRED WELD

SYMBOL

Figure 3.45 Designating root openings and included angle for groove welds
Source: Courtesy of the American Welding Society

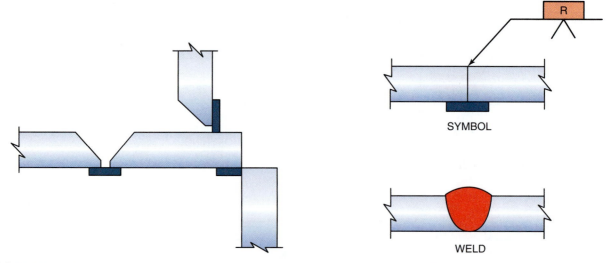

Figure 3.46 Backing strips

Figure 3.47 Butt weld with backing plate

Backing

A backing (strip) is a piece of metal placed on the back side of a weld joint. The backing must be thick enough to withstand the heat of the root pass as it is burned in. A backing strip may be used on butt joints, tee joints, and outside corner joints, Figure 3.46.

The backing may be left on the finished weld or removed after welding. If the backing is to be removed, the letter *R* is placed in the backing symbol, Figure 3.47. The backing is often removed for a finished weld because it can be a source of stress concentration and a crevice to promote rusting.

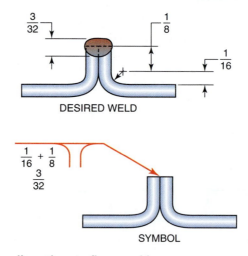

Figure 3.48 Applying dimensions to flange welds
Source: Courtesy of the American Welding Society

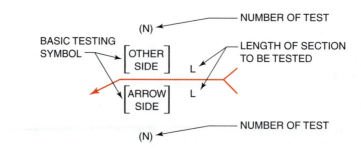

Figure 3.49 Basic nondestructive testing symbol

Flange Welds

The following welding symbols are used for light-gauge metal joints where the edges to be joined are bent to form flange or flare welds:

- Edge flange welds are shown by the edge flange weld symbol.
- Corner flange welds are indicated by the corner flange weld symbol.
- The dimensions of flange welds are shown on the same side of the reference line as the weld symbol and are placed to the left of the symbol, Figure 3.48. The radius and height above the point of tangency are indicated by showing both the radius and the height, separated by a plus sign.
- The size of the flange weld is shown by a dimension placed outward from the flanged dimensions.

Nondestructive Testing Symbols

The increased use of nondestructive testing (NDT) as a means of quality assurance has resulted in the development of standardized symbols. These symbols are used by the designer or engineer to indicate the area to be tested and the type of test to be used. The inspection symbol uses the same basic reference line and arrow as the welding symbol, Figure 3.49.

The symbol for the type of nondestructive test to be used, Table 3.3, is shown with a reference line. The location above, below, or on the line has the same significance as it does with a welding symbol: Symbols above the line indicate other side, symbols below the line indicate arrow side, and symbols on the line indicate no preference for the side to be tested, Figure 3.50. Some tests must be performed on both sides; in these cases, the symbol appears on both sides of the reference line.

Two or more tests may be required for the same section of weld. Figure 3.51 shows methods of combining testing symbols to indicate more than one type of test to be performed.

Module 3
Key Indicator 1, 2

Table 3.3 Standard Nondestructive Testing Symbols

Type of Nondestructive Test	Symbol
Visual	VT
Penetrant	PT
Dye penetrant	DPT
Fluorescent penetrant	FPT
Magnetic particle	MT
Eddy current	ET
Ultrasonic	UT
Acoustic emission	AET
Leak	LT
Proof	PRT
Radiographic	RT
Neutron radiographic	NRT

The length of weld to be tested and the number of tests to be made can be noted on the symbol. The length can be given to the right of the test symbol, usually in inches, or can be shown by the arrow line, Figure 3.52. The number of tests to be made is given in parentheses above or below the test symbol, Figure 3.53. The welding symbols and nondestructive testing symbols both can be combined into one symbol, Figure 3.54. The **combination symbol** may help both the welder and inspector to identify welds that need special attention. A special symbol can be used to show the direction of radiation used in a radiographic test, Figure 3.55.

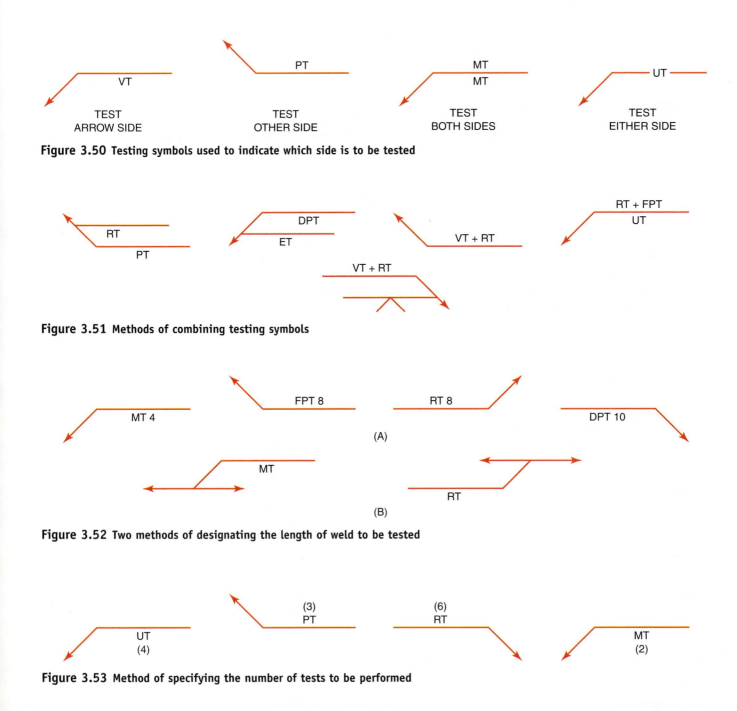

Figure 3.50 Testing symbols used to indicate which side is to be tested

Figure 3.51 Methods of combining testing symbols

Figure 3.52 Two methods of designating the length of weld to be tested

Figure 3.53 Method of specifying the number of tests to be performed

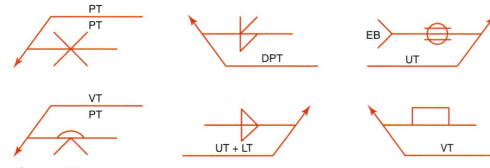

Figure 3.54 Combination weld and nondestructive testing symbols

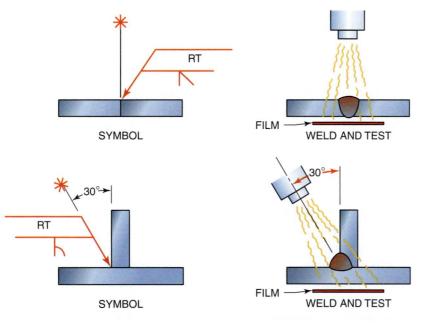

SYMBOL   FILM → WELD AND TEST

SYMBOL   FILM → WELD AND TEST

Figure 3.55 Combination symbol for weld and radiation source location for testing

SUMMARY

Mechanical drawings have been described as a universal language. In fact, Leonardo da Vinci's mechanical drawings of his ideas and inventions, dating back to the fifteenth century, are easily understood today, even though they were made more than 500 years ago and the notes were written in Italian. As a production welder you will be expected to follow simple or complex drawings in the fabrication of weldments. You must also be able to interpret the meaning of welding symbols. Understanding the significance of a welding symbol will prevent one of the most common problems in the field: overwelding. A weld that is made excessively large can cause a structural failure as easily as one that is undersized. Welded structures must often flex under load. Weldments must be flexible within limits so they can give, so they are not brittle, and so they will not break. Overwelding can cause a structure to be too rigid and subject to a brittle fracture. *Do not overweld.*

REVIEW

1. List the five joint types used in welding.
2. Sketch a V-grooved butt joint, and label all of the joint's dimensions.
3. Sketch a weld on plates in the 1G and 1F positions.
4. Sketch a weld on plates in the 2G and 2F positions.
5. Sketch a weld on plates in the 3G and 3F positions.
6. Sketch a weld on plates in the 4G and 4F positions.
7. Sketch a weld on a pipe in the 1G position.
8. Sketch a weld on a pipe in the 5G position.
9. Sketch a weld on a pipe in the 2G position.
10. Sketch a weld on a pipe in the 6G position.
11. Sketch a weld on a pipe in the 6GR position.
12. Why is it usually better to make a weld in the flat position?
13. Why are some joints back-gouged?
14. Why is cost a consideration in joint design?
15. What is contained in a set of drawings?
16. What information can be included in the title box of a drawing?
17. How is the front view of an object selected?
18. Why are sections and cut-aways used in drawings?
19. What types of information can be included on a welding symbol?
20. Why are welding symbols used?
21. What types of information may appear on the reference line of a welding symbol?
22. How is the reference line always drawn?
23. Why is a tail added to the basic welding symbol?
24. What is meant if the weld symbol is placed below the reference line?
25. How are the dimensions for a fillet weld given?
26. Sketch and dimension a V-groove weld symbol for a weld on the arrow side, with 1/8-in. root opening, 3/4 in. in size, and with a groove angle of 45°.
27. How is the removal of the backing strip noted on a welding symbol?
28. Using Figure 3.23 as a general example, prepare simple sketches of a welding table 2 ft wide 4 ft long 3 ft tall, using 1/4-in. mild steel plate and 1 1/2-in. mild steel angle iron material. Include front, top, right or left side, and bottom views.

Fabrication

OBJECTIVES

After completing this chapter, the student should be able to

- measure with a ruler or tape measure
- add and subtract whole numbers
- reduce simple fractions
- round numbers
- convert fractions to decimals
- convert decimals to fractions
- lay out a welding project

KEY TERMS

assembly	material shapes	tack welding
conversion	measuring	tolerance
fitting	part dimensions	

AWS SENSE EG2.0

Key Indicators Addressed in this Chapter:

Module 3: Drawing and Weld Symbol Interpretation

Key Indicator 1: Interprets basic elements of a drawing or sketch
Key Indicator 3: Fabricates parts from a drawing or sketch

INTRODUCTION

Fabricators must be able to communicate in the language of welding symbols, follow detailed directions, and sometimes express themselves by producing accurate technical drawings or sketches to represent weldments. Many people who previously found mathematics to be a struggle and an unpleasant chore discover fresh excitement when those numbers become part of a plan for something they are going to build with their own hands.

Fabrication shops and manufacturing facilities that want to compete in a global marketplace must be conversant with both standard and metric measuring techniques, because they may receive plans from anywhere in the world or may ship their products to anywhere in the world. This makes it necessary for you as a welder to be able to read numbers and fractions from your ruler or tape measure and convert them into decimals in the metric system or fractions in the standard system.

FABRICATION

Welders are often required to assemble parts to form a weldment. The weldment may form a completed project or may be only part of a larger structure. Some weldments are composed of two or three parts; others have hundreds or even thousands of individual parts, Figure 4.1. But even the largest weldments start by placing two parts together.

The number and type of steps required to take a plan and create a completed project vary depending on the complexity and size of the finished weldment. All welding projects start with a plan. This plan can range from a simple one that exists only in the mind of the welder to a very complex plan comprising a set of drawings. As a beginning welder, you must learn how to follow a set of drawings to produce a finished weldment.

Soon we will be fabricating large structures in space, Figures 4.2A and 4.2B. Work has already begun on the International Space Station, which

Figure 4.1 Large welded oil platform
Source: © BP p.l.c. 2003

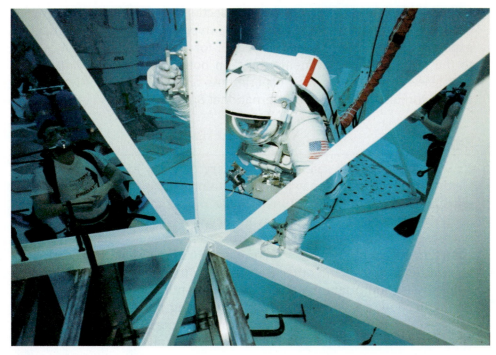

Figure 4.2A Welding in space
The neutral buoyancy tank allows divers to work in space suits underwater to simulate the microgravity of space
Source: Courtesy of NASA

Figure 4.2B Welding in space
Large structures could be fabricated in space someday
Source: Courtesy of NASA

will be assembled in space from large sections built here on Earth. Most of these assemblies will require some type of welding. Someday we expect to be welding in space. Research for welding in space dates back to the 1960s, with experiments done on board the *U.S. Sky Lab*. Today that research continues with experiments on the space shuttle program and in conjunction with the International Space Station.

Safety

As with any welding, safety is of primary concern in the fabrication of weldments. Fabrication may present certain safety problems not normally encountered in straight welding. Unlike most practice welding, a good proportion of larger fabrication work needs to be performed outside an enclosed welding booth. In addition, several welders may be working simultaneously on the same structure, so extra care must be taken to prevent burns to you or the other welders from the arc or hot sparks. Ventilation is also important, because the shop ventilation may not extend to the fabrication area. Often you will be working in an area with welding cables and torch hoses lying scattered on the floor. To prevent accidental tripping, these lines must be flat on the floor and should be covered if they are in a walkway.

These and other safety concerns are covered in Chapter 2, Safety in Welding. You should also read any safety booklets supplied with equipment before starting any project.

Shop Math

Measuring

Measuring for most welded fabrications does not require accuracies greater than those that can be obtained with a steel rule or a steel tape, Figure 4.3. Both steel rules and steel tapes are available in standard and metric units. Standard unit rules and tapes are available in fractional and decimal units, Figure 4.4.

Figure 4.3A Steel tape measures
Steel tape measures are available in lengths from 6 ft to 100 ft
Source: Courtesy of the Stanley Tool Div

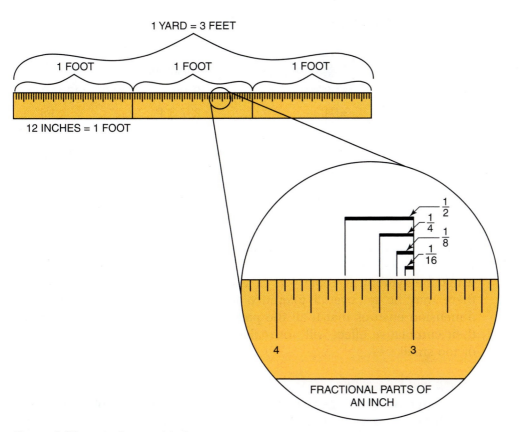

Figure 4.3B Yards, feet, and inches
The standard system of linear measurement is based on the yard. The yard is divided into 3 feet, each foot into 12 inches, and each inch into fractional parts
Source: Courtesy of Mark Huth

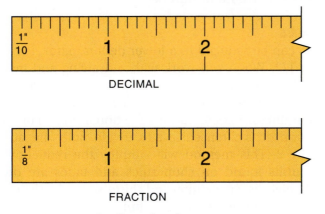

Figure 4.4 Two ways tapes can be dimensioned

Tolerances

All measuring, whether on a part or on the drawing, is in essence an estimate, because no matter how accurate the measurement is, there will always be a more accurate way of taking it. The more accurate the measurement, the more time it takes. To save time while still making an acceptable part, dimensioning **tolerances** have been established. Most drawings state a dimensioning tolerance, the amount by which the part can be larger or smaller than the stated dimensions and still be acceptable. Tolerances are usually expressed as plus (+) and minus (–). If the tolerance is the same for both the plus and the minus, it can be written

Table 4.1 Dimension Tolerances

		Acceptable Dimensions	
Dimension	**Tolerance**	**Minimum**	**Maximum**
10″	±1/8″	9 7/8″	10 1/8″
2′ 8″	±1/4″	2′ 7 3/4″	2′ 8 1/4″
10′	±1/8″	9′ 11 7/8″	10′ 1/8″
11″	±0.125	10.875″	11.125″
6′	±0.25	5′ 11.75″	6′ 0.25″
250 mm	±5 mm	245 mm	255 mm
300 mm	± 5 mm–0 mm	300 mm	305 mm
175 cm	±10 mm	174 cm	176 cm

using the symbol ±, Table 4.1. In addition to the tolerance for a part, there may be an overall tolerance for the completed weldment. This dimension ensures that if all the parts are either too large or too small, their cumulative effect will not make the completed weldment too large or too small.

Adding and Subtracting

Although most drawings give as many dimensions as possible, the welder may have to do some basic math to complete the project. Adding and subtracting fractions and mixed numbers can be accomplished quickly by following a simple rule:

RULE: Fractions that are to be added or subtracted must have the same denominator, or bottom number.

Reducing Fractions

Some fractions can be reduced to a lower denominator. For example, 2/4 is the same as 1/2. When you work with a drawing and make measurements, you can easily locate the fraction in either form on the scale. Usually, such reductions are necessary only when you are working with several different dimensions or various fractional units. For reducing fractions in the shop, it is often easiest to divide both the numerator and denominator by 2. This method will simplify the reduction because all the fractional units found on shop rules and tapes are divisible by 2: halves, fourths, eighths, sixteenths, and thirty-seconds. Using this method may require more than one reduction, but the simplicity of dividing by 2 offsets the time needed to repeat the reduction. Reduction of fractions will become easier with practice.

To reduce 4/8 in.:

$$\frac{4}{8} = \frac{4 \div 2}{8 \div 2} = \frac{2}{4}.$$

The new fraction is 2/4 in., and 2/4 can be reduced again:

$$\frac{2}{4} = \frac{2 \div 2}{4 \div 2} = \frac{1}{2}.$$

The new fraction is 1/2 in., the lowest form.

Rounding Numbers

When multiplying or dividing numbers, we often get a whole number followed by a long decimal fraction. When we divide 10 by 3, for example, we get 3.3333333. For all practical purposes, we need not lay out weldments to an accuracy greater than the second decimal place. We would therefore round off this number to 3.33, a dimension that would be easier to work with in the welding shop.

RULE: When rounding off a number, look at the number to the right of the last significant place to be used. If this number is less than 5, drop it and leave the remaining number unchanged. If this number is 5 or greater, increase the last significant number by 1 and record the new number.

Round off 15.6549 to the second decimal place:

Because the number in the third place is less than 5, the new number would be 15.65.

Round off 8.2764 to the second decimal place:

Because the number in the third place is 5 or more, the new number would be 8.28.

Round off 0.8539 to the third decimal place:

Because the number in the fourth place is 5 or more, the new number would be 0.854.

Round off 156.8244 to the first decimal place:

Because the number in the second place is less than 5, the new number would be 156.8.

Converting Fractions to Decimals

From time to time it may be necessary to convert fractional numbers to decimal numbers. A fraction-to-decimal **conversion** is needed before most calculators can be used to solve problems containing fractions, though some calculators allow the inputting of fractions without converting them to decimals.

RULE: To convert a fraction to a decimal, divide the numerator (top number in the fraction) by the denominator (bottom number in the fraction).

To convert 3/4 to a decimal:

$3 \div 4 = 0.75$.

To convert 7/8 to a decimal:

$7 \div 8 = 0.875$.

Converting Decimals to Fractions

This process is less exact than the conversion of fractions to decimals. Except for specific decimals, the conversion will leave a remainder unless a small enough fraction is selected. For example, if you are converting 0.765 to the nearest 1/4 in., 3/4 in. would be acceptable, and this conversion would leave a remainder of 0.015 in. (0.765 – 0.75 = 0.015). If you are working to a ±1/8-in. tolerance, which has up to a 1/4-in. difference from the minimum to maximum dimensions, a measurement of 3/4 is acceptable. More accurately, 0.765 can be converted to 49/64 in., a dimension that would be hard to lay out and impossible to cut using a hand torch.

RULE: To convert a decimal to a fraction, multiply the decimal by the denominator of the fractional units desired; that is, for eighths (1/8) use 8, for fourths (1/4) use 4, and so on. Place the whole number (dropping or rounding off the decimal remainder) over the fractional denominator used.

To convert 0.75 to fourths:

0.75 × 4 = 3.0 or 3/4.

To convert 0.75 to eighths:

0.75 × 8 = 6.0 or 6/8, which will reduce to 3/4.

To convert 0.51 to fourths:

0.51 × 4 = 2.04 or 2/4, which will reduce to 1/2.

To convert 0.47 to eighths:

0.47 × 8 = 3.76 or 3/8.

(Note that the 0.76 of the 3.76 is more than 0.5, so it could be rounded up, giving 4 or 4/8, which will reduce to 1/2.)

Conversion Charts

Occasionally a welder must convert the units used on a drawing to the type of units used on a layout rule or tape. Fortunately, charts are available that can easily be used to convert between fractions, decimals, and metric units, Table 4.2. To use these charts, locate the original dimension and then look at the dimension in the adjacent column(s) for the new units required.

To convert 1/16 in. to millimeters:

1/16 in. = 1.5875 mm.

To convert 0.5 in. to a fraction:

0.5 in. = 1/2 in.

To convert 0.375 in. to millimeters:

0.375 in. = 9.525 mm.

To convert 25 mm to a decimal inch:

25 mm = 0.98425 in.

To convert 19 mm to a fractional inch:

19 mm = 3/4 in. (approximately).

Both metric-to-standard conversions and standard-to-metric conversions often result in answers that contain long decimals number strings and cannot be easily located on the rule or tape. Most of the layout and fabrication work welders perform will not require such levels of accuracy. These small decimal fractions, in inches or millimeter scales, represent such a small difference that they cannot be laid out with a steel rule or tape. Such small differences can be important to some weldments, but in these cases some machining is required to obtain that level of accuracy. Because these small units are not normally included in a layout, they can be rounded off. Round off millimeter units to the nearest whole number; for example, 19.050 mm would be 19 mm, 1.5875 mm would be 2 mm, and so on. Round off decimal inch units to the nearest 1/16-in. fractional unit; for example, 0.47244 in. would become 0.5 in. (1/2″), and 0.23622 in. would become 0.25 in. (1/4″). In both cases of rounding, the whole number obtained is well within most welding layout and fabrication drawing tolerances, which are usually ±1/16 in. or ±1/8 in.

Table 4.2 Conversion of Decimal Inches to Millimeters and Fractional Inches to Decimal Inches and Millimeters

				Inches			Inches		
Inches dec	mm	Inches dec	mm	frac	dec	mm	frac	dec	mm
0.01	0.2540	0.51	12.9540	1/64	0.015625	0.3969	33/64	0.515625	13.0969
0.02	0.5080	0.52	13.2080	1/32	0.031250	0.7938	17/32	0.531250	13.4938
0.03	0.7620	0.53	13.4620						
0.04	1.0160	0.54	13.7160	3/64	0.046875	1.1906	35/64	0.546875	13.8906
0.05	1.2700	0.55	13.9700						
0.06	1.5240	0.56	14.2240	1/16	0.062500	1.5875	9/16	0.562500	14.2875
0.07	1.7780	0.57	14.4780	5/64	0.078125	1.9844	37/64	0.578125	14.6844
0.08	2.0320	0.58	14.7320						
0.09	2.2860	0.59	14.9860	3/32	0.093750	2.3812	19/32	0.593750	15.0812
0.10	2.5400	0.60	15.2400	7/64	0.109375	2.7781	39/64	0.609375	15.4781
0.11	2.7940	0.61	15.4940						
0.12	3.0480	0.62	15.7480	1/8	0.125000	3.1750	5/8	0.625000	15.8750
0.13	3.3020	0.63	16.0020						
0.14	3.5560	0.64	16.2560	9/64	0.140625	3.5719	41/64	0.640625	16.2719
0.15	3.8100	0.65	16.5100	5/32	0.156250	3.9688	21/32	0.656250	16.6688
0.16	4.0640	0.66	16.7640						
0.17	4.3180	0.67	17.0180	11/64	0.171875	4.3656	43/64	0.671875	17.0656
0.18	4.5720	0.68	17.2720	3/16	0.187500	4.7625	11/16	0.687500	17.4625
0.19	4.8260	0.69	17.5260						
0.20	5.0800	0.70	17.7800	13/64	0.203125	5.1594	45/64	0.703125	17.8594
0.21	5.3340	0.71	18.0340	7/32	0.218750	5.5562	23/32	0.718750	18.2562
0.22	5.5880	0.72	18.2880						
0.23	5.8420	0.73	18.5420	15/64	0.234375	5.9531	47/64	0.734375	18.6531
0.24	6.0960	0.74	18.7960						
0.25	6.3500	0.75	19.0500	1/4	0.250000	6.3500	3/4	0.750000	19.0500
0.26	6.6040	0.76	19.3040	17/64	0.265625	6.7469	49/64	0.765625	19.4469
0.27	6.8580	0.77	19.5580						
0.28	7.1120	0.78	19.8120	9/32	0.281250	7.1438	25/32	0.781250	19.8437
0.29	7.3660	0.79	20.0660	19/64	0.296875	7.5406	51/64	0.796875	20.2406
0.30	7.6200	0.80	20.3200						
0.31	7.8740	0.81	20.5740	5/16	0.312500	7.9375	13/16	0.812500	20.6375
0.32	8.1280	0.82	20.8280						
0.33	8.3820	0.83	21.0820	21/64	0.328125	8.3344	53/64	0.828125	21.0344
0.34	8.6360	0.84	21.3360	11/32	0.343750	8.7312	27/32	0.843750	21.4312
0.35	8.8900	0.85	21.5900						
0.36	9.1440	0.86	21.8440	23/64	0.359375	9.1281	55/64	0.859375	21.8281
0.37	9.3980	0.87	22.0980	3/8	0.375000	9.5250	7/8	0.875000	22.2250
0.38	9.6520	0.88	22.3520						
0.39	9.9060	0.89	22.6060	25/64	0.390625	9.9219	57/64	0.890625	22.6219
0.40	10.1600	0.90	22.8600	13/32	0.406250	10.3188	29/32	0.906250	23.0188
0.41	10.4140	0.91	23.1140						
0.42	10.6680	0.92	23.3680	27/64	0.421875	10.7156	59/64	0.921875	23.4156
0.43	10.9220	0.93	23.6220						
0.44	11.1760	0.94	23.8760	7/16	0.437500	11.1125	15/16	0.937500	23.8125
0.45	11.4300	0.95	24.1300	29/64	0.453125	11.5094	61/64	0.953125	24.2094
0.46	11.6840	0.96	24.3840						
0.47	11.9380	0.97	24.6380	15/32	0.468750	11.9062	31/32	0.968750	24.6062
0.48	12.1920	0.98	24.8920	31/64	0.484375	12.3031	62/64	0.984375	25.0031
0.49	12.4460	0.99	25.1460						
0.50	12.7000	1.00	25.4000	1/2	0.500000	12.7000	1	1.000000	25.4000

Using the rounding-off method with the conversion chart makes the converted units easier to locate on rules and tapes.

To convert 1/2 in. to millimeters:

1/2 in. = 13 mm.

To convert 0.625 in. to millimeters:

0.625 in. = 16 mm.

To convert 2 3/4 in. to millimeters:

$2 \times 25.4 = 50.8$

$3/4 \qquad = \underline{\underline{19.0}}$

$\qquad\qquad = 69.8$ rounded to 70 mm.

To convert 5.5 in. to millimeters:

$5 \times 25.4 = 127.0$

$0.5 \qquad = \underline{\underline{12.0}}$

$\qquad\qquad = 139.7$ rounded to 140 mm.

To convert 10 mm to fractional inches:

10 mm = 3/8 in.

To convert 14 mm to decimal inches:

14 mm = 0.5625 in.

To convert 300 mm to fractional inches:

$300 \div 25.4 = 11.81$ in. rounded to 11 13/16 in.

To convert 240 mm to decimal inches:

$240 \div 25.4 = 9.44$ in. rounded to 9 7/16 in.

Module 3
Key Indicator 1, 3

LAYOUT

The fabrication of parts may require that the welder lay out lines and locate points for cutting, bending, drilling, and assembling. Lines may be marked with a soapstone or a chalkline, scratched with a metal scribe, or punched with a center punch. If a piece of soapstone is used, it should be sharpened properly to increase accuracy, Figure 4.5. A chalk line will make a long, straight line on metal and is best used on large jobs, Figure 4.6. Both the scribe and punch can be used to lay out an accurate line, but the

Figure 4.6A Chalk lines
Pull the chalk line tight and then snap it
Source: Courtesy of Larry Jeffus

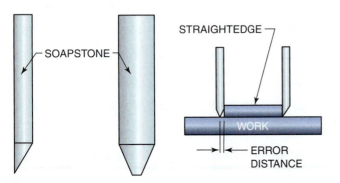

Figure 4.5 Proper method of sharpening a soapstone

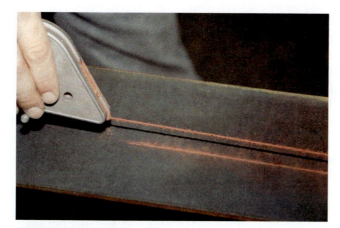

Figure 4.6C Chalk line reel
Source: Courtesy of Larry Jeffus

Figure 4.6B Chalk lines
Check that the line is dark enough to be clearly visible
Source: Courtesy of Larry Jeffus

punched line is easier to see when cutting. A punch can be held as shown in Figure 4.7, with the tip just above the surface of the metal. When the punch is struck with a lightweight hammer, it will make a mark. If you move your hand along the line and strike the punch in rapid succession, it will leave a series of punch marks for the cut to follow.

Always start a layout as close to a corner of the material as possible. By starting in a corner or along the edge, you can take advantage of the preexisting cut as well as reduce wasted material.

It is easy to cut the wrong line. In welding shops one person may lay out the parts and another make the cuts. Even when one person does both jobs, it is easy to cut the wrong line, either because of the restricted view through cutting goggles or because of the large number of lines on a part. To avoid making a cutting mistake, always identify whether lines are being used for cutting, for locating bends, as drill centers, or as assembly locations. The lines not to be cut may be marked with an *X*, or they may be identified by writing directly on the part. Mark the side of the line that is scrap so that when the kerf is removed from that side the part will be the proper size, Figure 4.8. Any lines that have been used for constructing

Figure 4.7 Punches
Holding the punch slightly above the surface allows the punch to be struck rapidly and moved along a line to mark it for cutting

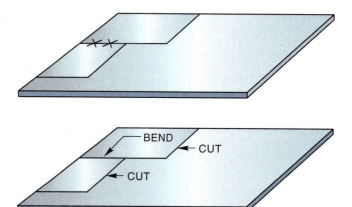

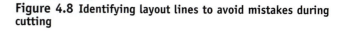

Figure 4.8 Identifying layout lines to avoid mistakes during cutting

the actual layout line or to locate points for drilling or are made in error must be erased completely or clearly marked to avoid confusion during cutting and assembly.

Some shops have their own shorthand methods for identifying layout lines, or you can develop your own system. Failure to develop and use a system for identifying lines will eventually result in a mistake. In a welding shop you will find only people who have made the wrong cut and people who will make the wrong cut. When it does happen, check with the welding shop supervisor to see what corrective steps can be taken. One advantage for most welding assemblies is that many cutting errors can be repaired by welding. Prequalified procedures are often established for just such an event, so check before deciding to scrap the part.

The process of laying out a part may be affected by the following factors:

- *material shape*—Figure 4.9 lists the most common metal shapes used for fabrication. Flat stock such as sheets and plates is easiest to lay out, and pipes and round tubing are the most difficult shapes to work with.
- *part shape*—Parts with square and straight cuts are easier to lay out than are parts with angles, circles, curves, and irregular shapes.
- *tolerance*—The smaller or tighter the tolerance that must be maintained, the more difficult the layout.
- *nesting*—The placement of parts together in a manner that will minimize the waste created is called *nesting*.

Parts with square or straight edges are the easiest to lay out. Simply measure the distance and use a square or straight edge to lay out the line to be cut, Figure 4.10. Straight cuts that are to be made parallel to an edge can be drawn by using a combination square and a piece of soapstone or scriber. Set the combination square to the correct dimension and drag it along the edge of the plate while holding the soapstone or scriber at the end of the combination square's blade, Figure 4.11.

Module 3
Key Indicator 1, 3

PRACTICE 4-1

Layout Square, Rectangular, and Triangular Parts

Using a piece of metal or paper, soapstone or pencil, tape measure, and square, lay out the parts shown in Figure 4.12. The parts must be laid out within ±1/16 in. of the dimensions. Convert the dimensions into S.I. metric units of measure.

Complete a copy of the Student Welding Report in Appendix I or provided by your instructor. ∎

PRACTICE 4-2

Laying Out Circles, Arcs, and Curves

Circles, arcs, and curves can be laid out using either a compass or a circle template, Figure 4.13. The diameter is usually given for a hole or round part, and the radius is usually given for arcs and curves, Figure 4.14. The center of the circle, arc, or curve may be located using dimension lines

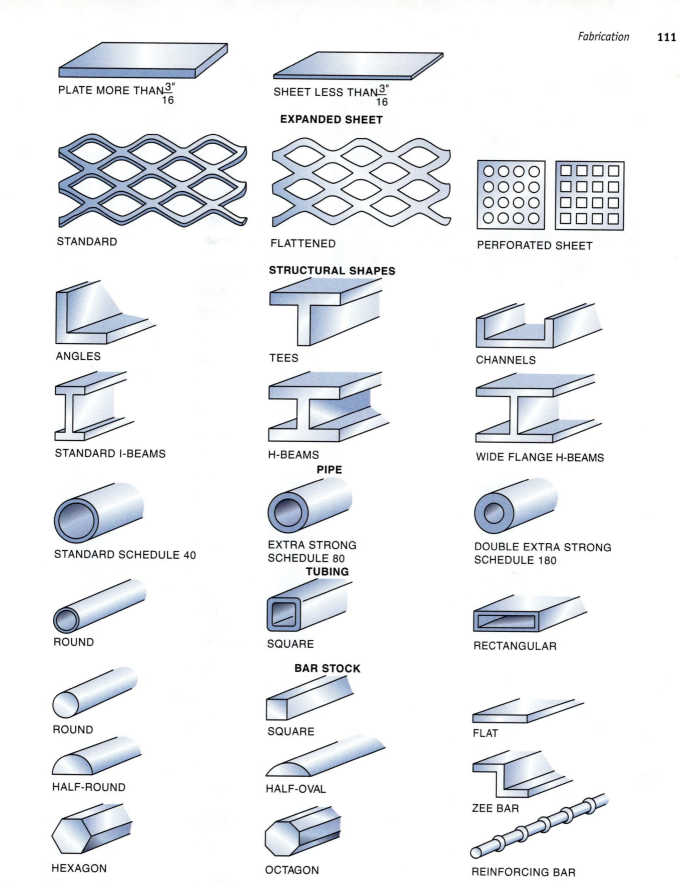

PLATE MORE THAN $\frac{3"}{16}$

SHEET LESS THAN $\frac{3"}{16}$

EXPANDED SHEET

STANDARD

FLATTENED

PERFORATED SHEET

STRUCTURAL SHAPES

ANGLES

TEES

CHANNELS

STANDARD I-BEAMS

H-BEAMS

WIDE FLANGE H-BEAMS

PIPE

STANDARD SCHEDULE 40

EXTRA STRONG SCHEDULE 80

DOUBLE EXTRA STRONG SCHEDULE 180

TUBING

ROUND

SQUARE

RECTANGULAR

BAR STOCK

ROUND

SQUARE

FLAT

HALF-ROUND

HALF-OVAL

ZEE BAR

HEXAGON

OCTAGON

REINFORCING BAR

Figure 4.9 Standard metal shapes
Most shapes are available with different surface finishes, such as hot-rolled, cold-rolled, or galvanized

Figure 4.10 Using a square to draw a straight line
Source: Courtesy of Larry Jeffus

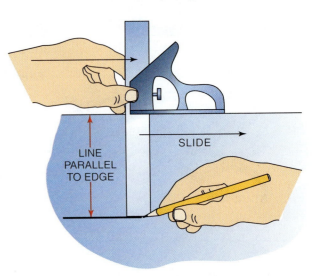

Figure 4.11 Using a combination square to lay out a strip of metal

and center lines. Curves and arcs that are to be made tangent to another line may be dimensioned with only their radiuses, Figure 4.14.

Using a piece of metal or paper, soapstone or pencil, tape measure, compass, or circle template and square, lay out the parts shown in

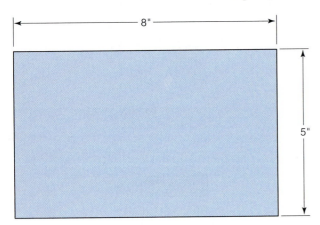

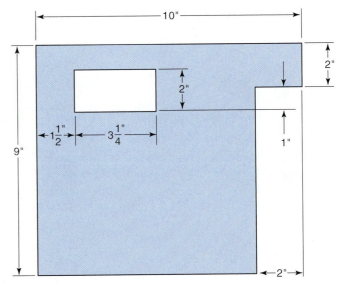

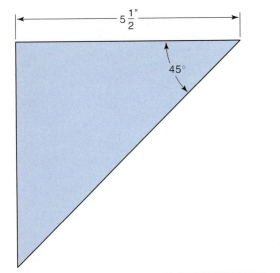

Figure 4.12 Layout parts for Practice 4-1

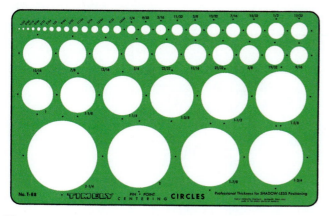

Figure 4.13A Circle template
Source: Courtesy of Timely Products Co., Inc.

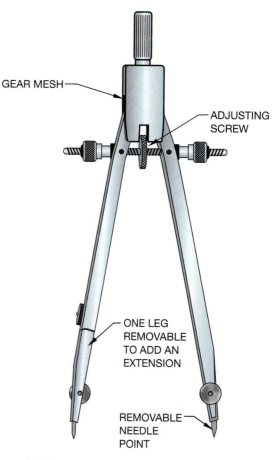

Figure 4.13B Compass
Source: Courtesy of J. S. Staedtler, Inc.

Figure 4.15. The parts must be laid out within ±1/16 in. of the dimensions. Convert the dimensions into S.I. metric units of measure.

Complete a copy of the Student Welding Report in Appendix I or provided by your instructor. ∎

Nesting

Laying out parts so that the least amount of scrap is produced is important. Odd-shaped and unusually sized parts often produce the largest amount of scrap. Computers can be used to lay out nested parts with a minimum of scrap. Some computerized cutting machines can also be programmed to nest parts.

Manual nesting of parts may require several tries at laying out the parts to achieve the least possible scrap.

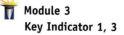 Module 3
Key Indicator 1, 3

PRACTICE 4-3

Nesting Layout

Using metal or paper that is 8 1/2 in. × 11 in., soapstone or pencil, tape measure, and square, lay out the parts shown in Figure 4.16 in a manner that results in the least scrap. Assume a 0-in. kerf width. Use as many 8 1/2-in. × 11-in. pieces of stock as necessary to produce the parts

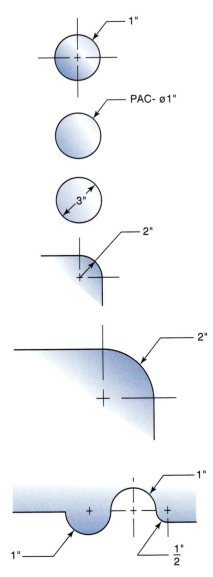

Figure 4.14 Dimensioning for arcs, curves, radii, and circles

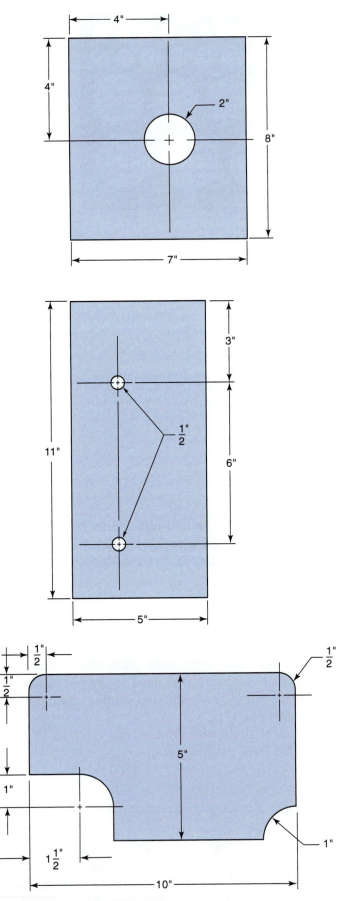

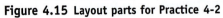

Figure 4.15 Layout parts for Practice 4-2

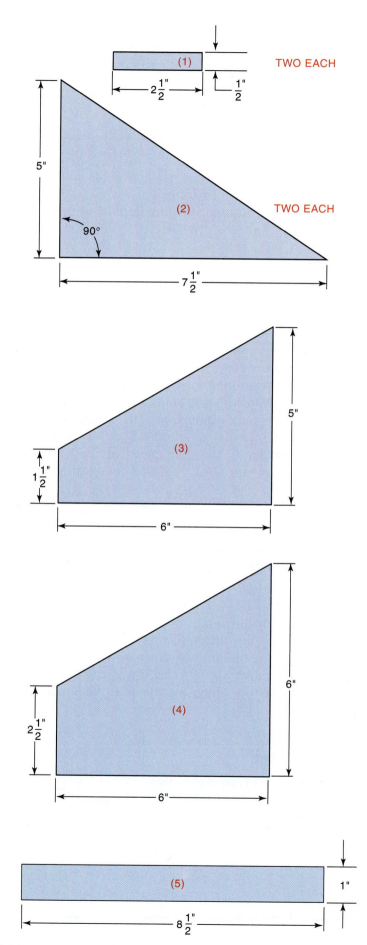

Figure 4.16 Parts to be nested

using your layout. The parts must be laid out within ±1/16 in. of the dimensions. Convert the dimensions into S.I. metric units of measure.

Complete a copy of the Student Welding Report in Appendix I or provided by your instructor. ▪

PRACTICE 4-4

Bill of Materials

Using the parts laid out in Practice 4-3 and paper and pencil, fill out the bill of materials form shown in Table 4.3.

Complete a copy of the Student Welding Report in Appendix I or provided by your instructor. ▪

Kerf Space

Because all cutting processes, except shearing, produce a kerf, this space must be included when parts are laid out side by side. The *kerf* is the space created as material is removed during a cut. The width of a kerf varies depending on the cutting process used. Of the cutting processes used in most shops, the metal saw produces one of the smallest kerfs and the hand-held oxyfuel cutting torch can produce one of the widest.

When only one or two parts are being cut, the kerf width may not need to be added to the **part dimension**. This space may be taken up during assembly by the root gap required for a joint. If a large number of parts are being cut out of a single piece of stock, the kerf width can add up and increase the stock required for cutting out the parts, Figure 4.17.

Module 3
Key Indicator 1, 3

PRACTICE 4-5

Allowing Space for the Kerf

Using a pencil, 8 1/2-in. × 11-in. paper, measuring tape or rule, and square, lay out four rectangles 2 1/2 in. × 5 1/4 in. down one side of the paper, leaving 3/32 in. for the kerf.

Two ways can be used to provide for the kerf spacing. One method is to draw a double line on the side of the part where the kerf is to be made, Figure 4.18. The other way is to lay out a single line and place an X on the side of the line on which the cut is to be made, Figure 4.19.

Table 4.3 Bill of Materials Form

Part	Number Required	Type of Material	Size	
			Standard Units	S. I. Units

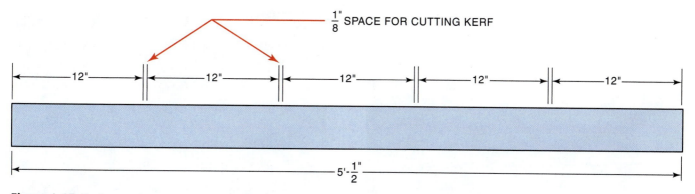

Figure 4.17 Kerf
Because of the kerf, an additional 1/2 in. of stock would be required to make these five 1-ft pieces

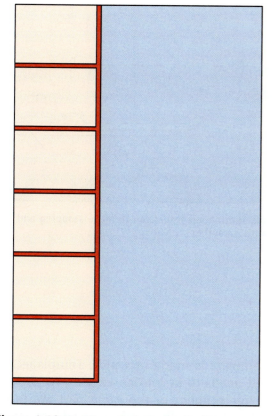

Figure 4.18 Kerf is made between the lines

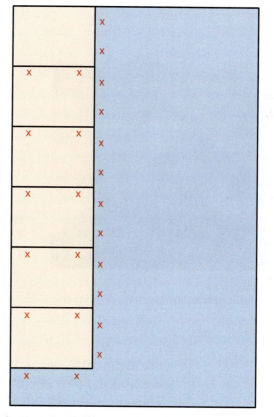

Figure 4.19 Xs mark the side of the line on which the kerf is to be made

Note that no kerf space need be left along the sides made next to the edge of the paper or next to the scrap. What is the total length and width of material needed to lay out these four parts?

Complete a copy of the Student Welding Report in Appendix I or provided by your instructor. ∎

Parts can be laid out by tracing either an existing part or a template, Figure 4.20. When using either process, be sure the line you draw is made as tight as possible to the part's edge, Figure 4.21. The inside edge of the line is the exact size of the part. Make the cut on the line or to the outside so that the part will be the correct size once it is cleaned

Figure 4.20A Tracing a part
Source: Courtesy of Larry Jeffus

Figure 4.20B Using a straight edge to make a line
Source: Courtesy of Larry Jeffus

Figure 4.21 Using a soapstone
Be sure that the soapstone is held tightly into the part being traced
Source: Courtesy of Larry Jeffus

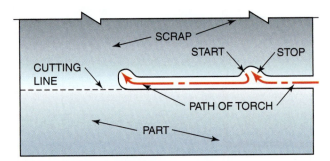

Figure 4.22 Turning out into scrap to make stopping and starting points smoother

up, Figure 4.22. Sometimes a template is made of a part. Templates are useful if the part is complex and needs to fit into an existing weldment. They are also helpful when a large number of the same part are to be made or when the part is only occasionally used. The advantage of using templates is that once the detailed layout work is completed, exact replicas can be made anytime they are needed. Templates can be made out of heavy paper, cardboard, wood, sheet metal, or other appropriate material. The sturdier the material, the longer the template will last.

Special tools have been developed to aid in laying out parts; one such tool is the *contour marker*, Figure 4.23. These markers are highly accurate when properly used, but they do require a certain amount of practice. A user familiar with this tool can lay out a huge variety of joints (within the limits of the tool being used). One advantage of tools like the contour marker is that all sides of a cut in structural shapes and pipe can be laid out from one side without relocating the tool, Figure 4.24.

Figure 4.23 Pipe lateral being laid out with contour marker
Source: Courtesy of Larry Jeffus

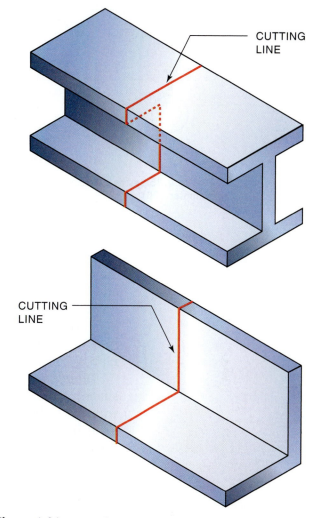

Figure 4.24 Layout for structural shapes

MATERIAL SHAPES

Metal stock can be purchased in a wide variety of shapes, sizes, and materials. Weldments may be constructed from combinations of **material shapes** and sizes. Only a single type of metal is used in most weldments, unless a special property such as corrosion resistance is required. In such cases dissimilar metals may be joined into the fabrication wherever they are needed. The most common metal used is carbon steel, and the most common shapes used are plate, sheet, pipe, tubing, and angles.

Plate is usually 3/16 in. (4.7 mm) or thicker and measured in inches and fractions of inches. Plates are available in widths from 12 in. (305 mm) up to 96 in. (2438 mm) and lengths from 8 ft (2.4 m) to 20 ft (6 m). Thickness ranges up to 12 in. (305 mm).

Sheets are usually 3/16 in. (4.7 mm) or less and measured in gauge or decimals of an inch. Several different gauge standards are used. The two most common are the Manufacturer's Standard Gauge for Sheet Steel, used for carbon steels, and the American Wire Gauge, used for most non-ferrous metals, such as aluminum and brass.

Pipe is dimensioned by its diameter and schedule, or strength. Pipe that is smaller than 12 in. (305 mm) is dimensioned by its inside diameter, and

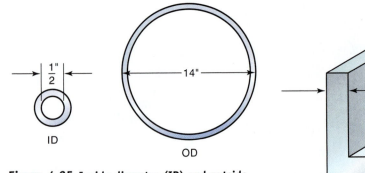

Figure 4.25 Inside diameter (ID) and outside diameter (OD)

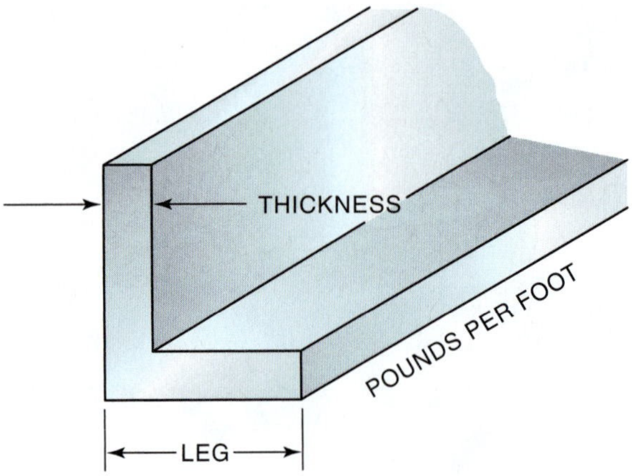

Figure 4.26 Specifications for sizing angles

the outside diameter is given for pipe that is 12 in. (305 mm) in diameter and larger, Figure 4.25. The strength of pipe is given as a schedule. Schedules 10 through 180 are available; schedule 40 is often considered a standard strength. The wall thickness for pipe is determined by its schedule (pressure range). The larger the diameter of the pipe, the greater its area. Pipe is available in welded (seamed) and extruded (seamless) forms.

Tubing sizes are always given as the outside diameter. The desired shape of tubing, such as square, round, or rectangular, must be listed with the ordering information. The wall thickness of tubing is measured in inches (millimeters) or as Manufacturer's Standard Gauge for Sheet Metal. Tubing should also be specified as rigid or flexible. The strength of tubing may also be specified as the ability of tubing to withstand compression, bending, or twisting loads.

Angles, sometimes referred to as angle iron, are dimensioned by giving the length of the legs of the angle and their thickness, Figure 4.26. Stock lengths of angles are 20 ft, 30 ft, 40 ft, and 60 ft (6 m, 9.1 m, 12.2 m, and 18.3 m).

Module 3
Key Indicator 1, 3

ASSEMBLY

The **assembly** process, bringing together all the parts of the weldment, requires proficiency in several areas. You must be able to read the drawing and interpret the information provided there to properly locate each part. An assembly drawing has the necessary information, both graphically and dimensionally, to allow the various parts to be properly located as part of the weldment. If the assembly drawing includes either pictorial or exploded views, this process is much easier for the beginning assembler; however, most assembly drawings are given as two, three, or more orthographic views, Figure 4.27. Orthographic views are more difficult to interpret until you have developed an understanding of their various elements.

On very large projects such as buildings or ships, a corner or centerline is established as a baseline. This is the point where all measurements for part location begin. With smaller weldments, a single part may be selected as a starting point. Often, the selection of the base part is automatic because all other parts are to be joined to this central part. On other weldments, however, the selection of a base part is strictly up to the assembler.

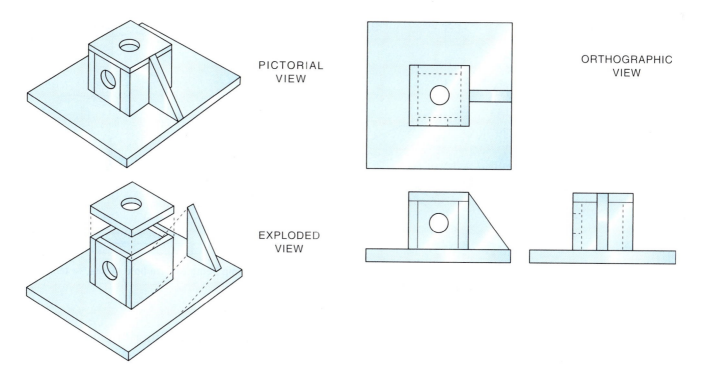

Figure 4.27 Types of drawings that can be used to show a weldment assembly

Select the largest or most central part to be the base for your assembly. All other parts will then be aligned to this one part. Using a base also helps to prevent location and dimension errors. Otherwise, a slight misalignment of one part, even within tolerances, will be compounded by the misalignment of other parts, resulting in an unacceptable weldment. Using a baseline or base part will result in a more accurate assembly.

Identify each part of the assembly and mark each piece for future reference. If it helps, you can hold the parts together and compare their orientation to the drawing. Locate points on the parts that can be easily identified on the drawing, such as holes and notches, Figure 4.28. Now

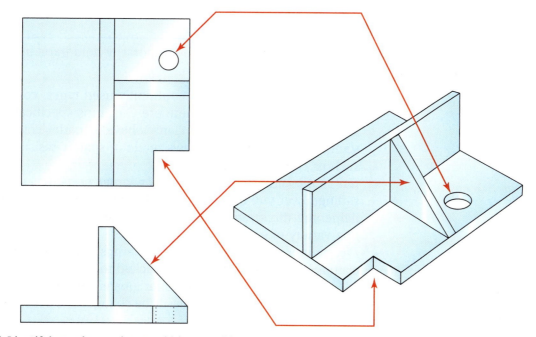

Figure 4.28 Identifying unique points to aid in assembly

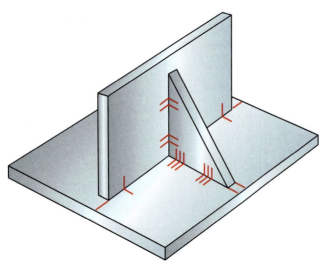

Figure 4.29 Layout markings to help locate the parts for tack welding

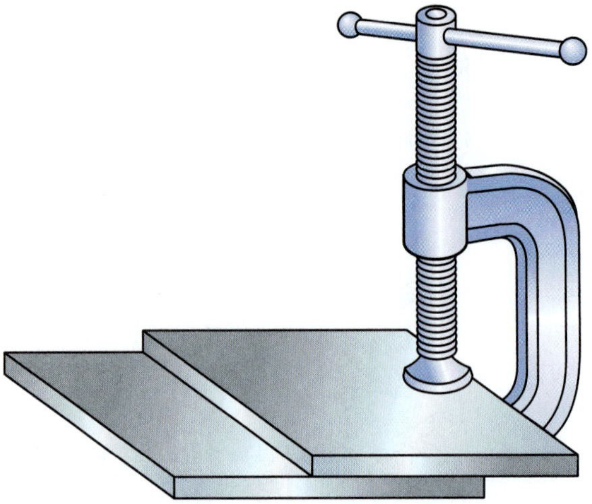

Figure 4.30 C-clamp being used to hold plates for tack welding
Source: Courtesy of Mike Gelleman

mark the location of these parts—top, front, or other orientation—so you can locate them during assembly.

Layout lines and other markings can be made on the base to locate other parts. Using a consistent method of marking helps prevent mistakes. One method is to draw parallel lines on both parts where they meet, Figure 4.29.

After the parts have been identified and marked, they can be either held by hand or clamped into place. Holding the parts in alignment by hand for tack welding is fast, but it often leads to errors and thus is not recommended for beginning assemblers. Experienced assemblers recognize that clamping the parts in place before tack welding is a much more accurate method, Figure 4.30.

ASSEMBLY TOOLS

Clamps

A variety of clamps can be used to temporarily hold parts in place so that they can be tack welded.

- *C-clamps*, among the most commonly used types, come in a variety of sizes, Figure 4.31. There are C-clamps specifically designed for welding. Some of these clamps have a spatter cover over the screw, and others have screws made of spatter-resistant materials such as copper alloys.
- *Bar clamps* are useful for clamping larger parts. Bar clamps have a sliding lower jaw that can be positioned against the part before tightening the screw clamping end, Figure 4.32. They are available in a variety of lengths.
- *Pipe clamps* are very similar to bar clamps. The advantage of pipe clamps is that the ends can be attached to a section of standard 1/2-in. pipe. This feature allows for greater flexibility in the length of the clamp, and the pipe can easily be changed if it becomes damaged.

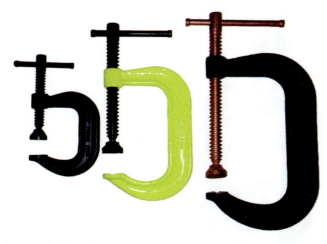

Figure 4.31 C-clamps
Source: Courtesy of Larry Jeffus

Figure 4.32 Bar clamps
Source: Courtesy of Woodworker's Supply Inc.

- *Locking pliers* are available in a range of sizes with a number of jaw designs, Figure 4.33. Their versatility and gripping strength make locking pliers very useful. Some locking pliers have a self-adjusting mechanism that allows them to be moved between parts of different thicknesses without the need to readjust them.
- *Cam-lock clamps*, also known as *toggle clamps*, are specialty clamps that are often used in conjunction with a jig or a fixture. They can be preset, allowing for faster work, Figure 4.34.
- *Specialty clamps* such as those for pipe welding, Figure 4.35, are available for many different types of jobs. Such specialty clamps make it possible to do faster and more accurate assembling.

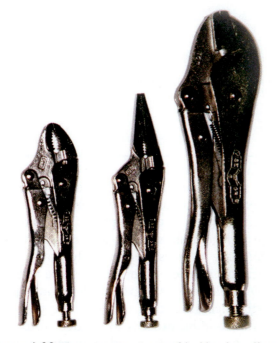

Figure 4.33 Three common types of locking jaw pliers
Source: Courtesy of Larry Jeffus

Figure 4.34 Toggle clamps
Source: Courtesy of Woodworker's Supply Inc.

Figure 4.35A Pipe alignment clamps and level
Source: Courtesy of Larry Jeffus

Figure 4.35B Pipe alignment clamp
Source: Courtesy of Larry Jeffus

Figure 4.35C Pipe gap adjustment tool
Source: Courtesy of Larry Jeffus

Fixtures

Fixtures are devices made to aid in assemblies and the fabrication of weldments. When a number of similar parts are to be made, fixtures are helpful. They can increase speed and accuracy in the assembly of parts. Fixtures must be strong enough to support the weight of the parts, be able to withstand the rigors of repeated assemblies, and remain within tolerance. They may have clamping devices permanently attached to speed up their use. Often, locating pins or other devices are used to ensure proper part location. A well-designed fixture allows adequate room for the welder to make the necessary tack welds. Some parts are left in the fixture throughout the welding process to reduce distortion. Making fixtures for every job is cost prohibitive and unnecessary for a skilled assembler.

FITTING

Not all parts fit exactly as they were designed to. There may be slight imperfections in cutting or distortion of parts as a result of welding, heating, or mechanical damage. Some **fitting** problems can be solved by grinding away the problem area. Hand grinders are most effective for this type of defect, Figure 4.36. Other situations may require that the parts be forced into alignment.

A simple way of correcting slight alignment problems is to make a small tack weld in the joint and then use a hammer and possibly an anvil to pound the part into place, Figure 4.37. Small tacks applied in this way become part of the finished weld. Be sure not to strike the part in a location that will damage the surface and render the finished part unsightly or unusable.

Greater aligning force can be applied using cleats or dogs with wedges or jacks. Cleats or dogs are pieces of metal that are temporarily attached to the weldment's parts to enable them to be forced into place. Jacks will do a better job if the parts must be moved more than about 1/2 in. (13 mm), Figure 4.38. Anytime cleats or dogs are used, they must be removed and the area ground smooth.

Some codes and standards do not allow cleats or dogs to be welded to the base metal. In these cases more expensive and time-consuming fixtures must be constructed to help align the parts if needed.

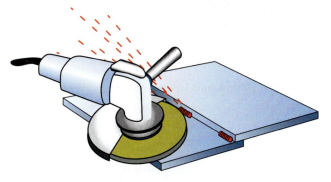

(A)

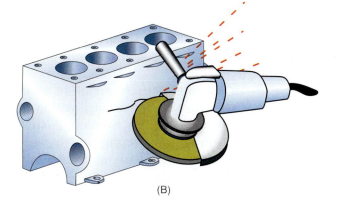

(B)

Figure 4.36 Abrasive grinding disk
Abrasive grinding disk can be used to grind a (A) weld or a (B) groove

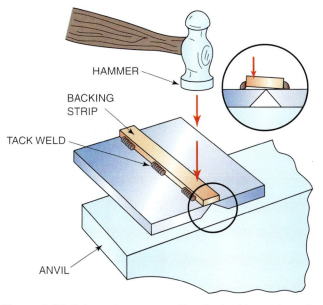

HAMMER

BACKING STRIP

TACK WELD

ANVIL

Figure 4.37 Using a hammer to align the backing strip and weld plates

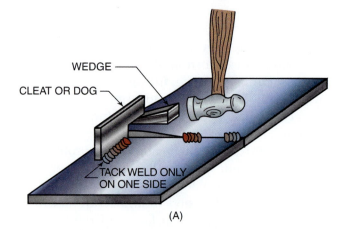

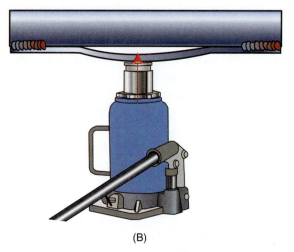

Figure 4.38 Alignment
(A) Cleat and wedge used in alignment and (B) hydraulic jack realignment

TACK WELDING

Tack welding is a temporary method of holding the parts in place until they can be completely welded. Usually, all of the parts of a weldment should be assembled before any finishing welding is started. This will help reduce distortion. Tack welds must be strong enough to withstand any pounding or forcing during assembly and any forces caused by weld distortion during final welding. They must also be small enough to be incorporated into the final weld without causing a discontinuity in its size or shape, Figure 4.39.

Tack welds must be made with an appropriate filler metal, in accordance with the welding procedure. They must be located well within the joint so that they can be fused again during the finish welding. Posttack welding cleanup is required to remove any slag or impurities that may cause flaws in the finished weld. Sometimes the ends of a tack weld must be ground down to a taper to improve their tie-in to the finished weld metal.

A good tack weld is one that does its job by holding parts in place yet is undetectable in the finished weld.

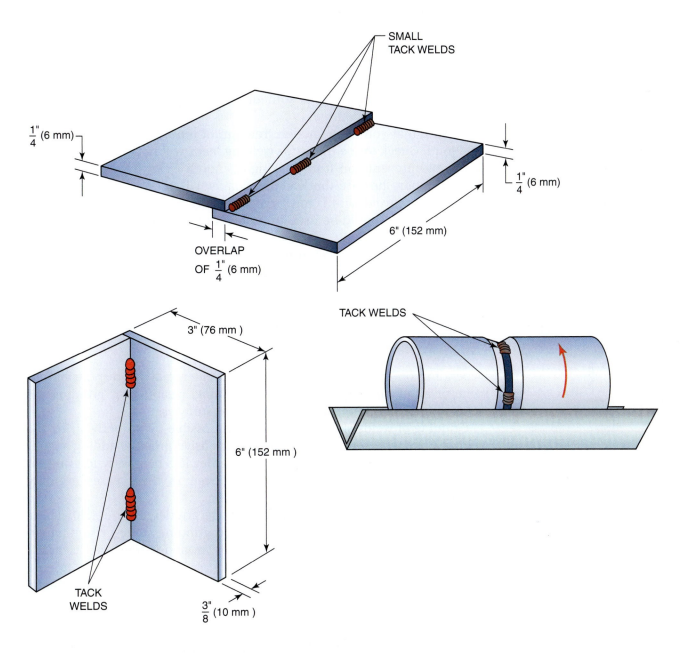

Figure 4.39 Tack welds
Make tack welds as small as possible

WELDING

Good welding requires more than just filling up the joints with metal. The order and direction in which welds are made can significantly affect distortion in the weldment. Generally, welding on an assembly should be staggered from one part to another. This allows both the welding heat and welding stresses to dissipate.

Keep the arc strikes in the welding joint so that they will be remelted as the weld is made. This will make the finished weldment look neater and reduce postweld cleanup. Some codes and standards do not allow arcs to be made outside of the welding joint.

Striking the arc in the correct location on an assembly is more difficult than working on a welding table, because you will often be in an awkward position. Several techniques will help you improve your arc starting accuracy. You can use your free hand to guide the electrode or weld gun in to the correct spot. Resting your arm, shoulder, or hip against the weldment can also help. It is sometimes helpful to practice starting the weld with the power off.

Be sure that you have enough freedom of movement to complete the weld joint. Check to see that your welding leads will not snag on anything that would prevent you from making a smooth weld. If you are welding out of position, be sure that welding sparks will not fall on you or other workers. If the weldment is too large to fit into a welding booth, portable welding screens should be used to protect other workers in the area from sparks and welding light.

Follow all safety and setup procedures for the welding process. Practice the weld to be sure that the machine is set up properly before starting on the weldment.

FINISHING

Depending on the size of the shop, the welder may be responsible for some or all of the finish work, from chipping, cleaning, or grinding the welds to applying paint or other protective surfaces.

Grinding of welds should be avoided if possible by properly sizing the weld as it is made. Grinding can be an expensive process, adding significant cost to the finished weldment. Sometimes it is necessary to grind for fitting purposes or for appearance, but even in these cases grinding should be minimized if possible.

Most grinding is done with a hand angle grinder, Figure 4.40. These grinders can be used with a flat or cupped grinding stone or sandpaper. As the grinder is used, the stone wears down, and it must be discarded once it has worn down to the paper backing. It is a good practice to hold the grinder at an angle so that if anything is thrown off the stone or metal

> **Caution**
>
> When using a portable grinder, be sure that it is properly grounded and that the sparks will not fall on others, cause damage, or start a fire. Always maintain control to prevent the stone from catching and gouging the part or yourself.

> **Caution**
>
> Be sure that any stone or sandpaper used is rated for a speed in revolutions per minute (RPM) that is equal to or greater than the speed of the grinder itself. Using a stone with a lower-rated RPM can result in its flying apart with an explosive force.

Figure 4.40 Wire brushes and grinding stones used to clean up welds
Source: Courtesy of Larry Jeffus

surface, it will not strike you or others in the area. Because of the speed of the grinding stone, any such impact can cause serious injury.

The grinder must be held securely so that there is a constant pressure on the work. If the pressure is too great, the grinder motor will overheat and may burn out. If the pressure is too light, the grinder may bounce, which could crack the grinding stone. Move the grinder in a smooth pattern along the weld. Watch the weld surface as it begins to take the desired shape and change your pattern as needed.

Painting and other finishes release fumes such as volatile organic compounds (VOCs), which are often regulated by local, state, and national governments. Special ventilation is required for most paints. A ventilation system should remove harmful fumes from the air before the air is released back into the environment. Check with your local, state, or national regulating authority before using such finishing products. Read and follow all the manufacturer's instructions for the safe use of its product.

> **Caution**
>
> Most paints are flammable and must be stored well away from any welding.

SUMMARY

Interpreting plans and joint designs is essential for the welder to recognize and follow simple as well as detailed instructions about a weldment in the shop or in the field. In this way you can work independently and require less supervision on the job. Engineers and designers from around the world may produce the plans, drawings, and sketches you will use on the job. A strong grasp of whole numbers, fractions, and decimals and the ability to convert them back and forth will allow the welder to quickly and accurately produce weldments from plans received from different sources.

REVIEW

1. Why do some parts have ± tolerances?
2. What is the rule that must be followed for adding or subtracting fractions?
3. Reduce the following fractions to their lowest denominator: 4/8, 16/32, 8/32, 10/16, 12/8, 8/4, and 10/2.
4. Round off the following numbers to two decimal places: 38.973, 7.976, 0.0137, 100.062, and 12.124.
5. Convert the following fractions to decimal equivalents: 1/2, 3/8, 9/16, 11/32, 15/16, and 1/32.
6. Convert the following decimals to the appropriate fractional units: 0.25 to fourths, 0.375 to eighths, 0.956 to sixteenths, 0.79 to fourths, and 1.29 to thirty-seconds.
7. Using the conversion chart in Table 4.2, convert the following standard dimensions to metric units: 1/8 in., 5/32 in., 11/16 in., 7/8 in., and 1 9/16 in.
8. Using the conversion chart in Table 4.2, convert the following metric dimensions to standard fractional units: 6.35 mm, 19.05 mm, 12.7 mm, 24.6062 mm, and 42.0688 mm.

9. Using the conversion chart in Table 4.2, convert the following metric dimensions to the nearest standard fractional units ±1/16 in.: 13 mm, 16 mm, 5 mm, 22 mm, and 3 mm.
10. How can layout lines be drawn on a metal surface?
11. Why should some layout lines be marked with an *X*?
12. How would a small or tight tolerance affect the layout?
13. How can circles, arcs, and curves be laid out?
14. Which cutting process does not produce a kerf space?
15. What can a template be made of?
16. What is the difference between plate and sheet material?
17. Why should you identify one part for assembly as the base?
18. What makes some C-clamps better for use in welding than others?
19. What precautions should be taken if a hammer is used to shift a part into alignment?
20. Why should the entire weldment be assembled and tack welded in place before finished welding is started?
21. Why is it important not to strike the arc outside of the weld joint?
22. What can happen if a grinding stone with a lower-rated RPM than the grinder is used?
23. Can an accurate and comprehensive set of drawings be a communication tool in intercultural projects?

CHAPTER

5

Welding Codes, Standards, and Cost

OBJECTIVES

After completing this chapter, the student should be able to

- distinguish between qualification and certification
- list four major considerations for selecting a code or standard
- write a welding procedure specification (WPS)
- identify the three most common codes and describe their major uses
- outline the steps required to certify a weld and welder
- explain how a tentative WPS becomes a certified WPS
- contrast cost differences in various processes and base metals

KEY TERMS

API Standard 1104

ASME Section IX

AWS D1.1

code

procedure qualification
   record (PQR)

specification

standard

welding procedure
   specification (WPS)

welding schedule

AWS SENSE EG2.0

Key Indicators Addressed in this Chapter:

Module 1: Occupational Orientation

Key Indicator 1: Prepares time or job cards, reports or records
Key Indicator 4: Follows written details to complete work assignment

Module 3: Drawing and Weld Symbol Interpretation

Key Indicator 1: Interprets basic elements of a drawing or sketch
Key Indicator 2: Interprets welding symbol information
Key Indicator 3: Fabricates parts from a drawing or sketch

INTRODUCTION

It is important to know that any weld produced is going to be the best one for the job. A method is also required to ensure that each weld made in the same plant or on the same type of equipment in another plant will be of the same quality.

To meet these requirements various agencies have established codes and standards. These detailed written outlines explaining exactly how a weld is to be laid out, performed, and tested have made consistent-quality welds possible. With the required information, skilled welders in shops all around the city, state, country, or world can make the same weld to the same level of safety, strength, and reliability.

A testing procedure to certify the welder ensures that the welder has the skills to make the weld. Passing a weld test is much easier when all of the detailed information is provided.

Selecting the code or standard to be used to judge a weld is as important as having a skilled welder. Not every product welded needs to be manufactured to the same level. The decision on the appropriate code or standard can be one of the most important aspects of welding fabrication. If the wrong one is selected, the cost of fabrication can be too high, or the parts might not stand up to in-service requirements.

For a welding business to operate profitably, the owner or manager must be able to make cost-effective welding decisions. A number of factors affect the cost of producing weldments. These factors include the following:

- material
- weld design
- welding processes
- finishing
- labor
- overhead

CODES, STANDARDS, PROCEDURES, AND SPECIFICATIONS

A welding **code** or **standard** is a detailed listing of the rules or principles that are to be applied to a specific classification or type of product.

A welding **specification** is a detailed statement of the legal requirements for a specific classification or type of weld to be made on a specific product. Products manufactured to code or specification requirements commonly must be inspected and tested to assure compliance.

Codes and specifications are intended to be guidelines only and must be qualified for specific applications by testing.

A number of organizations publish codes or specifications that cover a wide variety of welding conditions and applications. The specific code to be used is determined by the engineers, designers, or various other requirements:

- local, state, or federal government regulations—Many government agencies require that a specific code or standard be followed.
- bonding or insuring company—The weld must be shown to be fit for service requirements as established through testing. A bonding or insuring company must feel that the product is the safest that can be produced.
- end-user (customer) requirements—The manufacturer considers cost and reliability; that is, as stricter standards are applied to the welding, the cost of the weldments increases. The more lax the standard, the lower the cost, but also the reliability and possibly the safety decrease.
- standard industrial practices—The code or standard used is considered to be the standard one for the industry and has been in use for some time.

The three most commonly used codes are

- **API Standard 1104**, American Petroleum Institute—used for pipelines
- **ASME Section IX**, American Society of Mechanical Engineers—used for pressure vessels and nuclear components
- **AWS D1.1**, American Welding Society—used for bridges, buildings, and other structural steel

The following organizations publish welding codes and/or specifications. Most can be contacted for additional information and a current price list either directly or through the Internet.

AAR
Association of American Railroads
50 F Street, NW
Washington, DC 20001-1564
http://www.aar.org

AASHTO
American Association of State Highway and Transportation Officials
444 North Capitol Street, NW, Suite 249
Washington, DC 20001-1539
http://transportation1.org/aashtonew

AIA
Aerospace Industries Association of America
1000 Wilson Boulevard, Suite 1700
Arlington, VA 22209-3928
http://www.aia-aerospace.org

AISC
American Institute of Steel Construction
1 East Wacker Drive, Suite 700
Chicago, IL 60601-1802
http://www.aisc.org

ANSI
American National Standards Institute
25 West 43rd Street
New York, NY 10036-8007
http://webstore.ansi.org

API
American Petroleum Institute
1220 L Street, NW
Washington, DC 20005-4070
http://www.api.org

AREMA
American Railway Engineering and Maintenance of Way Association
10003 Derekwood Lane, Suite 210
Lanham, MD 20706-4875
http://www.arema.org

ASME
American Society of Mechanical Engineers
3 Park Avenue
New York, NY 10016-5990
http://www.asme.org

AWS
American Welding Society
550 NW LeJeune Road
Miami, FL 33126-5649
http://www.aws.org

AWWA
American Water Works Association
6666 West Quincy Avenue
Denver, CO 80235-3098
http://www.awwa.org

MIL
Department of Defense
Washington, DC 20301-0001
http://www.defenselink.mil

SAE
Society of Automotive Engineers
400 Commonwealth Drive
Warrendale, PA 15086-7511
http://www.sae.org

WELDING PROCEDURE QUALIFICATION

Welding Procedure Specification (WPS)

A welding procedure specification (WPS) is a set of written instructions by which a sound weld is made. Normally the procedure is written in compliance with a specific code, specification, or definition.

Welding procedure specification (WPS) is the standard terminology used by the American Welding Society (AWS) and the American Society of Mechanical Engineers (ASME). **Welding schedule** is the standard federal government, military, or aerospace terminology denoting a WPS. The shortened term *welding procedures* is the most common term used by industry to denote a WPS.

The WPS lists all of the parameters required to produce a sound weld to the specific code, specifications, or definition. Specific parameters such as welding process, technique, electrode or filler, polarity, amperage, voltage, preheat, and postheat should be included. The procedure should list a range or set of limitations on each, such as amps = 110–150, voltage = 17–22, with the more essential or critical parameters more closely defined or limited.

The WPS should give enough detail and specific information so that any qualified welder could follow it and produce the desired weld. The WPS should always be prepared as a tentative document until it is tested and qualified.

Qualifying the Welding Procedure Specification

The WPS must be qualified to prove or verify that the values assigned for the variables—amperage, voltage, filler, etc.—will provide a sound weld. Sample welds are prepared using the procedure and specifications listed in the tentative WPS. A record of all the parameters used to produce the test welds must be kept; be specific for the parameters with limits, such as voltage and amperage. This information should be recorded on a form called the **procedure qualification record (PQR)**.

In most cases, the inspection agency, inspector, client, or customer will request a copy of both the WPS and the PQR before allowing production welding to begin.

Qualifying and Certifying

The process of qualifying and then certifying both the WPS and welders has a number of specific requirements. The requirements may vary from one code or standard to another, but the general process is the same for most. Before you invest in the testing required to qualify and certify process and welders under a code, you must first obtain a copy of the code you are planning to use. The requirements of codes and standards change from time to time, and it is important that your copy is the most recent version.

The following generic schedule lists required activities you might follow when qualifying and certifying the welding process, the welder(s), and/or the welding operator.

1. A tentative welding procedure is prepared by a person with knowledge of the process and technique to be used and the code or specification to be satisfied.
2. Test samples are welded in accordance with the tentative WPS, and the welding parameters are recorded on the PQR. The test must be witnessed by an authorized person from an independent testing lab, the customer, an insurance company, or other individual(s) as specified by the code or listing agency.
3. The test samples are tested under the supervision of the same individuals or group who witnessed the test according to the applicable requirements, codes, or specifications.

4. If the test samples pass the applicable test, the procedure has completed qualification. It is then documented as qualified/finalized and is released for use in production.

5. If the test samples do not pass the applicable test, the tentative WPS value parameters are changed wherever feasible. Test samples are then rewelded and retested to determine whether they meet applicable requirements. This process is repeated until the test samples pass applicable requirements and the procedure is finalized and released.

6. The welder making the test samples to be used in qualifying the procedure is normally considered qualified and is then certified in the specific procedure.

7. Other welders to be qualified weld test samples per the WPS, and the samples are tested according to applicable requirements. If the samples pass, the welder is qualified to the specific procedure and certified accordingly.

8. A qualified WPS is usable for an indefinite length of time, usually until a process considered more efficient for a particular production weld is found.

9. The welder's qualification is normally considered effective for an indefinite period of time unless the welder is not engaged in the specific process of welding for which he or she is qualified for a period of more than six months; then the welder must requalify. Also a welder will need to requalify if for some reason the qualification is questioned.

Figures 5.1 and 5.2 are examples of WPS and PQR test records.

GENERAL INFORMATION

Normally, the format of the WPS is not dictated by the code or specification. Any format is acceptable provided it lists the parameters or variables (essential or nonessential, amps, volts, filler identification, etc.) listed by the code or specification. Most codes or specifications appear in an acceptable or recommended format.

Ideally, the WPS should include all of the information required to make the weld. A welder should be able to take the WPS without additional instructions and produce the weld. To this end, it is often a good idea to include supplementary information with the WPS. The information might be basic instructions for the process. With some WPSs you might include several pages as attachments to give the welder a review of the setup, operation, testing, inspecting, etc., that will help to ensure accuracy and uniformity in the welds.

Essential variables are those parameters where a change is considered to affect the mechanical properties of the weldment to the point of requiring requalification of the procedure. Nonessential variables are those parameters where a change can be made without requiring requalification of the procedure. However, a change in nonessential variables usually requires a revision to be made.

There are large differences among various codes. The AWS D1.1, *Structural Welding Code Steel,* allows some prequalified weld joints for specific processes (SMAW, SAW, FCAW, and GMAW). A written procedure

WELDING PROCEDURE SPECIFICATION (WPS)

Welding Procedures Specifications No: _____ Date: _____

TITLE:

Welding _____ of _____ to _____

SCOPE:

This procedure is applicable for _____

within the range of _____ through _____

Welding may be performed in the following positions _____

BASE METAL:

The base metal shall conform to _____

Backing material specification _____

FILLER METAL:

The filler metal shall conform to AWS classification No. _____ from AWS

specification _____. This filler metal falls into F-number _____

and A-number _____

SHIELDING GAS:

The shielding gas, or gases, shall conform to the following compositions and purity:

_____

JOINT DESIGN AND TOLERANCES:

PREPARATION OF BASE METAL:

ELECTRICAL CHARACTERISTICS:

The current shall be _____

The base metal shall be on the _____ side of the line.

PREHEAT:

BACKING GAS:

SAFETY:

WELDING TECHNIQUE:

INTERPASS TEMPERATURE:

CLEANING:

INSPECTION:

REPAIR:

SKETCHES:

BEND TEST: Specimen preparation

                Acceptance criteria for bend test:

Figure 5.1 Welding procedure specification (WPS)

PROCEDURE QUALIFICATION RECORD (PQR)

Welding Qualification Record No: _____(1)_____ WPS No: _____(2)_____ Date: _____(3)_____
Material specification _____(4)_____ to _____
P-No. _____(5)_____ to P-No. _____ Thickness and O.D. _____(6)_____
Welding process: Manual _____(7)_____ Automatic _____(8)_____
Thickness Range _____(9)_____

Filler Metal

Specification No. _____(10)_____ Classification _____(11)_____ F-number _____(12)_____
A-number _____(13)_____ Filler Metal Size _____(14)_____ Trade Name _____(15)_____
Describe filler metal (if not covered by AWS specification) _____(16)_____

Flux or Atmosphere

Shielding Gas _____(17)_____ Flow Rate _____(18)_____ Purge _____(19)_____
Flux Classification _____(20)_____ Trade Name _____(21)_____

Welding Variables

Joint Type _____(22)_____   Position _____(29)_____
Backing _____(23)_____   Preheat _____(30)_____
Passes and Size _____(24)_____   Bead Type _____(31)_____
No. of Arcs _____(25)_____   Current _____(32)_____
Ampere _____(26)_____   Volts _____(33)_____
Travel Speed _____(27)_____   Oscillation _____(34)_____
Interpass Temperature Range _____(28)_____

Weld Results

Appearance _____(35)_____ Weld Size _____(36)_____

Guided Bend Test

Type	Result	Type	Result
(37)	(38)		

Tensile Test

Specimen No.	Dimensions Width/ Thickness	Area	Ultimate Total Load, lb.	Ultimate Unit Stress, psi.	Character of Failure And Location
(39)	(40)	(41)	(42)	(43)	(44)

Welder's Name _____(45)_____ Identification No. _____(46)_____ Laboratory Test No. _____
By virtue of these test meets welder performance requirements.
Test Conducted by _____(47)_____ Address _____
            per _____(48)_____ Date _____(49)_____
We certify that the statements in this record are correct and that the test
welds performed and tested are in accordance with the WPS

Manufacture _____(50)_____
Signed by _____
Date _____

Figure 5.2 Procedure qualification record (PQR)

is required for these joints, but since the procedure is tentative, it does not require support via a written PQR, Figure 5.3.

The procedure qualification requirements regarding positions for groove welds in plate differ among codes. Some codes may require a written procedure for each position. The ASME Section IX, however, qualifies a welder for the 1G position when the welder qualifies for 2G, 3G, or 4G.

Ordinarily, the welder must be qualified/certified in accordance with a specific WPS. The welder's qualifying test plate may be examined radiographically or ultrasonically rather than through bend-tests. Specific codes or specifications must be referenced for details of the number of tensile, bend, or other types of test specimens and tests to be performed. For example, AWS D1.1 and ASME Section IX do not require a nick-break test specimen, but API Standard 1104 does require one.

PRACTICE 5-1

Writing a Welding Procedure Specification

Using the form provided and following the example, Figure 5.3, write a welding procedure specification. Figure 5.4 is a composite of sample WPS forms provided in the AWS, ASME, and API codes. You may want to obtain a copy of one of the codes or standards and compare a weld you make to the standard. Most of the unique information is provided in this short outline. Additional information that may be required for this form can be found in figures in this chapter. You may need to refer to books 2 or 3 in this series for specific welding process data or to your notes to establish the actual limits of the welding variables (voltage, amperage, gas flow rates, nozzle size, etc.).

Note that not all of the blanks will be filled in on the forms: The forms are designed to be used with a large variety of weld procedures, so they have spaces that will not be used every time.

1. The WPS number is usually made up following a system established by the company. This number may include coded information relating to the date the WPS was written, who wrote it, material or process data, and so on.
2. Date the WPS was written or effective
3. The welding process(es) that will be used to perform the weld: SMAW, GMAW, GTAW, and so on
4. The material type and thickness or pipe type and diameter and/or wall thickness. If all the material or pipe being joined is the same, then the same information will appear before and after "to."
5. Fillet or groove weld and the joint type: butt, lap, tee, and so on
6. Thickness range qualified or diameter range qualified: for both plate and pipe, a weld performed successfully on one thickness qualifies a welder to weld on material within that range. See Table 5.1 (page 143) for a list of thickness ranges.
7. Material position: 1G, 2G, 3G, 4G, 1F, 2F, 3F, 4F, 5G, 6G, 6GR, Figure 5.5 (page 145)
8. Base metal specification: the ASTM specification for the type and grade of material, including the P-number, Table 5.2 (page 143).
9. If a backing material is used, its ASTM or other specification information must be included here.

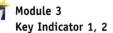

Module 1
Key Indicator 1, 4

Module 3
Key Indicator 1, 2

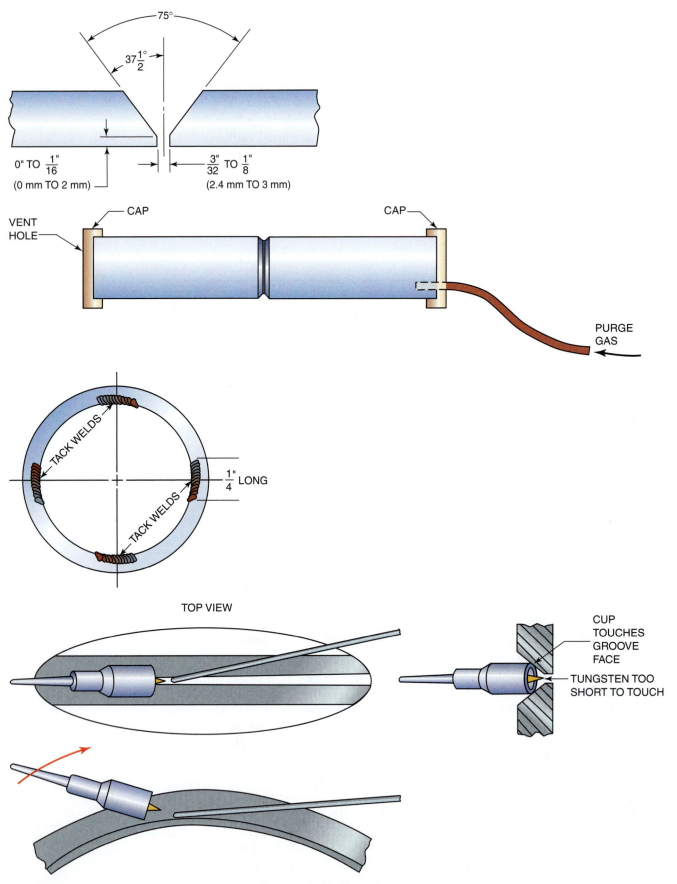

Figure 5.3 Welding procedure specification (WPS)

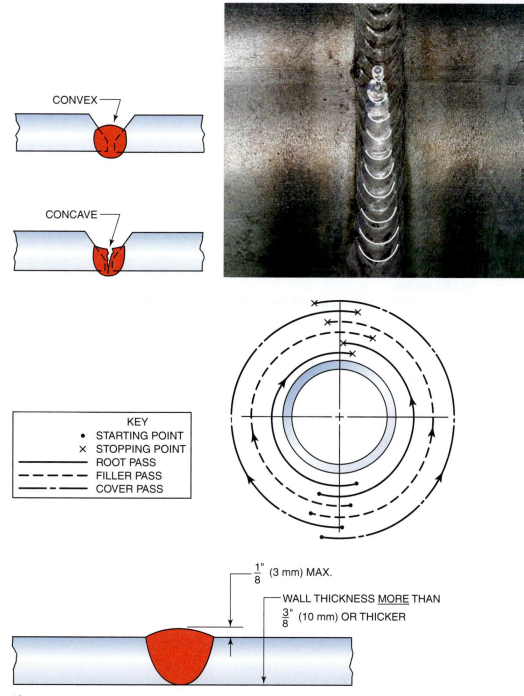

CONVEX

CONCAVE

KEY
• STARTING POINT
× STOPPING POINT
—— ROOT PASS
– – – FILLER PASS
–·–·– COVER PASS

$\frac{1}{8}"$ (3 mm) MAX.

WALL THICKNESS <u>MORE</u> THAN $\frac{3}{8}"$ (10 mm) OR THICKER

Figure 5.3 Continued

10. Classification number: the standard number found on the electrode or electrode box, such as E6010, E7018, E316-15, ER70S-3, or E70T-1.

11. Filler metal specification number: the AWS has specifications for chemical composition and physical properties for electrodes. Some of these specifications are listed in Table 5.3 (page 144).

12. F-number: a specific grouping number for several classifications of electrodes with similar composition and welding characteristics. Table 5.4 (page 144) lists the F-number corresponding to the electrode used.

WELDING PROCEDURE SPECIFICATION (WPS)

Welding Procedures Specifications No: _____(1)_____   Date: _____(2)_____

TITLE:

Welding _____(3)_____ of _____(4)_____ to _____(4)_____

SCOPE:

This procedure is applicable for _____(5)_____

within the range of _____(6)_____ through _____(6)_____

Welding may be performed in the following positions _____(7)_____

BASE METAL:

The base metal shall conform to _____(8)_____

Backing material specification _____(9)_____

FILLER METAL:

The filler metal shall conform to AWS classification No. _____(10)_____ from AWS

specification ____(11)____ . This filler metal falls into F-number ____(12)____

and A-number _____(13)_____

SHIELDING GAS:

The shielding gas, or gases, shall conform to the following compositions and purity:

_____(14)_____

JOINT DESIGN AND TOLERANCES:

(15)

PREPARATION OF BASE METAL:

(16)

ELECTRICAL CHARACTERISTICS:

The current shall be _____(17)_____

The base metal shall be on the _____(18)_____ side of the line.

PREHEAT: (19)

BACKING GAS: (20)

WELDING TECHNIQUE: (21)

INTERPASS TEMPERATURE: (22)

CLEANING: (23)

INSPECTION: (24)

REPAIR: (25)

SKETCHES: (26)

Figure 5.4 Welding procedure specification (WPS)

Table 5.1 Test Specimens and Range of Thickness Qualified

Plate Thickness (T) Tested in. (mm)	Plate Thickness (T) Qualified in. (mm)
1/8 ≤ T < 3/8*	1/8 to 2T
(3.1 ≤ T < 9.5)	(3.1 to 2T)
3/8 (9.5)	3/4 (19.0)
3/8 < T < 1	2T
(9.5 < T < 25.4)	2T
1 and over	Unlimited
(25.4 and over)	Unlimited

Pipe Size of Sample Weld	
Diameter in. (mm)	Wall Thickness, T
2 (50.8) or	Sch. 80
3 (76.2)	Sch. 40
6 (152.4) or	Sch. 120
8 (203.2)	Sch. 80

Pipe Size Qualified		
Diameter in. (mm)	Wall Thickness in. (mm)	
3/4 (19.0) through 4 (101.6)	Minimum 0.063 (1.6)	Maximum 0.674 (17.1)
4 (101.6) and over	0.187 (4.7)	Any

*Thickness (T) is equal to or greater than 1/8 in. (≤) and thickness (T) is less than 3/8 in. (≤).

Table 5.2 P-Numbers

	Type of Material
P-1	Carbon steel
P-3	Low alloy steel
P-4	Low alloy steel
P-5	Alloy steel
P-6	High alloy steel—predominantly martensitic
P-7	High alloy steel—predominantly ferritic
P-8	High alloy steel—austenitic
P-9	Nickel alloy steel
P-10	Specialty high alloy steels
P-21	Aluminum and aluminum-base alloys
P-31	Copper and copper alloy
P-41	Nickel

Table 5.3 Specification Numbers

A5.10	Aluminum—bare electrodes and rods
A5.3	Aluminum—covered electrodes
A5.8	Brazing filler metal
A5.1	Steel, carbon, covered electrodes
A5.20	Steel, carbon, flux cored electrodes
A5.17	Steel-carbon, submerged arc wires and fluxes
A5.18	Steel-carbon, gas metal arc electrodes
A5.2	Steel—oxyfuel gas welding
A5.5	Steel—low alloy covered electrodes
A5.23	Steel—low alloy electrodes and fluxes—submerged arc
A5.28	Steel—low alloy filler metals for gas shielded arc welding
A5.29	Steel—low alloy, flux cored electrodes

Table 5.4 F-Numbers

Group Designation	Metal Types	AWS Electrode Classification
F1	Carbon steel	EXX20, EXX24, EXX27, EXX28
F2	Carbon steel	EXX12, EXX13, EXX14
F3	Carbon steel	EXX10, EXX11
F4	Carbon steel	EXX15, EXX16, EXX18
F5	Stainless steel	EXXX15, EXXX16
F6	Stainless steel	ERXXX
F22	Aluminum	ERXXXX

13. A-number: the classification of weld metal analysis. Table 5.5 (page 146) lists the A-numbers.
14. Shielding gas(es) and flow rate for GMAW, FCAW, or GTAW.

Complete a copy of the Student Welding Report in Appendix I or provided by your instructor. ∎

PRACTICE 5-2

Procedure Qualification Record (PQR)

Module 1
Key Indicator 1, 4

Module 3
Key Indicator 1, 2, 3

Following the procedure you wrote in Practice 5-1, you are going to make the weld to see whether your tentative welding procedure and specification can be certified. Complete a copy of the form provided to record all of the appropriate information, Figure 5.2

1. The PQR number is usually made up following a system established by the company. This number may include coded information relating to the product being welded, material or process data, and the like.
2. The WPS number on which the PQR is based
3. The date on which the welding took place
4. Base metal specification: this is the ASTM specification for the type and grade of material.

5. Table 5.2 lists some commonly used metals and their P-numbers.
6. Test material thickness (or) test pipe outside diameter (OD) (and) wall thickness
7. Manual welding processes are used to qualify a welder. Specify the process: GMAW, FCAW, SMAW, GTAW, etc.
8. Automatic welding processes are used to qualify a welding operator. Specify the process: SAW, ESW, etc.
9. Thickness range qualified (or) diameter range qualified: for both plate and pipe, a weld performed successfully on one thickness qualifies a welder to weld on material within that range. See Table 5.1 for a list of thickness ranges.
10. Filler metal specification number: the AWS has specifications for chemical composition and physical properties for electrodes. Some of these specifications are listed in Table 5.3.
11. Classification number: this is the standard number found on the electrode or electrode box, such as E6010, E7018, E316-15, or ER1100.
12. F-number: a specific grouping number for several classifications of electrodes with similar composition and welding characteristics. See Table 5.4 for the F-number corresponding to the electrode used.
13. A-number: the classification of weld metal analysis. See Table 5.5 for a list of A-numbers.
14. Give the diameter of electrode used.
15. Give the manufacturer's identification name or number.
16. List the manufacturer's chemical composition and physical properties as provided if the filler metal is not covered by an AWS specification.
17. Shielding gas or gas mixture for GMAW, FCAW, or GTAW
18. Flow rate in cubic feet per hour (cfh)
19. The amount of time that the shielding gas is to flow to purge air from the welding zone
20. SAW flux classification
21. The manufacturer's identification name or number for the SAW flux

Table 5.5 A-Number Classification of Ferrous Metals

A No.	Types of Weld Deposit	Analysis					
		C %	Mn %	Si %	Mo %	Cr %	Ni %
1	Mild steel	0.15	1.6	1.0	–	–	–
2	Carbon-moly	0.15	1.6	1.0	0.4–0.65	0.5	–
3	Chrome (0.4 to 2%)-moly	0.15	1.6	1.0	0.4–0.65	0.4–2.0	–
4	Chrome (2 to 6%)-moly	0.15	1.6	2.0	0.4–1.5	2.0–6.0	–
5	Chrome (6 to 10.5%)-moly	0.15	1.2	2.0	0.4–1.5	6.0–10.5	–
6	Chrome-martensitic	0.15	2.0	1.0	0.7	11.0–15.0	–
7	Chrome-ferritic	0.15	1.0	3.0	1.0	11.0–30.0	–
8	Chromium-nickel	0.15	2.5	1.0	4.0	14.5–30.0	7.5–15.0
9	Chromium-nickel	0.30	2.5	1.0	4.0	25.0–30.0	15.0–37.0
10	Nickel to 4%	0.15	1.7	1.0	0.55	–	0.8–4.0
11	Manganese-moly	0.17	1.25–2.25	1.0	0.25–0.75	–	0.85
12	Nickel-Chrome-moly	0.15	0.75–2.25	1.0	0.25–0.8	1.5	1.25–2.25

22. Butt, lap, tee, or other joint type
23. Backing strip material specification: this is the ASTM specification number
24. The number of passes and the size
25. Usually one (1) except for some automatic SAW process that may use multiple electrodes with multiple arcs
26. The current (in amps) used to make the weld. If the machine being used for the weld does not have an amp meter, a meter must be attached to the welding lead, within 2 feet (1.5 m) of the electrode holder, to get this reading.
27. The travel speed in inches per minute is usually given for machine or automatic welds.
28. This is the maximum temperature that the base metal is allowed to reach during the weld. Welding must stop and the part allowed to cool if this temperature is reached.
29. Test position: 1G, 2G, 3G, 4G, 1F, 2F, 3F, 4F, 5G, 6G, 6GR. See Figure 5.5.
30. This is the minimum temperature that the base metal must reach before welding can start.
31. Groove or fillet weld
32. AC, DCEP, or DCEN
33. The voltage is included for all welding processes.
34. The type of electrode movement used when making the weld
35. Visually inspect the weld and record any flaws.
36. Record the legs and reinforcement dimensions. Measure and record the depth of the root penetration.
37. Four (4) test specimens are used for 3/8-in. or thinner metal. Two (2) will be root bent and two (2) face bent. For thicker metal all four (4) will be side bent.
38. Visually inspect the specimens after testing and record any discontinuities.
39. Identification number that was marked on the specimen
40. Width and thickness of test section of the specimen
41. Cross-sectional area of specimen in the test area
42. The load at which the specimen failed
43. The maximum load divided by the specimen's original area converts the ultimate total load for the specimen to the measure of pounds per square inch (psi) that was required to break the material.
44. The type of failure, whether it was ductile or brittle, and where the failure occurred relative to the weld
45. Welder's name: the person who performed the weld
46. Identification number: on a welding job, every person has an identification number that is used on the time card and paycheck. In this space, you can write the class number or section number, since you do not have a clock number.
47. The name of the person who interpreted the results
48. Qualifications of the test interpreter: this is usually a certified welding inspector or other qualified person.
49. Date the results of the test were completed
50. The name of the company that requested the test

Complete a copy of the Student Welding Report in Appendix I or provided by your instructor. ▪

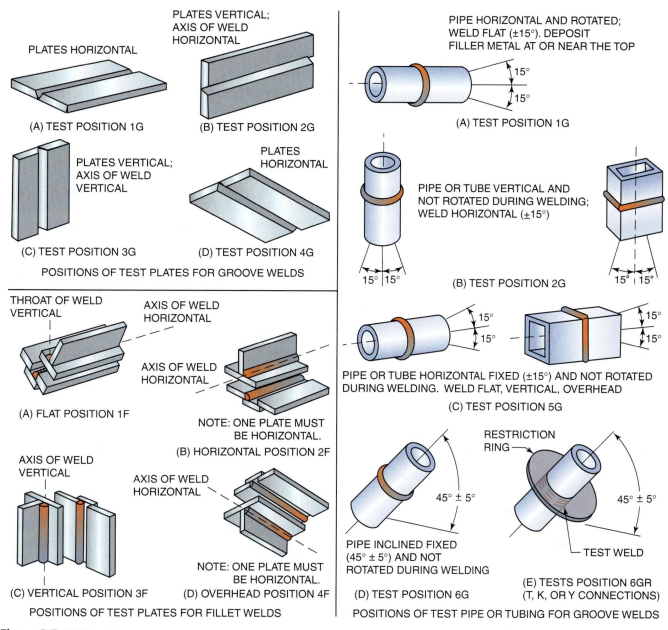

PLATES HORIZONTAL

(A) TEST POSITION 1G

PLATES VERTICAL;
AXIS OF WELD
HORIZONTAL

(B) TEST POSITION 2G

PLATES VERTICAL;
AXIS OF WELD
VERTICAL

(C) TEST POSITION 3G

PLATES
HORIZONTAL

(D) TEST POSITION 4G

POSITIONS OF TEST PLATES FOR GROOVE WELDS

THROAT OF WELD
VERTICAL

AXIS OF WELD
HORIZONTAL

(A) FLAT POSITION 1F

AXIS OF WELD
HORIZONTAL

NOTE: ONE PLATE MUST
BE HORIZONTAL.

(B) HORIZONTAL POSITION 2F

AXIS OF WELD
VERTICAL

(C) VERTICAL POSITION 3F

AXIS OF WELD
HORIZONTAL

NOTE: ONE PLATE MUST
BE HORIZONTAL.

(D) OVERHEAD POSITION 4F

POSITIONS OF TEST PLATES FOR FILLET WELDS

PIPE HORIZONTAL AND ROTATED;
WELD FLAT (±15°). DEPOSIT
FILLER METAL AT OR NEAR THE TOP

15°
15°

(A) TEST POSITION 1G

PIPE OR TUBE VERTICAL AND
NOT ROTATED DURING WELDING;
WELD HORIZONTAL (±15°)

15° 15°
15° 15°

(B) TEST POSITION 2G

15°
15°
15°
15°

PIPE OR TUBE HORIZONTAL FIXED (±15°) AND NOT ROTATED
DURING WELDING. WELD FLAT, VERTICAL, OVERHEAD

(C) TEST POSITION 5G

RESTRICTION
RING

45° ± 5°
45° ± 5°

PIPE INCLINED FIXED
(45° ± 5°) AND NOT
ROTATED DURING WELDING

TEST WELD

(D) TEST POSITION 6G

(E) TESTS POSITION 6GR
(T, K, OR Y CONNECTIONS)

POSITIONS OF TEST PIPE OR TUBING FOR GROOVE WELDS

Figure 5.5 Weld positions

WELDING COSTS

Estimating the costs of welding can be a difficult task because of the many variables involved. One approach is to have a welding engineer specify the type and size of weld to withstand the loads that the weldment must bear. Then the welding engineer must select the welding process and filler metal that will provide the required welds at the least possible cost. This method is used in large shops, where welding expenditures on a product can range from a few thousand dollars to well over a million dollars. With competition for work resulting in small profit margins, each job must be analyzed carefully.

A second approach, used by smaller shops, is to get a price for the materials and then estimate the production time. Process and filler metal costs are considered, but little thought is given to the hidden costs of equipment depreciation, joint efficiency, power, and so on. With the majority of the welding jobs in these shops taking from a few hours to a few days, and with costs ranging from a few hundred dollars to a few thousand dollars, extensive cost analysis cannot be justified. Detailed estimating may take more time than the job itself. However, to remain profitable *and* competitive, some cost estimation is required. Only the costs that a small shop should consider when estimating a job will be covered in this section.

Cost Estimation

A number of factors affect welding cost. These factors can be divided into two broad categories: fixed and variable. *Fixed costs* are those expenses that must be paid each and every day, week, month, or year, regardless of work or production. Examples of fixed costs include rent, taxes, insurance, and advertising spend. *Variable costs* are those expenses that change with the quantity of work being produced. Examples of variable costs include supplies, utilities, labor, and equipment leasing costs. Expenses within both categories must be considered when making welding job estimates. These cost areas include the following:

- *material cost*—The cost of new stock required to produce the weldment is fixed by the supplier. It is often possible to help control these costs by getting bids from several suppliers and combining as many jobs as possible to get any discount for bulk purchases.
- *scrap cost*—Scrap is an inevitable part of any project. Scrap costs a company in two ways: by wasting expensive resources and by requiring cleanup and removal. Reducing scrap production through proper planning will result in a direct saving for any project.
- *process cost*—The major welding processes (SMAW, GMAW, and FCAW) differ widely in their cost of equipment, operating supplies, and production efficiency. SMAW has the lowest initial cost and has excellent flexibility but also a higher total cost for large jobs.
- *filler metal*—The cost of filler metal per pound is only a small part of its actual cost. The major welding processes (SMAW, GMAW, and FCAW) have widely varying deposition and efficiency rates.
- *labor cost*—Total labor cost includes wages and benefits. Insurance, sick leave, vacation, social security, retirement, and other benefits can range from 25% to 75% of the total labor cost. Because labor costs are figured on an hourly basis, they can be controlled only by increasing productivity.
- *overhead costs*—Overhead costs are often intangible costs related to doing business. These costs include building rent or mortgage, advertising, insurance, utilities, taxes, licenses, governmental fees, accounting, loan payments, and property upkeep.
- *finishing cost*—Postwelding cleanup, painting, or other finishing add to the weldment's final production cost. Many of these finishing processes can produce some level of health hazard, and a major concern to everyone is the environment. Complying with local, state, and federal environmental laws can add significantly to the cost of painting, dipping, and plating. New environmentally friendly paints, low-pressure spray guns, and water-based products are a few of the advances in finishing that have helped reduce environmental compliance costs.

Joint Design

Joint design is an important consideration when estimating weld cost. The root opening, root face thickness, and bevel angles must be studied carefully when making design decisions. All these factors affect the weld dimensions, Figure 5.6, which in turn determine the amount of weld metal needed to fill the joint. Plate thickness is a major factor that, in most cases, cannot be changed. Increasing the root opening increases cost. But larger root openings often allow the bevel angle to be reduced while maintaining good access to the weld root.

To keep welding costs down, the joints should have the smallest possible root opening and the smallest reasonable bevel angle. These conditions are more easily achieved with welding processes that provide deep penetration. The other benefit of deep penetration is the option of a deeper root face and its significant effect on the volume of filler metal needed. In addition, the amount of reinforcement affects welding costs. Some reinforcement is unavoidable to ensure a full-thickness weld; however, too much reinforcement requires extra time and material, which can reduce the weld's strength and fatigue life.

The weld should be approximately the same size as the metal is thick. Welds that are undersized or oversized can cause joint failure. Welds that are undersized do not have enough area to hold the parts under load. Welds that are oversized can make the joint too stiff. The lack of flexibility of the weld causes the metal near the joint to be highly stressed, Figure 5.7. A piece of wire bent between two pliers will break, but a piece of wire bent between your fingers will withstand more bending before it breaks.

Welds made on parts of unequal size should allow enough joint flexibility to prevent a crack from forming along the edge of the weld. It is possible to taper the thicker metal to reduce the thickness or to build up the thinner metal.

Overwelding also contributes to welding costs. Welders often believe that "if a little is good, a lot is better, and too much is just right." Too often it is assumed that a large reinforcement means greater strength, but that is never true.

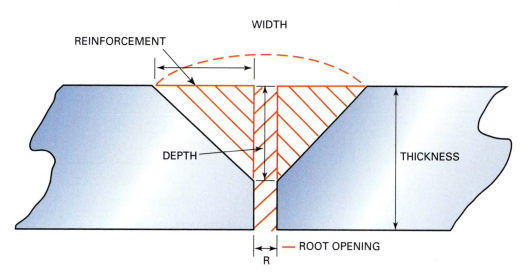

Figure 5.6 Calculation of weld requirements depending on joint design

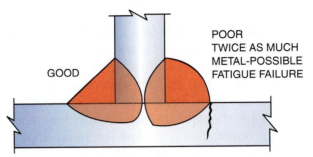

Figure 5.7 Overwelding
Overwelding can be harmful as well as costly

Cutting weld volume also reduces the labor cost. It should be remembered that the labor content of welds almost always exceeds the cost of the consumables. Reducing the amount of filler metal needed also cuts the time needed to make the welds. This significant reduction in labor costs, Figure 5.8, justifies the price of more expensive filler metals or welding processes. This assumes that the metals and processes provide enough increase either in penetration (to allow joint designs requiring less filler metal) or in deposition rates (to reduce welding times).

Groove Welds

For groove welds the bevel angle greatly affects the filler metal volume. As the groove angle increases, more filler metal is required to fill the groove

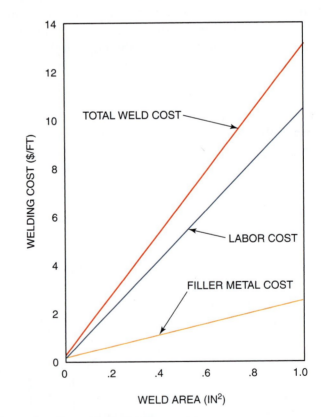

Figure 5.8 Increasing the weld size increases cost
Note that labor is 80% of the total cost (based on typical modern welding rates and efficiencies)

during the weld. Because the volume is in proportion to the angle, the change in volume can be calculated. To determine the weld volume, begin by finding the cross-sectional area of the weld groove. First find the rectangular area formed by the root opening:

$$\text{Root area} = \text{root opening} \times \text{plate thickness}$$

Next find the area of the triangular space of a bevel, which is equal to 1/2 of the width times the depth. The area of a V-groove (if both sides are at the same bevel angle) is equal to the width times the depth:

$$\text{Root area} = \frac{\text{bevel width} \times \text{bevel depth}}{2};$$

$$\text{V-groove area} = \text{bevel width} \times \text{bevel depth}$$

The cross-sectional area in Figure 5.6 is the sum of the small rectangular area formed by the root opening and plate thickness, plus the area of the triangular space formed by each of the bevel angles:

$$\text{Total cross-sectional area} = \text{root area} + \text{bevel area(s)}$$

The total groove volume is then determined by multiplying the groove area by the weld length:

$$\text{Total groove volume} = \text{cross-sectional area} \times \text{weld length}$$

PRACTICE 5-3

Finding Weld Groove Volume

Using a pencil, paper, and calculator, determine the total volume of the following groove welds, Figure 5.9.

Module 1
Key Indicator 1, 4

1. V-groove joint with the following dimensions:
   - width, 3/8 in.
   - depth, 3/8 in.
   - root opening, 1/8 in.
   - thickness, 1/2 in.
   - weld length, 144 in.
2. Bevel joint with the following dimensions:
   - width, 0.25 in.
   - depth, 0.375 in.
   - root opening, 0.062 in.
   - thickness, 0.5 in.
   - weld length, 96 in.
3. V-groove joint with the following dimensions:
   - width, 12 mm
   - depth, 12 mm
   - root opening, 2 mm
   - thickness, 15 mm
   - weld length, 3600 mm

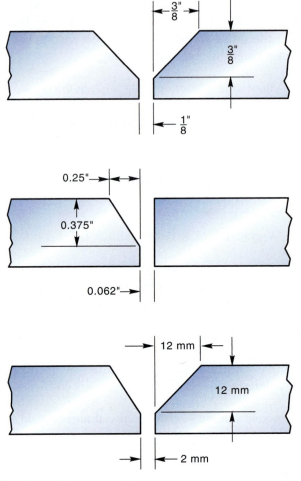

Figure 5.9 Finding the weld groove volume

4. Bevel joint with the following dimensions:
   - width, 15 mm
   - depth, 15 mm
   - root opening, 3 mm
   - thickness, 18 mm
   - weld length, 3 m

Complete a copy of the Student Welding Report in Appendix I or provided by your instructor. ∎

Fillet Welds

The deep penetration processes of fillet welds offer lower costs and improved weld quality. Smaller fillet welds with deeper penetration also have the potential for yielding much stronger welds, Figure 5.10. The cross-sectional area of a fillet weld is equal to 1/2 of the weld leg height times the weld leg width:

$$\frac{2800}{7.00} \div (7.4 \times .50)5m \text{ an-hours}$$

$$400 \div 3.7 = 108m \text{ an-hours}$$

The fillet weld volume is determined in the same manner as the groove weld volume, by multiplying the area by the length:

Total fillet volume = cross-sectional area × weld length

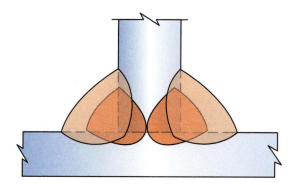

Figure 5.10 Smaller fillet welds with deeper penetration
The deeper penetrating and smaller 1/4-inch (6-mm) fillet weld is stronger than the larger fillet weld and contains about one-half the amount of filler metal

Weld Metal Cost

In their technical data sheets, manufacturers of filler metal provide information regarding the welding metal. The number of electrodes per pound or the length of wire per pound can be used to determine the pounds of electrodes needed to produce a weld. To make this determination, the weight of filler metal required to fill the groove or make the fillet weld must be determined. The weight of filler metal is determined by multiplying the weld volume by the density of the metal, Table 5.6:

Weight of filler metal = volume × metal density

PRACTICE 5-4

Calculate Deposition Rates

Using a pencil, paper, and calculator, determine the weight of metal required for each of the welds described in Practice 5-3. Calculate the weight for both steel and aluminum base and filler metals, Table 5.6.

Using the weight of weld metal deposited allows for better comparisons when a number of different welds are being made. The weight of weld metal can either be determined for the welding prints or measured as the welder uses up supplies, Figure 5.11.

Complete a copy of the Student Welding Report in Appendix I or provided by your instructor. ▪

Cost of Electrodes, Wires, Gases, and Flux

You must obtain the current cost per pound of electrode or welding wire plus the cost of shielding gas or flux (if applicable) from the supplier. The shielding gas flow rate varies slightly with the kind of gas used. The flow rates in Table 5.7 are average values whether the shielding gas is an

Module 1
Key Indicator 1, 4

Table 5.6 Density of Metals

Material	Weight lb/in.^{3}	Weight g/cm.^{3}
Aluminum	0.096	2.73
Steel	0.287	7.945

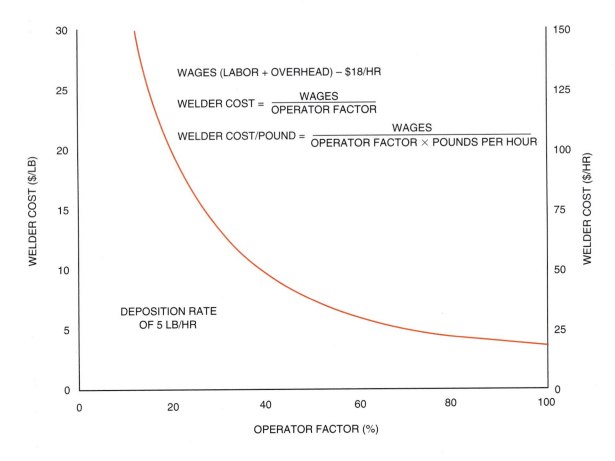

Figure 5.11 Operating factors
Low operating factors are costly

Table 5.7 Approximate Shielding Gas Flow Rate, Cubic Feet Per Hour

	GMAW		FCAW		
Wire diameter	.035″	.045″–1/16″	.045″	1/16″	5/64″–1/8″
CFH	30	35	35	40	45

argon mixture or pure CO_2. Use these rates in your calculations if the actual flow rate is not available.

In the submerged arc process (SAW), the ratio of flux to wire consumed in the weld is approximately 1 to 1 by weight. When the loss due to flux handling and flux recovery systems is considered, the average ratio of flux to wire is approximately 1.4 lb of flux for each pound of wire consumed. If the actual flux-to-wire ratio is unknown, use 1.4 for cost estimating.

Deposition Efficiency

Not every pound of electrode filler metal purchased is converted into weld metal. Some portion of every electrode is lost as slag, spatter, and/ or fume. Some material, such as SMAW electrode stub ends, is unused. The amount of raw electrode deposited as weld metal is measured as deposition efficiency. If all is deposited, the deposition efficiency is 100%. If half is lost, the deposition efficiency is 50%. The argon-shielded

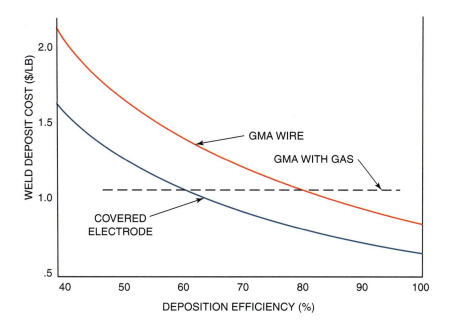

Figure 5.12 Deposition efficiency
Weld metal cost is affected by the process deposition efficiency

GMA process is about 95% efficient. The SMAW process is between 40% and 60% efficient, depending on the electrode and the welder using it. Thus, a relatively costly filler metal with high efficiency can be as cost-effective as one that appears to be cheaper, Figure 5.12. The efficiency can be calculated using the following formula:

$$\text{Deposition efficiency} = \frac{\text{weight of weld metal}}{\text{weight of electrode used}} \text{ or } \frac{\text{deposition rate (lb/hr)}}{\text{burn-off rate (lb/hr)}}$$

The deposition efficiency tells us how many pounds of weld metal can be produced from a given weight of the electrode of welding wire. As an example, 100 lb of a flux cored electrode with an efficiency of 85% will produce approximately 85 lb of weld metal. One hundred pounds of coated electrode with an efficiency of 65% will produce approximately 65 lb of weld metal, less the weight of the stubs discarded.

Note that electrodes priced at $0.65 actually cost about $1.30 per pound as weld metal if about 50% is wasted. The more expensive GMA wire at $0.85 costs about $0.90 per pound as weld metal when deposited because only 5% is lost as spatter and fume. Even with the additional $0.15 for shield gas, the final cost of $1.05 is less than that of welds made with covered electrodes.

Deposition Rate

The deposition rate is the rate at which weld metal can be deposited by a given electrode or welding wire, expressed in pounds per hour. It is based on continuous operation, with no time allowed for stops and starts for inserting a new electrode, cleaning slag, terminating the weld, or other reasons. The deposition rate increases as the welding current is increased. When solid or flux cored wires are used, the deposition rate increases as the electrical stickout is increased and the same welding current is maintained. True deposition rates for each welding filler metal, whether it is a coated electrode or a solid or flux cored wire, can only be established by an actual test. The weldment is weighed before and after welding and the whole process is timed. Tables 5.8, 5.9, 5.10A, and 5.10B contain average values for the deposition rate of the various welding filler metals based on welding laboratory tests and published data.

Coated Electrodes

The deposition efficiency of coated electrodes according to the AWS definition and in published data does not subtract the unused electrode stub that is discarded. This is understandable, as the stub length can vary with the operator and the application. Long, continuous welds are usually conducive to short stubs, whereas on short, intermittent welds the stub length tends to be longer. Figure 5.13 illustrates how the stub loss influences the electrode efficiency when using coated electrodes.

In Figure 5.13, a 14-in.-long, 5/32-in.-diameter E7018 electrode at 140 amperes is considered. It is 75% efficient, and a 2-in. stub loss is assumed. The 75% efficiency applies only to the 12 inches of the electrode consumed in making the weld, not to the 2-in. stub. When the 2-in. stub loss and the 25% loss to slag, spatter, and fumes are considered, the efficiency minus stub loss is lowered to 64.3%. This means that for every 100 lb of electrodes, you can expect an actual deposit of approximately 64.3 lb of weld metal if all electrodes are used to a 2-in. stub length.

The formula for efficiency including stub loss is important. It must always be used when estimating the cost of depositing weld metal by the SMAW method. Table 5.11 shows the formula used to establish the efficiency of coated electrodes including stub loss. It is based on the electrode length and is slightly inaccurate. That is, it does not consider that electrode weight is not evenly distributed because of flux removed from the electrode holder end (indicated by the dotted lines in Figure 5.13). Use of the formula will result in a 1.5% to 2.3% error that varies with electrode size, coating thickness, and stub length. The formula is acceptable for estimating purposes, however.

For the values given in Figure 5.13 the formula is

$$
\begin{aligned}
\text{Efficiency} - \text{stub loss} &= \frac{(14 - 2) \times 0.75}{14} \\
&= \frac{(12 \times 0.75)}{14} \\
&= \frac{9}{14} \\
&= 0.6429 \text{ or } 64.3\%
\end{aligned}
$$

In this example, the electrode length is known, the stub loss must be estimated, and the efficiency is taken from Tables 5.8 and 5.9. Use an average stub loss and 3 in. for coated electrodes if the shop practices concerning stub loss are not known.

Efficiency of Flux Cored Wires

Flux cored wires have a lower flux-to-metal ratio than coated electrodes and therefore a higher deposition efficiency. Stub loss need not be considered, as the wire is continuous. The E70T-1 and E70T-2 types of

Table 5.8 Deposition Data

E6010			
Electrode Diameter	**Amperes**	**Deposition Rate, lb/hr**	**Efficiency %**
1/8	100	2.1	76.3
	130	2.3	68.8
5/32	140	2.8	73.6
	170	2.9	64.1
3/16	160	3.3	74.9
	190	3.5	69.7
7/32	190	4.5	76.9
	230	5.1	73.1
1/4	220	5.9	77.9
	260	6.2	76.2

E6011			
Electrode Diameter	**Amperes**	**Deposition Rate, lb/hr**	**Efficiency %**
1/8	120	2.3	70.7
5/32	150	3.7	77.0
3/16	180	4.1	73.4
7/32	210	5.0	74.2
1/4	250	5.6	71.9

E6012			
Electrode Diameter	**Amperes**	**Deposition Rate, lb/hr**	**Efficiency %**
1/8	130	2.9	81.8
5/32	165	3.2	78.8
	200	3.4	69.0
3/16	220	4.0	77.0
	250	4.2	74.5
7/32	320	5.6	69.8
1/4	320	5.6	70.0
	360	6.6	67.7
	380	7.1	66.0
5/16	400	8.1	70.2

E6013			
Electrode Diameter	**Amperes**	**Deposition Rate, lb/hr**	**Efficiency %**
5/32	140	2.6	75.6
	160	3.0	74.1
	180	3.5	71.2
3/16	180	3.2	73.9
	200	3.8	71.1
	220	4.1	72.9

Continued

Table 5.8 (Continued)

E6013			
Electrode Diameter	**Amperes**	**Deposition Rate, lb/hr**	**Efficiency %**
7/32	250	5.3	71.3
	270	5.7	73.0
	290	6.1	72.7
1/4	290	6.2	75.0
	310	6.5	73.5
	330	7.1	72.1
5/16	360	8.6	70.7
	390	9.4	71.8
	450	10.3	71.3

E7014			
Electrode Diameter	**Amperes**	**Deposition Rate, lb/hr**	**Efficiency %**
1/8	120	2.4	63.9
	150	3.1	61.1
5/32	160	3.0	71.9
	200	3.7	67.0
3/16	230	4.5	70.9
	270	5.5	73.2
7/32	290	5.8	67.2
	330	7.1	70.3
1/4	350	7.1	68.7
	400	8.7	69.9
5/16	440	8.9	62.2
	500	11.1	65.4

Table 5.9 Deposition Data

E7016			
Electrode Diameter	**Amperes**	**Deposition Rate, lb/hr**	**Efficiency %**
5/32	140	3.0	70.5
	160	3.2	69.1
	190	3.6	66.0
3/16	175	3.8	71.0
	200	4.2	71.0
	225	4.4	70.0
	250	4.8	65.8
1/4	250	5.9	74.5
	275	6.4	74.1
	300	6.8	73.2
	350	7.6	71.5

Continued

Table 5.9 (Continued)

E7016			
Electrode Diameter	**Amperes**	**Deposition Rate, lb/hr**	**Efficiency %**
5/16	325	8.0	77.3
	375	9.0	76.3
	425	10.2	76.7

E7024			
Electrode Diameter	**Amperes**	**Deposition Rate, lb/hr**	**Efficiency %**
1/8	140	4.2	71.8
	180	5.1	70.7
5/32	180	5.3	71.3
	210	6.3	72.5
	240	7.2	69.4
3/16	245	7.5	69.2
	270	8.3	70.5
	290	9.1	68.0
7/32	320	9.4	72.4
	360	11.6	69.1
1/4	400	12.6	71.7

Low Alloy, Iron Powder Electrodes of the Types E7018, E8018, E9018, E10018, E11018, and E12018			
Electrode Diameter	**Amperes**	**Deposition Rate, lb/hr**	**Efficiency %**
3/32	70	1.37	70.5
	90	1.65	66.3
	110	1.73	64.4
1/8	120	2.58	71.6
	140	2.74	70.9
	160	2.99	68.1
5/32	140	3.11	75.0
	170	3.78	73.5
	200	4.31	73.0
3/16	200	4.85	76.4
	250	5.36	74.6
	300	5.61	70.3
7/32	250	6.50	75.0
	300	7.20	74.0
	350	7.40	73.0
1/4	300	7.72	78.0
	350	8.67	77.0
	400	9.04	74.0

Table 5.10A Deposition Data

Flux Cored Electrodes—Gas Shielded Types E70T-1, E71T-1, E70T-2, and All Low Alloy Types			
Electrode Diameter	**Amperes**	**Deposition Rate, lb/hr**	**Efficiency %**
0.45	180	5.3	85.0
	200	5.5	86.0
	240	6.9	84.0
	280	13.0	83.0
0.52	190	4.8	85.0
	210	5.3	83.5
	270	7.6	83.0
	300	9.8	85.0
1/16	200	5.2	85.0
	275	10.1	85.0
	300	11.5	85.0
	350	13.3	86.0
5/64	250	6.4	85.0
	350	10.5	85.0
	450	14.8	85.0
3/32	400	12.7	85.0
	450	15.0	86.0
	500	18.5	86.0
7/64	550	17.1	85.0
	625	19.6	86.0
	700	23.0	86.0
1/8	600	16.2	86.0
	725	22.5	86.0
	850	29.2	85.0

Flux Cored Electrodes—Self-Shielded			
Type and Diameter	**Amperes**	**Deposition Rate, lb/hr**	**Efficiency %**
E70T-3			
3/32	450	14.0	88
E70T-4			
3/32	400	18.0	85
.120	450	20.0	81
E70T-6			
5/64	350	11.9	86
3/32	480	14.7	81
E70T-7			
3/32	325	11.4	80
7/64	450	18.0	86

Continued

Table 5.10A (Continued)

Flux Cored Electrodes—Self-Shielded			
Type and Diameter	Amperes	Deposition Rate, lb/hr	Efficiency %
E71T-7			
.068	200	4.2	76
5/64	300	8.0	84
E71T-8			
5/64	220	4.4	77
3/32	300	6.7	77
EG1T8-K6			
5/64	235	4.3	76
E71T8-Nil			
5/64	235	4.3	77
3/32	345	8.2	84
E70T-10			
3/32	400	13.0	69
E71T-11			
5/64	240	4.5	87
3/32	250	5.0	91
E70T4-K2			
3/32	300	14.0	83

NOTE: Values shown are optimum for each type and size.

Table 5.10B Deposition Data

Gas Metal Arc Welding Solid Wires			
Diameter, in.	Amperes	Melt-Off Rate, lb/hr	Efficiency %
.030	75	2.0	
	100	2.7	98% A
	150	4.2	2% O_2
	200	7.0	98%
.035	80	3.2	
	100	2.8	
	150	4.3	75% Argon
	200	6.3	25% CO_2
	250	9.2	96%
.045	100	2.1	
	125	2.9	
	150	3.7	
	200	5.7	
	250	7.4	Straight
	300	10.4	CO_2
	350	13.5	93%

Continued

Table 5.10B (Continued)

Gas Metal Arc Welding Solid Wires			
Diameter, in.	Amperes	Melt-Off Rate, lb/hr	Efficiency %
1/16	250	6.7	
	275	8.6	
	300	9.2	
	350	11.5	
	400	14.3	
	450	17.8	

NOTE: For GMAW and SAW, melt-off rate may be used as the deposition rate in the cost formulas. Using the proper deposition efficiency will account for losses due to spatter, clipping the wire end, etc.

Submerged Arc Welding (1″ Stickout)			
Diameter, in.	Amperes	Melt-Off Rate, lb/hr	Efficiency %
5/64	300	7.0	Assume
	400	10.2	99%
	500	15.0	Efficiency.
3/32	400	9.4	Assume
	500	13.0	99%
	600	17.2	Efficiency.
1/8	400	8.5	Assume
	500	11.5	99%
	600	15.0	Efficiency.
	700	19.0	
5/32	500	11.3	Assume
	600	14.6	99%
	700	18.4	Efficiency.
	800	22.0	
	900	26.1	
3/16	600	13.9	Assume
	700	17.5	99%
	800	21.0	Efficiency.
	900	25.0	
	1,000	29.2	
	1,100	34.0	

NOTE: Values for 1″ stickout.

gas-shielded wires have efficiencies of 83% to 88%. The gas-shielded basic slag wire (E70T-5) is 85% to 90% efficient with CO_2 as the shielding gas. The efficiency can reach 92% when a 75% argon–25% CO_2 gas mixture is used. Use the efficiency figures in Table 5.10A for your calculations if the actual values are not known.

The efficiency of self-shielded flux cored wires is subject to greater variation because of the large assortment of available types designed for

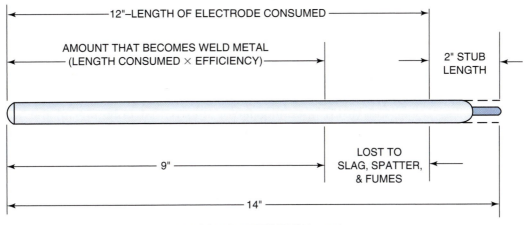

DEPOSITION EFFICIENCY = 75%
ACTUAL EFFICIENCY, INCLUDING STUB LOSS = 9 ÷ 14 = 64.3%

Figure 5.13 Deposition efficiency and stub loss

Table 5.11 Efficiency Minus Stub Loss Formula

$$\text{Efficiency minus stub loss} = \frac{(\text{Electrode length} - \text{Stub length}) \times \text{Deposition efficiency}}{\text{Electrode length}}$$

specific applications. The efficiency of the high-deposition, general-purpose type, such as E70T-4, is 81% to 86%, depending on wire size and electrical stickout. Table 5.10A shows the optimum conditions for each wire size and may be used in your calculations.

Efficiency of Solid Wires for GMAW

The efficiency of solid wires in GMAW is very high and will vary with the shielding gas or gas mixture used, Table 5.10B. Using CO_2 will produce the most spatter, and the average efficiency will be about 93%. Using a 75% argon–25% CO_2 gas mixture will result in somewhat less spatter and an efficiency of approximately 96%. A 98% argon–2% oxygen mixture will produce even less spatter, and the average efficiency will be about 98%. Stub loss need not be considered, since the wire is continuous. Table 5.12 shows the average efficiencies to use in your calculations if the actual efficiency is not known.

Efficiency of Solid Wires for SAW

In submerged arc welding there is no spatter loss, and an efficiency of 99% may be assumed. The only loss during welding is the short piece the operator must clip off the end of the wire to remove the fused flux that forms at

Table 5.12 Deposition Efficiencies—Gas Metal Arc Welding Carbon and Low Alloy Steel Wires

Shielding Gas	Efficiency Range	Average Efficiency
Pure CO_2	88% to 95%	93%
75% A–25% CO_2	94% to 98%	96%
98% A–2% O_2	97% to 98.5%	98%

the termination of each weld. This is done to ensure a good start on the next weld.

Operating Factor

Operating factor is the percentage of a welder's working day actually spent on welding. It is the arc time in hours divided by the total hours worked. A 45% (0.45) operating factor means that only 45% of the welder's day is actually spent on welding. The rest of the time is spent installing a new electrode or wire, cleaning slag, positioning the weldment, cleaning spatter from the welding gun, and so on.

When using coated electrodes (SMAW), the operating factor can range from 15% to 40% depending on material handling, fixturing, and operator dexterity. If the operating factor is not known, an average of 30% may be used for cost estimates involving the shielded metal arc welding process.

When welding with solid wires (GMAW) using the semiautomatic method, operating factors ranging from 45% to 55% are easily attainable. For cost estimating purposes, use a 45% operating factor. The estimated operating factor of FCAW should be about 5% lower than that of GMAW to allow for slag removal time.

In semiautomatic submerged arc welding, slag removal and loose flux handling must be considered. A 40% operating factor is typical for this process.

Automatic welding using the GMAW, FCAW, and SAW processes requires that each application be studied individually. Operating factors ranging from 50% to 100% may be obtained, depending on the degree of automation.

Table 5.13 shows average operating factor values for the various welding processes. These figures may be used for cost estimating when the actual operating factor is not known.

It is necessary to know the productivity of welders to determine the cost of the finished part. In some shops this cost is passed directly on to the customer in the form of cost plus. It may also be used to determine at what level the shop can bid on new work and still make a profit. It is not often used to promote or penalize welders.

Knowing the number of parts produced is useful if there are a number of welders making the same or similar parts in a production shop. A comparison of the productivity can be made because there should be little difference in the average time compared with each welder's actual time.

The length of weld produced is a useful tool when a lot of the same type of welding is required. Welding on items such as ships, tanks, and large vessels may take weeks, months, or years. The length that a welder produces in this type of production can be recorded on a regular basis.

The deposition rates of the process also affect costs. Processes with high deposition rates can be very cost-effective, Figure 5.14. A compelling reason for replacing covered electrodes with small-diameter cored wires

Table 5.13 Approximate Operating Factor

Welding Process			
SMAW	GMAW	FCAW	SAW
30%	50%	45%	40%

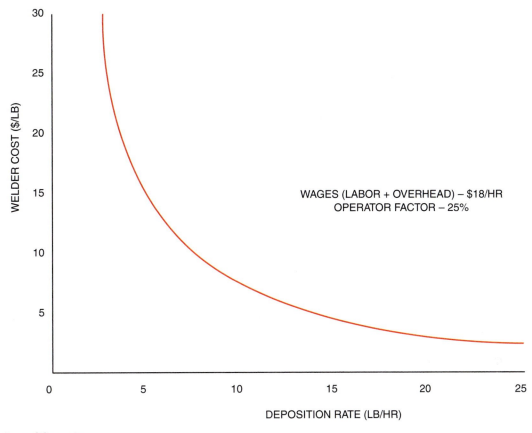

Figure 5.14 Deposition rate
High deposition rates are desirable

is that the deposition rates in the vertical position can be increased from about 2 lb per hour to more than 5 lb per hour. Thus, the welder cost in this example drops from more than $30 per pound to about $12 per pound of weld metal deposited.

Since operating factors and deposition rates are strongly related, their effects on weld costs are examined together. The time spent preparing and positioning weld joints for submerged arc welding is costly, but the high deposition rates of that process justify the time, Figure 5.15. Covered electrodes, however, cannot compete with most other processes unless the setup time and other factors can be reduced. The speed with which alloys and electrode types can be changed explains why covered electrodes have remained competitive in small job shops, especially when typical welds are quite short. Changeover times with GMAW processes can be lengthy, and if the welding jobs are small, the operating factor can drop below 15%. Excessively high deposition rates may be needed to compensate for this deficit.

Factors for Cost Formulas

Labor and Overhead

Labor and overhead may be considered jointly in your calculations. Labor is the welder's hourly rate of pay, including wages and benefits. Overhead includes allocated portions of plant operating and maintenance costs. Weld shops in manufacturing plants normally have established labor and

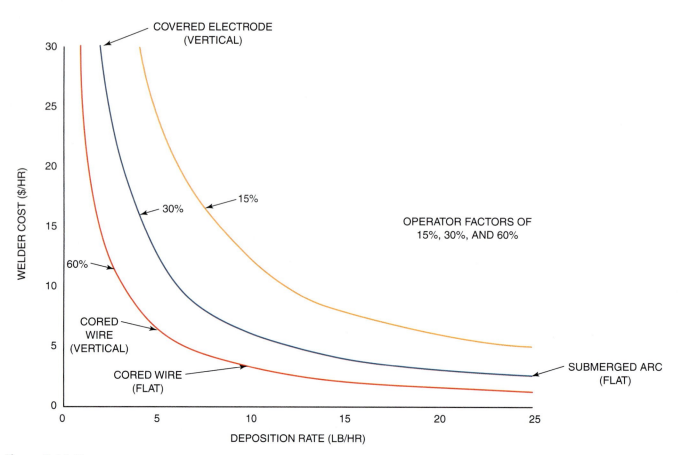

Figure 5.15 New processes
Newer processes can reduce welder costs

overhead rates for each department. Labor and overhead rates can vary greatly from plant to plant and with location. Table 5.14 shows how labor and overhead can vary and suggests an average value to use in your calculations when the actual value is unknown.

Cost of Power

The cost of electrical power is a very small part of the cost of depositing weld metal and in most cases is less than 1% of the total. It will be necessary for you to know the power cost expressed in dollars per kilowatt-hour ($/kWh) if required for a total cost estimate.

Calculating the Cost per Pound of Deposited Weld Metal

Example 1

Calculate the cost of welding 1280 feet of a single-bevel butt joint, as shown in Figure 5.16, using the following data:

- electrode, 13/16-in. diameter, 14 in. long, E7018, operated at 25 volts, 250 amps
- stub loss, 2 in.
- labor overhead, $45.00/hr
- electrode cost, $1.45/lb
- power cost, $0.045/kWh

Table 5.14 Approximate Hourly Labor and Overhead Rates

Small shops	$7.50 to $15.00
Large shops	$15.00 to $35.00
Average	$20.00

The formula for the calculations is shown on the weld metal cost worksheet in Figure 5.17. The following text explains each step in the calculations:

Line 1: Labor and Overhead—$45.00/hr (given)

- deposition rate: from shielded metal arc welding deposition data chart in Table 5.9, 5.36 lb/hr
- operating factor: Since it is not stated, use the average value of 30% (0.30) shown in Table 5.13.
- The cost of labor and overhead per pound of deposited weld metal can now be calculated as $27.99/lb.

Line 2: Electrode Cost per Pound—$1.45 (given)

- deposition efficiency: From shielded metal arc welding deposition table in Table 5.9, 74.6%. Since this is a coated electrode, the efficiency must be adjusted for stub loss by the formula shown in Table 5.11. We know that the electrode length is 14 in. and the stub loss is 2 in. (given). The formula becomes

$$\text{Efficiency} - \text{stub loss} = \frac{(14 - 2) \times 0.746}{14} \\ = 0.639 \text{ or } 63.9\%$$

- 63.9% is the adjusted efficiency to be used in line 2. The cost of the electrode per pound of deposited weld metal can now be calculated as $.89/lb.

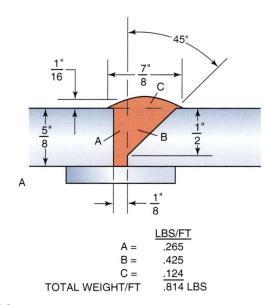
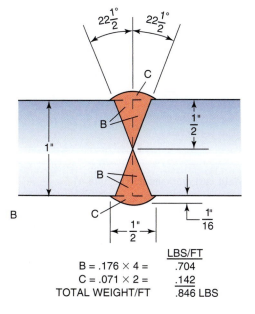

Figure 5.16 Estimating weld metal weight

EXAMPLE 1
WELD METAL COST WORKSHEET
COST PER POUND OF DEPOSITED WELD METAL

1	LABOR & OVERHEAD	$\dfrac{\text{LABOR \& OVERHEAD COST/HR}}{\text{DEPOSITION RATE (LB/HR)} \times \text{OPERATING FACTOR}}=$	$\dfrac{45.00}{5.36 \times .30} = \dfrac{45.00}{1.608} = \underline{27.99}$
2	ELECTRODE	$\dfrac{\text{ELECTRODE COST/LB}}{\text{DEPOSITION EFFICIENCY}}=$	$\dfrac{1.45}{.639} = \underline{2.27}$
3	GAS	$\dfrac{\text{GAS FLOW RATE (CU FT/HR)} \times \text{GAS COST/CU FT}}{\text{DEPOSITION RATE (LB/HR)}}=$	$\dfrac{\quad \times \quad}{\quad} = \dfrac{}{} = \underline{NA}$
4	FLUX	$\dfrac{\text{FLUX COST/LB} \times 1.4}{\text{DEPOSITION EFFICIENCY}}=$	$\dfrac{\quad \times 1.4}{\quad} = \dfrac{}{} = \underline{NA}$
5	POWER	$\dfrac{\text{COST/KWH} \times \text{VOLTS} \times \text{AMPS}}{1000 \times \text{DEPOSITION RATE}}=$	$\dfrac{.045 \times 25 \times 250}{1000 \times 5.36} = \dfrac{281.25}{5,360} = \underline{.05}$
6	TOTAL COST PER LB OF DEPOSITED WELD METAL	SUM OF 1 THROUGH 5, ABOVE	$ \underline{30.31}$

COST PER FOOT OF DEPOSITED WELD METAL

7	COST PER POUND OF DEPOSITED WELD METAL $\times$ POUNDS PER FOOT OF WELD JOINT $=$	$\underline{30.31} \times \underline{.814} = \$ \underline{24.67}$

COST OF WELD METAL – TOTAL JOB

8	TOTAL FEET OF WELD $\times$ COST PER FOOT $=$	$\underline{1,280} \times \underline{24.67} = \$ \underline{31,577.60}$

Figure 5.17 Weld cost worksheet (Example 1)

Lines 3 and 4: Not applicable for coated electrodes
Line 5: Cost of Power—$.045/kWh (given)

- volts and amperes: 25 V and 250 A (given)
- constant: The 1000 already entered is a constant necessary to convert to watt-hours.

- deposition rate: 5.36 lb/hr are used in line 1. The cost of electrical power to deposit 1 pound of weld metal can now be calculated as $0.05.

Line 6: Total lines 1, 2, and 5 to find the total cost of depositing 1 pound of weld metal.

- The total is $30.31.

Calculating the Cost per Foot of Deposited Weld Metal

Calculating the weight of weld metal requires that we consider the following items:

1. area of the weld's cross section
2. length of the weld
3. volume of the weld in cubic inches
4. weight of the weld metal per cubic inch

In the fillet weld shown in Figure 5.18, we know that the area of the cross section (the triangle) is equal to one-half the base times the height. The volume of the weld is equal to the area times the length. The weight of the weld is then the volume times the weight of the material (steel) per cubic inch.

We can then write the formula:

Weight of weld metal = 1/2 × base × height × length × weight of material

Substituting the values from Figure 5.18, we have

Wt/ft = 0.5 × 0.5 × 0.5 × 12 × 0.283 = 0.4245 lb

Table 5.15 eliminates the need for these calculations for steel fillet and butt joints, as it lists the weight per foot directly.

Estimating the weight per foot of a weld using the table requires that you make a drawing of the weld joint to exact scale. Dimension the leg lengths, root gap, thickness, angles, and other pertinent measurements as shown in Figure 5.16. Divide the cross section of the weld into right triangles and rectangles as shown. Where required, sketch in the reinforcement, which is the domed portion above or below the surface of the plate. The

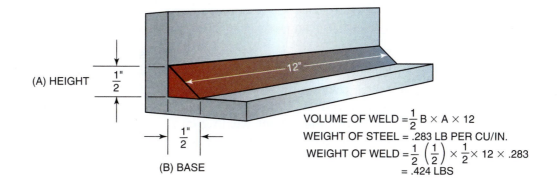

Figure 5.18 Calculating the weight per foot of a fillet weld

reinforcement should extend slightly beyond the edges of the joint. Measure the length and height of the reinforcement and note them on your drawing. The reinforcement is only an approximation because the contour cannot be exactly controlled during welding.

Refer to Table 5.15 for the weights per foot of each of the component parts of the weld as sketched. The sum of the weights of all the components is the total weight of the weld per foot, as shown in Figure 5.16A.

Line 7: The total cost per pound as determined in line 6 is entered and multiplied by the weight per foot as determined in Figure 5.16A.

Line 8: The cost of the weld for the total job is determined by multiplying the total feet of weld (given) by the cost per foot as determined in line 7.

Example 2

Calculate the total cost of depositing 1280 feet of weld metal using the CO_2 shielded, flux cored welding process in the double V-groove joint shown in Figure 5.16B using the following data:

1. electrode: 3/32 in., E70T-1 at 31 volts, 450 amps
2. labor and overhead: $45.00/hour
3. deposition rate: 15 lb/hour, from Table 5.10A
4. operating factor: 45% (0.45) average, from Table 5.13
5. electrode cost: $1.22/pound (from supplier)
6. deposition efficiency: 86% (0.86) from Table 5.10A
7. gas flow rate: 45 cu ft per hour, from Table 5.7
8. gas cost: $0.17 per cubic foot (from supplier)
9. cost of power: $0.045/kWh
10. weight per foot of weld: 0.846 lb/ft, from Figure 5.16B.

These values are shown inserted into the formulas on the weld metal cost worksheet in Figure 5.19.

Comparing Weld Metal Costs

Note that the amount of weld metal deposited in Examples 1 and 2 is almost the same, but the total cost of depositing the weld metal is three times higher in Example 1, as shown here:

Example 1: 1280 ft × 0.814 lb/ft = 1041.9 lb at $31,577.60
Example 2: 1280 ft × 0.846 lb/ft = 1082.9 lb at $9356.80

This is because the flux cored process has a higher deposition rate, efficiency, and operating factor, and it allows a tighter joint because of the deep penetrating characteristics of the process.

When you are comparing welding processes, all efforts should be made to use the proper welding current for the electrode or wire in the position in which the weld must be made. As an example, consider depositing a given-size fillet weld in the vertical up position using the GMAW process and FCAW process semiautomatically. In both processes the welding current and voltage must be lowered to weld out of position. In GMAW, the short-circuiting arc transfer must be used. Example 3 compares the weld metal cost per pound deposited by these processes,

Table 5.15 Weight Per Foot of Weld Metal for Fillet Welds and Elements of Common Butt Joints in Steel

WEIGHT PER FOOT OF WELD METAL FOR FILLET WELDS
AND ELEMENTS OF COMMON BUTT JOINTS (LB/FT) STEEL

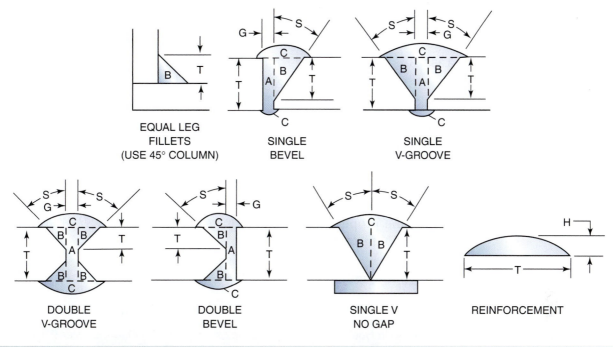

EQUAL LEG FILLETS (USE 45° COLUMN)   SINGLE BEVEL   SINGLE V-GROOVE

DOUBLE V-GROOVE   DOUBLE BEVEL   SINGLE V NO GAP   REINFORCEMENT

T	Weight/Ft of Rectangle AG						Weight per Foot of Triangle BS						Weight/Ft Reinforcement CH			
Inches	1/16″	1/8″	3/16″	1/4″	3/8″	1/2″	5°	10°	15°	22 1/2°	30°	45°	1/16″	1/8″	3/16″	1/4″
1/8	.027	.053	.080	.106	.159	.212	.002	.005	.007	.011	.015	.027				
3/16	.040	.080	.119	.159	.239	.318	.005	.011	.016	.025	.035	.060	.027			
1/4	.053	.106	.159	.212	.318	.425	.009	.019	.028	.044	.061	.106	.035			
5/16	.066	.133	.199	.265	.390	.531	.015	.029	.044	.069	.096	.166	.044	.084		
3/8	.080	.159	.239	.318	.478	.637	.021	.042	.064	.099	.138	.239	.053	.106		
7/16	.091	.186	.279	.371	.557	.743	.028	.057	.087	.129	.188	.325	.062	.124		
1/2	.106	.212	.318	.425	.637	.849	.037	.075	.114	.176	.245	.425	.071	.141	.212	
9/18	.119	.239	.358	.478	.716	.955	.047	.095	.144	.223	.311	.451	.080	.159	.239	
5/8	.133	.265	.398	.531	.796	1.061	.058	.117	.178	.275	.383	.664	.088	.177	.265	.354
11/16	.146	.292	.438	.584	.876	1.167	.070	.142	.215	.332	.464	.804	.097	.195	.292	.389
3/4	.159	.318	.478	.637	.995	1.274	.084	.169	.256	.396	.552	.956	.106	.212	.318	.424
13/16	.172	.345	.517	.690	1.035	1.380	.098	.198	.301	.464	.648	1.121	.115	.230	.345	.460
7/8	.186	.371	.557	.743	1.114	1.486	.114	.230	.349	.538	.751	1.300	.124	.248	.371	.495
15/18	.199	.398	.597	.796	1.194	1.592	.131	.263	.400	.618	.863	1.493	.133	.266	.398	.530
1	.212	.425	.637	.849	1.274	1.698	.149	.300	.456	.703	.981	1.698	.141	.283	.424	.566
11/8	.239	.478	.716	.955	1.433	1.910	.188	.379	.577	.890	1.241	2.149	.159	.318	.477	.637
11/4	.265	.531	.796	1.061	1.592	2.123	.232	.468	.712	1.099	1.532	2.653	.177	.354	.531	.707
13/8	.292	.584	.876	1.167	1.751	2.335	.281	.567	.861	1.330	1.853	3.210	.195	.389	.584	.777
11/2	.318	.637	.955	1.274	1.910	2.547	.334	.674	1.023	1.582	2.206	3.821	.212	.424	.637	.849

Table 5.15 Weight Per Foot of Weld Metal for Fillet Welds and Elements of Common Butt Joints in Steel (Continued)

T	Weight/Ft of Rectangle AG						Weight per Foot of Triangle BS						Weight/Ft Reinforcement CH			
Inches	1/16″	1/8″	3/16″	1/4″	3/8″	1/2″	5°	10°	15°	22 1/2°	30°	45°	1/16″	1/8″	3/16″	1/4″
15/8	.345	.690	1.035	1.380	2.069	2.759	.393	.792	1.201	1.857	2.589	4.484	.230	.460	.690	.920
13/4	.371	.743	1.114	1.486	2.229	2.972	.455	.918	1.393	2.154	3.002	5.200	.248	.495	.743	.990
17/8	.390	.796	1.194	1.592	2.388	3.184	.523	1.053	1.599	2.473	3.447	5.970	.266	.531	.796	1.061
2	.425	.849	1.274	1.698	2.547	3.396	.594	1.197	1.820	2.813	3.921	6.792	.283	.566	.849	1.132
21/4	.478	.955	1.433	1.910	2.865	3.821	.752	1.516	2.303	3.561	4.963	8.596	.318	.637	.955	1.273
21/2	.530	1.061	1.592	2.123	3.184	4.245	.928	1.871	2.844	4.396	6.127	10.613	.354	.707	1.061	1.415
23/4	.584	1.167	1.751	2.335	3.502	4.669	1.123	2.264	3.441	5.319	7.414	12.841	.389	.778	1.167	1.556
3	.636	1.274	1.910	2.547	3.821	5.094	1.337	2.695	4.095	6.330	8.823	15.282	.424	.849	1.273	1.698

using the proper current and voltage for depositing a 1/4-in. fillet weld on 1/4-in. plate, vertically up.

Example 3

	FCAW	GMAW
Electrode type	0.045-in.-diameter E71T-1	0.045-in.-diameter ER70S-3
Labor and overhead	$45.00/hour	$45.00/hour
Welding current	180 amp	125 amp
Deposition rate	5.3 lb/hr (Table 5.10A)	2.9 lb/hr (Table 5.10B)
Operation factor	45% (Table 5.13)	50% (Table 5.13)
Electrode cost	$2.19/lb	$1.05/lb
Deposition efficiency	85% (Table 5.10A)	96% (Table 5.12)
Gas flow rate	35 cfh (Table 5.7)	35 cfh (Table 5.7)
Gas cost per cu ft	$0.11 CO_2	$0.17 75% Ar–25% CO_2

We can eliminate the cost of electrical power when comparing processes because the difference is very small. The tabulated data are shown in Table 5.16.

As you can see, the cost of depositing the weld metal is about 33% less using the flux cored arc welding process. Since there is no slag to help hold the vertical weld puddle in the GMAW process, the welding current with solid wire must be lowered considerably. This, of course, lowers the deposition rate, and since labor plus overhead is the largest factor involved, the lowering of the deposition rate substantially raises deposition costs. In the flat or horizontal position, where the welding current on the solid wire would be much higher, the cost difference would be considerably less.

EXAMPLE 2
WELD METAL COST WORKSHEET
COST PER POUND OF DEPOSITED WELD METAL

1		
LABOR & OVERHEAD	$\dfrac{\text{LABOR \& OVERHEAD COST/HR}}{\underset{\text{RATE (LB/HR)}}{\text{DEPOSITION}} \times \underset{\text{FACTOR}}{\text{OPERATING}}}$ =	$\dfrac{45.00}{15 \times .45} = \dfrac{45.00}{6.75} = 6.67$

2		
ELECTRODE	$\dfrac{\text{ELECTRODE COST/LB}}{\text{DEPOSITION EFFICIENCY}}$ =	$\dfrac{1.22}{.86} = 1.42$

3		
GAS	$\dfrac{\underset{\text{(CU FT/HR)}}{\text{GAS FLOW RATE}} \times \text{GAS COST/CU FT}}{\text{DEPOSITION RATE (LB/HR)}}$ =	$\dfrac{45 \times .17}{15} = \dfrac{7.65}{15} = .51$

4		
FLUX	$\dfrac{\text{FLUX COST/LB} \times 1.4}{\text{DEPOSITION EFFICIENCY}}$ =	$\dfrac{\times 1.4}{} = \dfrac{}{} = NA$

5		
POWER	$\dfrac{\text{COST/KWH/} \times \text{VOLTS} \times \text{AMPS}}{1000 \times \text{DEPOSITION RATE}}$ =	$\dfrac{.045 \times 31 \times 450}{1000 \times 15} = \dfrac{627.75}{15,000} = .04$

6		
TOTAL COST PER LB OF DEPOSITED WELD METAL	SUM OF 1 THROUGH 5, ABOVE	\$ 8.64

COST PER FOOT OF DEPOSITED WELD METAL

7		
COST PER POUND OF DEPOSITED WELD METAL ×	POUNDS PER FOOT OF WELD JOINT =	$8.64 \times .846 = \$ 7.31$

COST OF WELD METAL – TOTAL JOB

8		
TOTAL FEET OF WELD × COST PER FOOT =		$1,280 \times 7.31 = \$ 9,356.80$

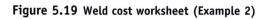

Figure 5.19 Weld cost worksheet (Example 2)

Table 5.16 Formulas for Calculating Cost Per Pound Deposited of Weld Metal

FORMULAS FOR CALCULATING COST PER POUND DEPOSITED WELD METAL	*Flux cored arc welding E71T .045" diameter 180 amps.*	*Gas metal arc welding ER70S-3 .045" diameter 125 amps*
LABOR & OVERHEAD = LABOR & OVERHEAD COST/HR / (DEPOSITION RATE (LB/HR) × OPERATING FACTOR) =	$\dfrac{45}{5.3 \times .45} = \dfrac{45}{2.385} = 18.87$	$\dfrac{45}{2.9 \times .50} = \dfrac{45}{1.45} = 31.03$
ELECTRODE = ELECTRODE COST/LB / DEPOSITION EFFICIENCY =	$\dfrac{2.19}{.85} = 2.58$	$\dfrac{1.05}{.96} = 1.09$
GAS = (GAS FLOW RATE (CU FT/HR) × GAS COST/CU FT) / =	CO₂ $\dfrac{35 \times 11}{5.3} = \dfrac{3.85}{5.3} = .13$	15/25 $\dfrac{30 \times 11}{2.9} = \dfrac{5.1}{2.9} = 1.16$
SUM OF THE ABOVE	TOTAL VARIABLE COST/LB DEPOSITED WELD METAL 22.18	TOTAL VARIABLE COST/LB DEPOSITED WELD METAL 33.88

Other Useful Formulas

The following formulas will assist you in making other useful calculations:

$$\text{Total pounds} = \frac{\text{wt/ft of weld} \times \text{no. of ft of weld}}{\text{deposition efficiency}}$$

Substracting the values from Example 1:

$$\frac{0.814 \times 1280}{0.639} = 1631 \text{lb}$$

Welding Time Required (Ref. Example 1)

$$\text{Welding time} = \frac{\text{wt/ft of weld} \times \text{ft of weld}}{\text{deposition rate} \times \text{operating factor}}$$

Substracting the values in Example 1 :

$$\frac{0.814 \times 1280}{5.36 \times 0.30} = \frac{1042}{1608} = 648 \text{ hr}$$

Amortization of Equipment Costs

Calculations show that you can save $7.00 per pound of deposited weld metal by switching from E7018 electrodes and the SMAW process to an ER70S-3 solid wire using the GMAW process. However, the cost of the necessary equipment (power source, wire feeder, and gun) is $2800. How long will it take to amortize or regain the cost of the equipment knowing that the deposition rate of the ER70S-3 is 7.4 lb/hr and the operating factor of the GMAW process is 50%? The formula is

$$\text{Equipment cost} \div (\text{deposition rate} \times \text{operation factor})$$
$$= \text{man-hour savings/lb in dollars.}$$

Subtracting the values in the formula:

$$\frac{2800}{7.00} \div (7.4 \times .50) \text{ 5m an} - \text{hours}$$
$$400 \div 3.7 = 108\text{m an} - \text{hours}$$

If we divide 108 into eight-hour days (108/8 = 13.5), the deposited weld metal savings from one person working an eight-hour day for 13 1/2 days will pay for the cost of the equipment.

SUMMARY

Over the years, often through trial and error, welders and welding engineers have developed standards, codes, and specifications that, when followed, will produce welds that are sound and provide years of service. It is important to know that under such codes and standards, not all welds must be perfect. Some levels of imperfection are acceptable and, through years of experience, such minor flaws have been determined not to be critical or to result in structural failure.

Being able to follow codes and standards is important in that it helps control welding costs. The more precise the weld produced, the more expensive the weld is to produce. As a welder attempts perfection, welding preparation time, welding time, postweld cleanup time, and unnecessary rewelding time all increase, thus increasing the cost of the weld. Knowing what is "fit for service" as established by the code or standard is essential. Likewise, failing to produce a weld to the required code or standard can cause structural failure. Therefore, it is important that you familiarize yourself with the standards for your company's requirements.

Controlling the cost of a weldment is as important as producing a quality welded product, because if you spend too much time in preparing and producing the weld to a quality standard far above that required by the industry, the end product may be excessively expensive and unmarketable. An example of a product requiring a relatively low level of welding skills is yard art. For such decorative or ornamental pieces, the customer is most frequently looking at cost. Welding on these items must merely hold them together to meet market demands. An example of a product that requires a high level of welding skills is the space shuttle rocket engines, where all the welds must be precise at any expense. Most welding requirements obviously fall somewhere between these two extremes. For both your own benefit and your company's, you must learn to meet their needs and standards in the most cost-effective manner.

REVIEW

1. What are codes and standards?
2. Why is it important to select the correct welding code or standard?
3. What is the difference between welding codes or standards and welding specifications?
4. What might influence the selection of a particular code or specification for welding?
5. What information should be included in a WPS?
6. What is the purpose of the PQR?
7. Who should witness the test welding being performed for a tentative WPS?
8. Ideally, a WPS should be written with enough information to allow a good welder to _____.
9. List examples of fixed and variable costs that must be considered when estimating the costs for a job.

10. List examples of overhead costs that a welding shop might have.
11. What effect on the weld cost does increasing the groove angle have?
12. What potential problems can be caused by having too small or too large a weld bead?
13. What is the cross-sectional area of a single-bevel groove weld that is 1/2 in. wide and 5/8 in. deep on a 3/4-in.-thick plate with a 1/8-in. root opening? What is the SI area?
14. What is the cross-sectional area of a V-groove weld that is 6 mm wide and 8 mm deep on a 10-mm-thick plate with a 2-mm root opening? What is the area in square inches?
15. What is the cross-sectional area of a fillet weld that has equal legs of 1/2 in.? What is the SI area?
16. What is the cross-sectional area of a fillet weld that has a 10-mm leg and an 8-mm leg? What is the area in square inches?
17. How many pounds of steel electrode are required to make a weld that has a volume of 18 cu in.? What is the SI weight?
18. What are the advantages of calculating the weight of filler metal needed to make a weld?
19. Using Table 5.7, what would be the flow rate for a 1/16-in. FCAW electrode?
20. What is the approximate ratio of flux to filler wire for SAW?
21. Why must deposition efficiency be used when determining how much electrode will be needed for a job?
22. Approximately how many pounds of weld will a 30-pound spool of 3/32-in. E70T-4 FCA welding wire produce? (Refer to Table 5.10A.)
23. What is the melt-off rate for 0.035 GMAW filler wire at the 100-ampere setting? (Refer to Table 5.10B.)
24. What is the approximate length of the unused electrode stub of a 1/8-in. E7018 SMAW electrode?
25. What does the flux-to-metal ratio have to do with FCAW deposition efficiency?
26. Which shielding gas or gas mixture provides GMAW with the highest average efficiency? (Refer to Table 5.12.)
27. What is the operating factor?
28. Of the major processes, which has the highest and which has the lowest operating factor, and what are their percentages? (Refer to Table 5.13.)
29. With reference to Figure 5.14, what would be the welder cost in dollars per pounds if the deposition rate were 15 lb/hr?
30. Why is SMAW still used in many shops on small jobs if its operating factors and deposition rates are so low compared with those of most other processes?

Flame Cutting

After completing this chapter, the student should be able to

- describe the oxyfuel gas cutting process and list three of the most commonly used fuel gases
- list two metals that can be cut using the oxyfuel gas cutting process and three metals that cannot
- select the eye protection that must be used for flame cutting
- determine the correct size and type of cutting tip for a specific job
- set up, light, and clean the tip of a cutting torch
- list four actions that can be taken to ensure a smooth cut when using a hand torch
- lay out a line to be cut
- explain the chemical process that takes place during the burning away of the metal when an oxyfuel gas cutting torch is used
- explain what the kerf surface can reveal about what was correct or incorrect with the preheat flame, cutting speed, and oxygen pressure
- make a machine cut and then evaluate the results
- describe soft slag and hard slag and what causes them
- make a manual flat, straight cut in thin plate, thick plate, and sheet metal
- make a flame-cut hole
- describe the two major methods of controlling distortion of the metal during the heating or cutting process
- make a straight line cut in the vertical and overhead positions
- cut out internal and external shapes

KEY TERMS

coupling distance	injector torch	preheat flame
cutting lever	kindling point	preheat hole
cutting tip	machine cutting torch	slag
drag	MPS gases	soapstone
drag line	orifice	soft slag
equal-pressure torch	oxyacetylene hand torch	tip cleaner
hard slag	oxyfuel gas cutting (OFC)	venturi
high-speed cutting tip		

AWS SENSE EG2.0

Key Indicators Addressed in this Chapter:

Module 8: Thermal Cutting Processes

Unit 1: Manual Oxyfuel Gas Cutting (OFC) Key Indicators

Key Indicator 1: Performs safety inspections of manual OFC equipment and accessories

Key Indicator 2: Makes minor external repairs to manual OFC equipment and accessories

Key Indicator 3: Sets up for manual OFC operations on carbon steel

Key Indicator 4: Operates manual OFC equipment on carbon steel

Key Indicator 5: Performs manual OFC straight, square edge, cutting operations, in all positions, on limited thickness range of carbon steel

Key Indicator 6: performs manual OFC shape, square edge, cutting operations in all positions, on limited thickness range of carbon steel

Key Indicator 7: Performs manual OFC straight, bevel edge, cutting operations, in all positions, on limited thickness range of carbon steel

Key Indicator 8: Performs manual OFC scarfing and gouging operations to remove base and weld metal, in 1G and 2G positions, on carbon steel

Unit 2: Mechanized Oxyfuel Gas Cutting (OFC)

Key Indicator 1: Performs safety inspections of mechanized OFC equipment and accessories

Key Indicator 2: Makes minor external repairs to mechanized OFC equipment and accessories

Key Indicator 3: Sets up for mechanized OFC operations on carbon steel

Key Indicator 4: Operates mechanized OFC equipment on carbon steel

Key Indicator 5: Performs mechanized OFC straight, square edge, cutting operations, in 1G and 2G positions, on limited thickness range of carbon steel

Key Indicator 6: Performs mechanized OFC straight, bevel edge, cutting operations, in 1G and 2G positions, on limited thickness of carbon steel

Module 9: Welding Inspection and Testing Principles

Key Indicator 1: Examines cut surfaces and edges of prepared base metal parts

INTRODUCTION

Oxyfuel gas cutting (OFC) describes a group of oxygen cutting processes that use an oxyfuel gas flame to heat metal to its kindling temperature before a high-pressure stream of oxygen is directed onto the metal to cut it. The kindling temperature of a material is the temperature at which rapid oxidation (combustion) can begin. The kindling temperature of steel in pure oxygen is 1600°F to 1800°F (870°C to 900°C). The OFC processes are identified by the type of fuel gas used

Table 6.1 Fuel Gases Used for Flame Cutting

Fuel Gas	Flame (Fahrenheit)	Temperature* (Celsius)
Acetylene	5589°	3087°
MAPP®	5301°	2927°
Natural gas	4600°	2538°
Propane	4579°	2526°
Propylene	5193°	2867°
Hydrogen	4820°	2660°

*Approximate neutral oxyfuel flame temperature.

with oxygen to produce the preheat flame. Oxyfuel gas cutting is most commonly performed with oxyacetylene (OFC-A). Table 6.1 lists a number of fuel gases used for OFC. MPS (MAPP®) gas is increasingly being used today for cutting and rivals acetylene's popularity in some areas of the United States.

More welders use the oxyfuel cutting torch than any other welding process. The cutting torch is used by workers in virtually all areas, including manufacturing, maintenance, automotive repair, railroad work, farming, and more. Unfortunately, it is one of the most misused processes. Most workers know how to light the torch and make a cut, but their cuts are of very poor quality. Often, in addition to making bad cuts, they use unsafe torch techniques. A good oxyfuel cut should be straight and square and should require little or no postcut cleanup. Excessive postcut cleanup results in extra cost, which is an expense that cannot be justified.

Manual, mechanized, and automatic OFC processes are used in industry. Hand-controlled, manual cutting is used for short-run production, one-of-a-kind fabrication, and demolition and scrapping operations. Manual cutting is also used in the field for steel construction. Mechanized or automatic cutting is widely used in production work where a large number of identical cuts must be made or where very precise cuts are required. In mechanized or automatic cutting, more than one cutting head may be mounted so that several cuts can be made at the same time.

Various oxyfuel cutting specialties are found on the job, including flame cutting, gouging, beveling, scarfing, and the operation of an automated cutting machine. In addition to these cutting jobs, some welders work with scrap metal, such as scrap autos, or in construction demolition.

METALS CUT USING THE OXYFUEL PROCESS

Oxyfuel gas cutting is used to cut iron base alloys. Low-carbon steels (up to 0.3% carbon) are easy to cut. Any metal that requires preheating for welding, such as high strength and high alloy carbon steels, should also be preheated before cutting. Failure to preheat some high-strength alloys before cutting can result in a very thin hardness zone on the cut surface.

This hardness zone can cause cracks to start in the finished part. If high-strength steel flame-cut parts are to be bent or formed, the hardened edge may cause cracks to form. These surface cracks can cause the part to fracture and fail. High-nickel steels, cast iron, and stainless steel are difficult to cut. Most nonferrous metals—such as brass, copper, and aluminum—cannot be cut using oxyfuel cutting. A few reactive nonferrous metals, such as titanium and magnesium, can be cut. However, these metals seldom are cut with the OFC process because of the extensive postcut cleanup required.

EYE PROTECTION FOR FLAME CUTTING

The National Bureau of Standards has identified appropriate filter plates to be used for eye protection in flame cutting. The recommended filter plates are identified by shade number and are related to the type of cutting operation being performed.

Goggles or other suitable eye protection must be worn for flame cutting. Goggles should have vents near the lenses to prevent fogging. Cover lenses or plates should be provided to protect the filter lens. All lens glass should be ground properly so that the front and rear surfaces are smooth. Filter lenses must be marked so that the shade number can be readily identified, Table 6.2.

CUTTING TORCHES

The **oxyacetylene hand torch** is the most common type of oxyfuel gas cutting torch used in industry. The hand torch, as it is often called, may be either a part of a combination welding and cutting torch set or a cutting torch only, Figure 6.1. The combination welding–cutting torch offers more flexibility because a cutting head, welding tip, or heating tip can be attached quickly to the same torch body, Figure 6.2. Combination torch sets are often used in schools, automotive repair shops, auto body shops,

Table 6.2 A General Guide for the Selection of Eye and Face Protection Equipment

Type of Cutting Operation	Hazard	Suggested Shade Number
Light cutting, up to 1 in.	Sparks, harmful	3 or 4
Medium cutting, 1–6 in.	rays, molten metal,	4 or 5
Heavy cutting, over 6 in.	flying particles	5 or 6

Figure 6.1 Dedicated oxyfuel cutting torch
Source: Courtesy of Victor Equipment, a Thermadyne Company

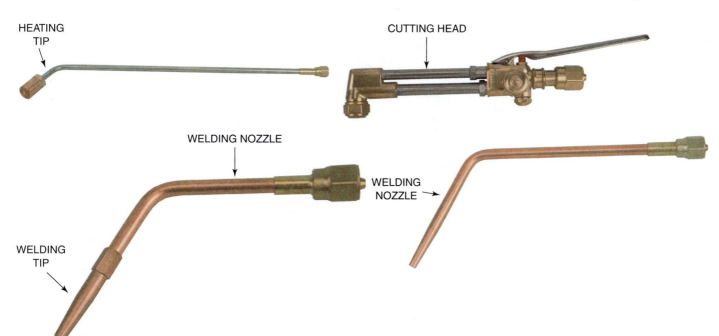

HEATING TIP

CUTTING HEAD

WELDING NOZZLE

WELDING NOZZLE

WELDING TIP

Figure 6.2 Torch attachments
The flexible combination torch set has attachments for heating, cutting, welding, and brazing
Source: Courtesy of Victor Equipment, a Thermadyne Company

small welding shops and for any job where multipurpose equipment is needed. A cut made with either type of torch has the same quality; however, the dedicated cutting torches are usually longer and have larger gas flow passages than the combination torches. The added length of the dedicated cutting torch helps keep the operator farther away from the heat and sparks and allows thicker material to be cut.

Oxygen is mixed with the fuel gas to produce a high-temperature preheating flame. The two gases must be completely mixed before they leave the tip and burn. Two methods are used to mix the gases. One method uses a mixing chamber, and the other method uses an injector chamber.

The mixing chamber may be located in the torch body or in the tip, Figure 6.3. Torches that use a mixing chamber are known as

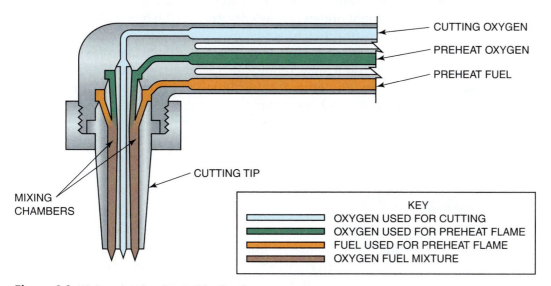

CUTTING OXYGEN

PREHEAT OXYGEN

PREHEAT FUEL

CUTTING TIP

MIXING CHAMBERS

KEY		
	OXYGEN USED FOR CUTTING	
	OXYGEN USED FOR PREHEAT FLAME	
	FUEL USED FOR PREHEAT FLAME	
	OXYGEN FUEL MIXTURE	

Figure 6.3 Mixing chamber located in the tip

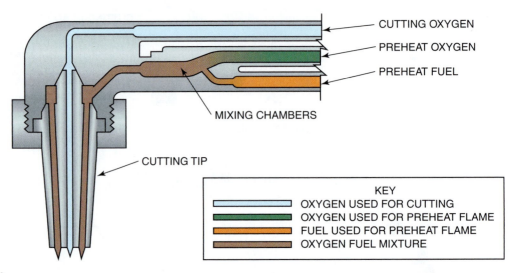

CUTTING OXYGEN
PREHEAT OXYGEN
PREHEAT FUEL
MIXING CHAMBERS
CUTTING TIP

KEY
OXYGEN USED FOR CUTTING
OXYGEN USED FOR PREHEAT FLAME
FUEL USED FOR PREHEAT FLAME
OXYGEN FUEL MIXTURE

Figure 6.4 Injector mixing torch

equal-pressure torches, because the gases must enter the mixing chamber under the same pressure. The mixing chamber is larger than both the gas inlet and the gas outlet. This larger size causes turbulence in the gases, resulting in the gases mixing thoroughly.

Injector torches work with both equal gas pressures and low-fuel gas pressures, Figure 6.4. The injector allows the oxygen to draw the fuel gas into the chamber even if the fuel gas pressure is as low as 6 oz/in.^{2} (26 g/cm^2). The injector works by passing the oxygen through a **venturi**, which creates a low-pressure area that pulls the fuel gases in and mixes them. An injector-type torch must be used if a low-pressure acetylene generator or low-pressure residential natural gas is used as the fuel gas supply.

The cutting head may hold the cutting tip at a right angle or a slight angle to the torch body. Torches with the tip slightly angled make it easier for the welder to cut flat plate. Torches with a right-angle tip make it easier for the welder to cut pipe, angle iron, I-beams, and other uneven material shapes. Both types of torches can be used to cut any type of material, but practice is needed to keep the cut square and accurate.

The location of the **cutting lever** varies from one torch to another, Figure 6.5. Most cutting levers pivot from the front or back end of the torch body. Personal preference will determine which one a welder uses.

A **machine cutting torch**, sometimes referred to as a *line burner* or *track torch*, operates in a similar manner to a hand cutting torch. The machine cutting torch may require two oxygen regulators, one for the preheat oxygen and the other for the cutting oxygen stream. The addition of a separate cutting oxygen supply allows the flame to be more accurately adjusted. It also allows the pressures to be adjusted during a cut without disturbing the other parts of the flame. Various machine cutting torches are shown in Figures 6.6–6.8.

CUTTING TIPS

Most **cutting tips** are made of copper alloy, but some tips are chrome. Chrome plating prevents spatter from sticking to the tip, thus prolonging its useful life. Tip designs change for the different types of uses and gases, and from one torch manufacturer to another, Figure 6.9.

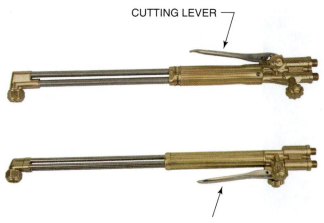

CUTTING LEVER

CUTTING LEVER

Figure 6.5 Cutting lever
The cutting lever may be located on the front or back of the torch body
Source: Courtesy of Victor Equipment, a Thermadyne Company

Figure 6.7 Multiple-head cutting machine
Source: Courtesy of ESAB Welding & Cutting Products

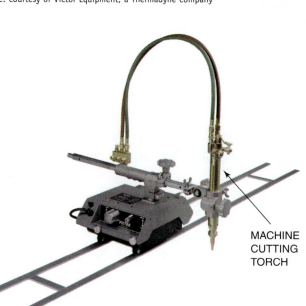

MACHINE CUTTING TORCH

Figure 6.6 Portable oxyfuel cutting machine
Source: Courtesy of Victor Equipment, a Thermadyne Company

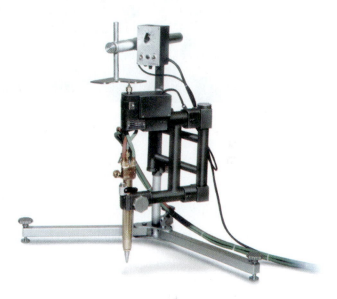

Figure 6.8 Portable cutting machine for highly complex shapes
Source: Courtesy of ESAB Welding & Cutting Products

Tips for straight cutting are either standard or high-speed, Figure 6.10. The **high-speed cutting tip** is designed for a higher cutting oxygen pressure, which allows the torch to travel faster. High-speed tips are also available for different types of fuel gases.

The diameter of the center cutting orifice determines the thickness of the metal that can be cut. A larger-diameter oxygen orifice is required for cutting thick metal. There is no standard numbering system for sizing cutting tips. Each manufacturer uses its own system, though there are similarities among some systems. Table 6.3 lists several manufacturers' tip numbering systems. The center hole diameter (in inches) is given below the tip number to allow size comparisons. For example, in Table 6.3 Airco's tip number 00 has a center orifice size equal to a number 70 drill

Figure 6.9 Cutting torch seals
Source: Courtesy of American Torch Tip

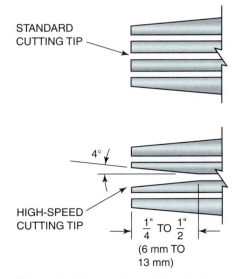

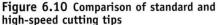

Figure 6.10 Comparison of standard and high-speed cutting tips

Table 6.3 Comparison of Manufacturers' Oxyacetylene Cutting Tip Identifications

| Manufacturer | Metal Thickness, Inches (mm) | | | | | | | | | | |
	1/8 (3)	1/4 (6)	1/2 (13)	3/4 (19)	1 (24)	1 1/2 (37)	2 (49)	2 1/2 (61)	3 (74)	4 (98)	5 (123)
Cutting orifice drill number	70	68	60	56	54	53	50	47	45	39	31
Airco	00	0	1	1	2	2	3	4	4	5	6
ESAB	1/4	1/4	1/2	1 1/2	1 1/2	1 1/2	4	4	4	4	8
Harris	000	00	0	1	1	2	2	3	3	3	4
Oxweld	2	3	4	6	6	6	8	8	8	8	8
Purox	3	3	4	4	5	5	7	7	7	7	9
Smith	00	0	1	2	2	3	3	4	4	4	5
Victor	000	00	0	1	2	2	3	4	4	5	6

size. This cutting tip is designed for cutting metal approximately 1/8 in. (3 mm) thick. Other manufacturers' tips designed for this thickness have the following numbers: 000, 00, 1/4, 2, and 3.

Finding the correctly sized tip for a job can be confusing, especially if you are using the cutting unit for the first time. To make it easier to select a tip, you can use a standard set of tip cleaners to find the size of the center cutting orifice. Table 6.4 lists the material thickness that can be cut with each size of tip cleaner.

If the manufacturer's recommendations for gas pressure are not available, you can use Table 6.4 to find the approximate pressures to be used with the tip. A number of factors determine gas pressures, including the equipment manufacturer, the condition of the equipment, hose length, hose diameter, regulator size, and operator skill. In all cases, start out

Table 6.4 Center Cutting Orifice Size, Metal Thickness, and Gas Pressures for Oxyacetylene Cutting

	Metal Thickness, Inches (mm)										
Tip Size	1/8 (3)	1/4 (6)	1/2 (13)	3/4 (19)	1 (24)	1 1/2 (37)	2 (49)	2 1/2 (61)	3 (74)	4 (98)	5 (123)
Cutting orifice drill number	70	68	60	56	54	53	50	47	45	39	31
WYPO tip cleaner number*	10	10	15	18	22	24	26				
Campbell Hausfeld tip cleaner number*	3	3	6	9	10	11	12				
Oxygen pressure, psi**	20	20	25	30	35	35	40	40	40	45	45
	25	25	30	35	40	40	45	45	45	55	55
Oxygen pressure, kPa**	140	140	170	200	240	240	275	275	275	310	310
	170	170	200	240	275	275	310	310	310	380	380
Acetylene pressure, psi**	3	3	3	3	3	3	4	4	5	6	8
	5	5	5	5	5	5	8	8	11	13	14
Acetylene pressure, kPa**	20	20	20	20	20	20	30	30	35	40	55
	35	35	35	35	35	35	55	55	75	90	95

*There is no standard numbering system for tip cleaners, so numbers can differ from one manufacturer to another.

**Tip size and pressures are approximate. Use the manufacturer's specification for equipment being used when available.

with the pressure recommended by the manufacturer of the equipment being used. Adjust the pressure to fit the particular job.

A wide variety of tip shapes are available for specialized cutting jobs. Each tip, of course, also comes in several sizes. Some tips are specialized for the kind of fuel gas being used. Different means are used to attach the cutting tip to the torch head. Some tips screw in; others have a push fitting.

Different designs are used for manual and for mechanized and automated cutting tips. Mechanized and automated cutting tips are designed for high-speed cutting with high-speed oxygen flow.

Always choose the correct type and size of tip for each cutting job. Check the manufacturer's literature for recommendations. Make sure the tip is designed for the type of fuel gas being used. Inspect the tip before using it. If the tip is clogged or dirty, clean it and clean out the orifices with an appropriate-size drill. Check to ensure there is no damage to the threads. If the threads or the tapered seat are damaged, do not use the tip.

The amount of **preheat flame** required to make a perfect cut is determined by the type of fuel gas used and by the material thickness, shape, and surface condition. Materials that are thick, are round, or have surfaces covered with rust, paint, oil, and so on require more preheat flame, Figure 6.11.

Different cutting tips are available for each of the major fuel gases. The type or number of **preheat holes** determines the fuel gas to be used in the tip. Table 6.5 lists fuel gases and the range of preheat holes or tip designs used with each gas. Acetylene is used in tips with from one

(A)

(B)

(C)

(D)

Figure 6.11 Special cutting tips
Special cutting tips come in a variety of shapes, for many purposes. They can have different sizes and numbers of preheat holes. (A) 10-in.-long cutting tip; (B) water-cooled cutting tip; (C) two-piece cutting tip; (D) sheet metal cutting tip
Source: Courtesy of ESAB Welding & Cutting Products

Table 6.5 Fuel Gas and Number of Preheat Holes Needed in the Cutting Tip

Fuel Gas	Number of Preheat Holes
Acetylene	One to six
MPS (MAPP®)	Eight- to two-piece tip
Propane and natural gas	Two-piece tip

CAUTION

Acetylene must be used in tips that are designed to be used with acetylene. If acetylene is used in tips designed for other fuel gases, the tip may overheat, causing a backfire or even causing the tip to explode.

to six preheat holes. Some large acetylene cutting tips may have eight or more preheat holes.

MPS gases are used in tips with eight preheat holes or in a two-piece tip that is not recessed, Figure 6.12. These gases have a slower flame combustion rate than acetylene. For tips with fewer than eight preheat

Figure 6.12 Parts of a two-piece cutting tip
Source: Courtesy of ESAB Welding & Cutting Products

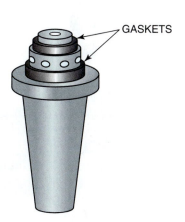

Figure 6.13 Gaskets
Some cutting tips use gaskets to make a tight seal

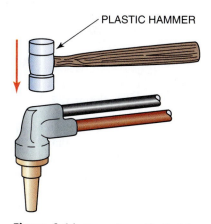

Figure 6.14 Removing a tip that is stuck
Tap the back of the torch head. The tip itself should never be tapped

holes, there may not be enough heat to start a cut, or the flame may pop out when the cutting lever is pressed.

Propane and natural gas should be used in two-piece tips that are typically deeply recessed, Figure 6.12. The flame burns at such a slow rate that it may not stay lit on any other tip.

Some cutting tips have metal-to-metal seals. When they are installed in the torch head, a wrench must be used to tighten the nut. Other cutting tips have fiber packing seats to seal the tip to the torch. If a wrench is used to tighten the nut for this type of tip, the tip seat may be damaged, Figure 6.13. A torch owner's manual should be checked or a welding supplier should be asked about the best way to tighten various torch tips.

When removing a cutting tip, if the tip is stuck in the torch head, tap the back of the head with a plastic hammer, Figure 6.14. Any tapping on the side of the tip may damage the seat.

To check the assembled torch tip for a good seal, open the oxygen valve and spray the tip with a leak-detecting solution, Figure 6.15.

If the cutting tip seat or the torch head seat is damaged, it can be repaired using a reamer designed for the specific torch tip and head, Figure 6.16, or it can be sent out for repair. New fiber packings are available for tips with packings. The original leak-checking test should be repeated to be sure the new seal is good.

OXYFUEL CUTTING, SETUP, AND OPERATION

Setting up a cutting torch system is exactly like setting up oxyfuel welding equipment, except for the adjustment of gas pressures. This section covers gas pressure adjustments and cutting equipment operations.

> **CAUTION**
>
> Handle and store tips carefully to prevent damage to the tip seats and to keep dirt from becoming stuck in the small holes.

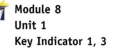

Module 8
Unit 1
Key Indicator 1, 3

Module 8
Unit 2
Key Indicator 1, 3

Figure 6.15 Checking a cutting tip for leaks
Source: Courtesy of Larry Jeffus

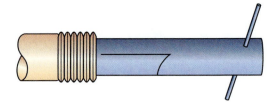

Figure 6.16 Reamer
Damaged torch seats can be repaired using a specially designed reamer

PRACTICE 6-1

Setting Up a Cutting Torch

Module 8
Unit 1
Key Indicator 1, 2, 3

Module 8
Unit 2
Key Indicator 1, 2, 3

Demonstrate to other students and your instructor the proper method of setting up cylinders, regulators, hoses, and the cutting torch.

1. The oxygen and acetylene cylinders must be securely chained to a cart or wall before the safety caps are removed.
2. After removing the safety caps, stand to one side and crack (open and quickly close) the cylinder valves, making sure there are no sources of possible ignition nearby. Cracking the cylinder valves blows out any dirt they may contain.
3. Visually inspect all of the parts for any damage and repair or cleaning requirements.
4. Attach the regulators to the cylinder valves and tighten them securely with a wrench.
5. Attach a reverse-flow valve or flashback arrestor, if the torch does not have them built in, to the hose connection on the regulator or to the hose connection on the torch body, depending on the type of reverse-flow valve in the set. Occasionally, test each reverse-flow valve by blowing through it to make sure it works properly.
6. If the torch you will be using is a combination torch, attach the cutting head at this time.
7. Install a cutting tip on the torch.
8. Before you open the cylinder valves, back out the pressure-regulating screws so that when the valves are opened the gauges will show 0 lb working pressure.
9. Stand to one side of the regulators' faces as you open the cylinder valves slowly.
10. The oxygen valve is opened all the way until it becomes tight, but do not over tighten; the acetylene valve is opened no more than one half turn.

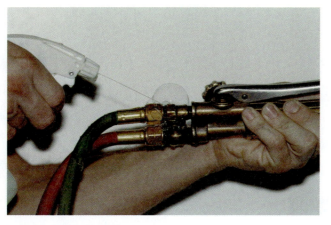

Figure 6.17 Leak-checking gas fittings
Source: Courtesy of Larry Jeffus. See Welding Principles and Practices on DVD.DVD 4—Oxyacetylene Welding

11. Open one torch valve and then turn the regulating screw in slowly until 2 psig to 4 psig (14 kPag to 30 kPag) shows on the working pressure gauge. Allow the gas to escape so that the line is completely purged.

12. If you are using a combination welding and cutting torch, the oxygen valve nearest the hose connection must be opened before the flame-adjusting valve or cutting lever will work.

13. Close the torch valve and repeat the purging process with the other gas.

14. Be sure there are no sources of possible ignition nearby.

15. With both torch valves closed, spray a leak-detecting solution on all connections, including the cylinder valves. Tighten any connection that shows bubbles, Figure 6.17.

Complete a copy of the Student Welding Report in Appendix I or provided by your instructor. ■

PRACTICE 6-2

Cleaning a Cutting Tip

Using a cutting torch set that is assembled and adjusted as described in Practice 6-1 and a set of **tip cleaners**, you will clean the cutting tip.

1. Turn on a small flow of oxygen, Figure 6.18. This blows out any dirt loosened during the cleaning.

2. File the end of the tip flat, using the file provided in the tip cleaning set, Figure 6.19.

3. Try several sizes of tip cleaners in a preheat hole until the correct size cleaner is determined. It should easily go all the way into the tip, Figure 6.20.

4. Push the cleaner in and out of each preheat hole several times. Tip cleaners are small, round files; excessive use of them will greatly increase the **orifice** (hole) size.

5. Depress the cutting lever and, by trial and error, select the correct size tip cleaner for the center cutting orifice. A tip cleaner should never be forced.

 Module 8
Unit 1
Key Indicator 2

Module 8
Unit 2
Key Indicator 2

Figure 6.18 Opening the oxygen valve
Source: Courtesy of Larry Jeffus

Figure 6.19 Filing the end of the tip flat
Source: Courtesy of Larry Jeffus

Figure 6.20 Tip cleaner
A tip cleaner should be used to clean the flame and center cutting holes

Complete a copy of the Student Welding Report in Appendix I or provided by your instructor. ■

PRACTICE 6-3

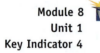

Module 8
Unit 1
Key Indicator 4

Lighting the Torch

Wearing welding goggles, gloves, and any other required personal protective clothing, and with a cutting torch set that is safely assembled, you will light the torch.

1. Set the regulator working pressure for the tip size. If you do not know the correct pressure for the tip, start with the fuel set at 5 psig (35 kPag) and the oxygen set at 25 psig (170 kPag).
2. Point the torch tip upward and away from any equipment or other students.
3. Open just the acetylene valve and use only a sparklighter or striker to ignite the acetylene. The torch may not stay lit. If it goes out, close the valve slightly and try to relight the torch.
4. If the flame is small, it will produce heavy black soot and smoke. In this case, turn the flame up to stop the soot and smoke. You

need not be concerned if the flame jumps slightly away from the torch tip.

5. With the acetylene flame burning smoke-free, slowly open the oxygen valve and, using only the oxygen valve, adjust the flame to a neutral setting, Figure 6.21.

6. When the cutting oxygen lever is depressed, the flame may become slightly carbonizing. This can occur because the high flow of oxygen through the cutting orifice causes a drop in line pressure.

7. With the cutting lever depressed, readjust the preheat flame to a neutral setting.

The flame will become slightly oxidizing when the cutting lever is released. Since an oxidizing flame is hotter than a neutral flame, the metal being cut will be preheated faster. When the cut is started by depressing the lever, the flame automatically returns to the neutral setting and does not oxidize the top of the plate. Extinguish the flame by first turning off the oxygen and then the acetylene.

> **CAUTION**
>
> Sometimes a large cutting tip will pop when the acetylene is turned off first. If that happens, turn the oxygen off first. Always refer to the manufacturer's instructions when determining the shutdown sequence.

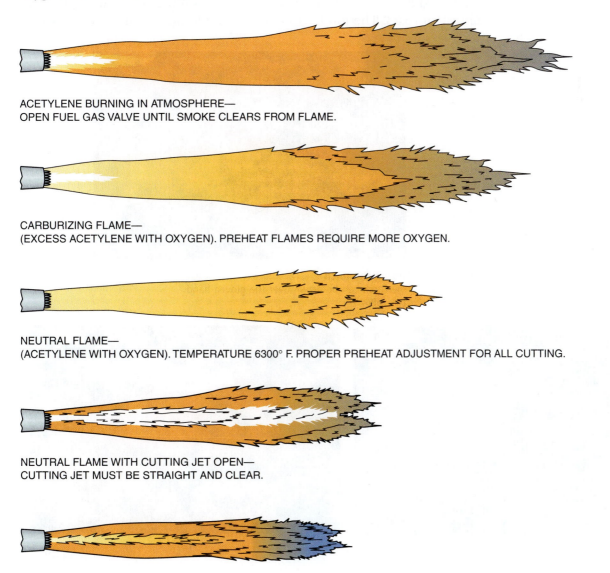

ACETYLENE BURNING IN ATMOSPHERE—
OPEN FUEL GAS VALVE UNTIL SMOKE CLEARS FROM FLAME.

CARBURIZING FLAME—
(EXCESS ACETYLENE WITH OXYGEN). PREHEAT FLAMES REQUIRE MORE OXYGEN.

NEUTRAL FLAME—
(ACETYLENE WITH OXYGEN). TEMPERATURE 6300° F. PROPER PREHEAT ADJUSTMENT FOR ALL CUTTING.

NEUTRAL FLAME WITH CUTTING JET OPEN—
CUTTING JET MUST BE STRAIGHT AND CLEAR.

OXIDIZING FLAME—
(ACETYLENE WITH EXCESS OXYGEN). NOT RECOMMENDED FOR AVERAGE CUTTING.

Figure 6.21 Oxyacetylene flame adjustments for the cutting torch

Complete a copy of the Student Welding Report in Appendix I or provided by your instructor. ∎

HAND CUTTING

When a cut is made with a hand torch, it is important for the welder to be steady to make the cut as smooth as possible. A welder must also be comfortable and free to move the torch along the line to be cut. It is a good idea for a welder to get into position and practice the cutting movement a few times before lighting the torch. Even when the welder and the torch are braced properly, even the most subtle physical movements will cause a slight ripple in the cut. Attempting a cut without leaning on the work, to brace oneself, is tiring and causes inaccuracies.

The torch should be braced with the left hand if the welder is right-handed or with the right hand if the welder is left-handed. The torch may be moved by sliding it toward you over your supporting hand, Figure 6.22 and Figure 6.23. The torch can also be pivoted on the supporting hand. If

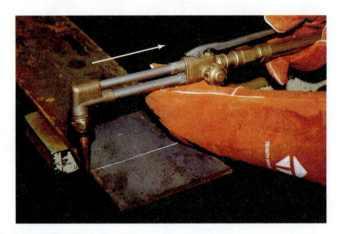

Figure 6.22 Moving the torch
For short cuts, the torch can be drawn over the gloved hand
Source: Courtesy of Larry Jeffus

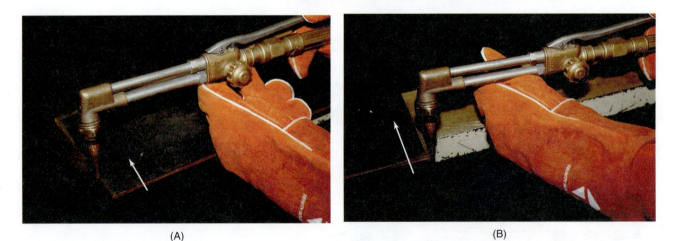

(A)   (B)

Figure 6.23 Moving the torch
For longer cuts, the torch can be moved by sliding your gloved hand along the plate parallel to the cut: (A) start and (B) finish. Always check for free and easy movement before lighting the torch
Source: Courtesy of Larry Jeffus. See Welding Principles and Practices on DVD.DVD 4—Oxyacetylene Welding

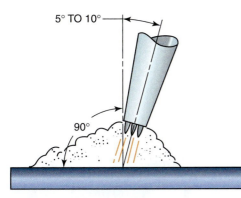

Figure 6.24 Forward torch angle
A slight forward angle helps when cutting thin material

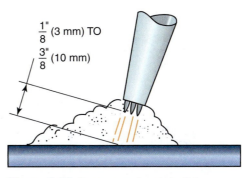

Figure 6.25 Inner cone to work distance (coupling distance)

the pivoting method is used, care must be taken to prevent the cut from becoming a series of arcs.

A slight forward torch angle helps the flame preheat the metal, keeps some of the reflected flame heat off the tip, aids in blowing dirt and oxides away from the cut, and keeps the tip clean for a longer time because slag is less likely to be blown back onto it, Figure 6.24. The forward angle can be used only for a straight-line square cut. If shapes are cut using a slight angle, the part will have beveled sides.

When a cut is made, the inner cones of the flame should be kept 1/8 in. (3 mm) to 3/8 in. (10 mm) from the surface of the plate, Figure 6.25. This distance is known as the **coupling distance**.

To start a cut on the edge of a plate, hold the torch at a right angle to the surface or pointed slightly away from the edge, Figure 6.26. The torch must also be pointed so that the cut is started at the very edge. The edge of the plate heats up more quickly, allowing the cut to be started sooner. Also, fewer sparks will be blown around the shop. Once the cut is started, the torch should be rotated back to a right angle to the surface or to a slight leading angle.

If a cut is to be started other than at the edge of the plate, the inner cones should be held as close as possible to the metal. Touching the

CAUTION

Never use a cutting torch to cut open a used can, drum, tank, or other sealed container. The heat, sparks, and oxygen cutting stream may cause even nonflammable residue inside to burn or explode. If a used container must be cut, one end must be removed first and all residue cleaned out. In addition to the possibility of a fire or an explosion, you might be exposing yourself to hazardous fumes. Before making a cut, check the material safety data sheet (MSDS) for safety concerns.

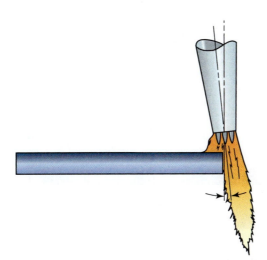

Figure 6.26 Starting a cut on the edge of a plate
Notice how the torch is pointed at a slight angle away from the edge

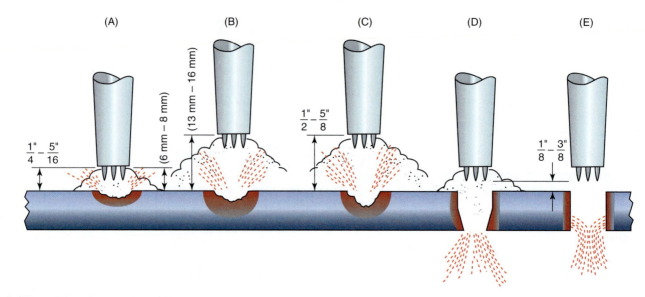

Figure 6.27 Sequence for piercing plate

metal with the inner cones will speed up the preheat time. When the metal is hot enough to start cutting, the torch should be raised as the cutting lever is slowly depressed. When the metal is pierced, the torch should be lowered again, Figure 6.27. By raising the torch tip away from the metal, the number of sparks blown into the air is reduced, and the tip is kept cleaner. If the metal being cut is thick, it may be necessary to move the torch tip in a small circle as the hole goes through the metal. If the metal is to be cut in both directions from the spot where it was pierced, the torch should be moved backward a short distance and then forward, Figure 6.28. This prevents slag from refilling the kerf at the starting point and making it difficult to cut in the other direction. The kerf is the space produced during any cutting process.

Starts and stops can be made more easily and better if one side of the metal being cut is scrap. When it is necessary to stop and reposition oneself before continuing the cut, the cut should be turned out a short distance into the scrap side of the metal, Figure 6.29. The extra space

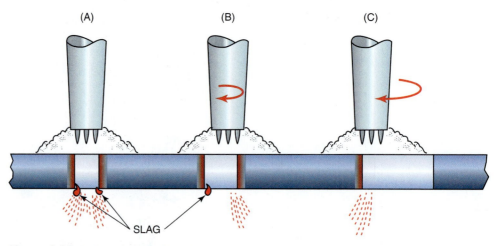

Figure 6.28 Cutting in both directions
A short, backward movement (A) before the cut is carried forward (B) clears the slag from the kerf (C). Slag left in the kerf may cause the cutting stream to gouge into the base metal, resulting in a poor cut

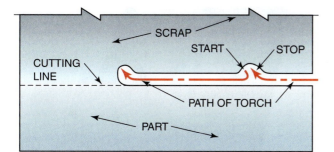

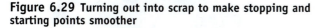

Figure 6.29 Turning out into scrap to make stopping and starting points smoother

Figure 6.30 Drag
Drag is the distance by which the bottom of a cut lags behind the top

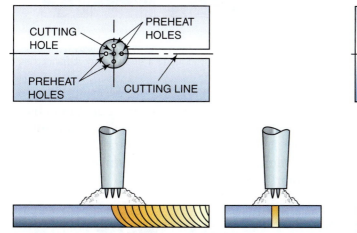

Figure 6.31 Tip alignment for a square cut

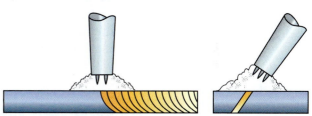

Figure 6.32 Tip alignment for a bevel cut

that this procedure provides will allow a smoother and more even start with less chance that slag will block the cut. If neither side of the cut is to be scrap, the forward movement should be stopped for a moment before releasing the cutting lever. This action will allow the **drag**, or the distance that the bottom of the cut is behind the top, to be reduced before stopping, Figure 6.30. To restart, use the procedure for starting a cut at the edge of the plate.

Proper alignment of the preheat holes will speed up and improve the cut. The holes should be aligned so that one is directly on the line ahead of the cut and another is aimed down into the cut when making a straight-line square cut, Figure 6.31. The flame is directed toward the smaller piece and the sharpest edge when cutting a bevel. For this reason, the tip should be changed so that at least two of the flames are on the larger plate and none of the flames is directed onto the sharp edge, Figure 6.32. If the preheat flame is directed at the edge, it will be rounded off as it is melted off.

LAYOUT

A line to be cut can be laid out with a piece of **soapstone** or a chalk line. To obtain an accurate line, a scribe or a punch can be used. If a piece of soapstone is used, it should be sharpened properly to increase accuracy, Figure 6.33. A chalk line will make a long, straight line on metal and is best used on large jobs. The scribe and punch can both be used to lay out an

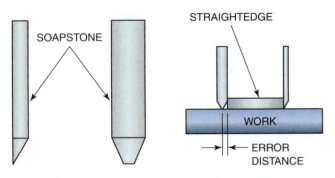

Figure 6.33 Proper method of sharpening a soapstone

Figure 6.34 Punching
Holding the punch slightly above the surface allows the punch to be struck rapidly and moved along a line to mark it for cutting
Source: Courtesy of Larry Jeffus

accurate line, but a punched line is easier to see when cutting. A punch can be held as shown in Figure 6.34, with the tip just above the surface of the metal. When the punch is struck with a lightweight hammer, it will make a mark. If you move your hand along the line and rapidly strike the punch, it will leave a series of punch marks for the cut to follow.

SELECTING THE CORRECT TIP AND SETTING THE PRESSURE

Each welding equipment manufacturer uses its own numbering system to designate tip size. It would be impossible to memorize every system. Each manufacturer, however, does relate the tip number to the numbered drill size used to make the holes. On the back of most tip cleaning sets, the manufacturer lists the equivalent drill size for each tip cleaner. By remembering approximately which tip cleaner was used on a particular tip for a metal thickness range, a welder can easily select the correct tip when using a new torch set. Using the tip cleaner that you are familiar with, try it in the various torch tips until you find the correct tip that the tip cleaner fits. Table 6.6 lists the tip drill size, pressure range, and metal thickness range for which the tip can be used.

PRACTICE 6-4

Setting the Gas Pressures

The working pressure of the regulators can be set by following a table or by watching the flame.

1. To set the regulator by watching the flame, first set the acetylene pressure at 2 psig to 4 psig (14 kPag to 30 kPag) and then light the acetylene flame.
2. Open the acetylene torch valve one to two turns and reduce the regulator pressure by backing out the setscrew until the flame starts to smoke.
3. Increase the pressure until the smoke stops. This is the maximum fuel gas pressure the tip needs. With a larger tip and a longer hose, the pressure must be set higher. This lowest-possible setting is the

Table 6.6 Cutting Pressure and Tip Size

| Metal Thickness in. (mm) | Center Orifice Size | | Oxygen Pressure lb/in.^{2}(kPa) | Acetylene lb/in.^{2} (kPa) |
	No. Drill Size	Tip Cleaner No.*		
1/8 (3)	60	7	10 (70)	3 (20)
1/4 (6)	60	7	15 (100)	3 (20)
3/8 (10)	55	11	20 (140)	3 (20)
1/2 (13)	55	11	25 (170)	4 (30)
3/4 (19)	55	11	30 (200)	4 (30)
1 (25)	53	12	35 (240)	4 (30)
2 (51)	49	13	45 (310)	5 (35)
3 (76)	49	13	50 (340)	5 (35)
4 (102)	49	13	55 (380)	5 (35)
5 (127)	45	**	60 (410)	5 (35)

*The tip cleaner number when counted from the small end toward the large end in a standard tip cleaner set.
**Larger than normally included in a standard tip cleaner set.

best one, and it is the safest one to use. There is less chance of a leak. If the hose is damaged, the resulting fire will be much smaller than a fire burning from a hose with a higher pressure. There is also less chance of a leak with the lower pressure.

4. With the acetylene adjusted so that the flame just stops smoking, slowly open the torch oxygen valve.
5. Adjust the torch to a neutral flame. When the cutting lever is depressed, the flame will become carbonizing, because it will not have enough oxygen pressure.
6. Holding the cutting lever down, increase the oxygen regulator pressure slightly. Readjust the flame, as needed, to a neutral setting using the oxygen valve on the torch.
7. Increase the pressure slowly and readjust the flame as you watch the length of the clear cutting stream in the center of the flame, Figure 6.35A. The center stream will stay fairly long until a pressure is reached that causes turbulence and disrupts the cutting stream. This turbulence will cause the flame to shorten in length considerably, Figure 6.35B.
8. With the cutting lever still depressed, reduce the oxygen pressure until the flame lengthens once again. This is the maximum oxygen pressure that this tip can use without turbulence in the cutting stream, which would cause a very poor cut. The lower pressure also will prevent the sparks from being blown a longer distance from the work, Figure 6.36.

Complete a copy of the Student Welding Report in Appendix I or provided by your instructor. ■

THE CHEMISTRY OF A CUT

The oxyfuel gas cutting torch works when the metal being cut rapidly oxidizes or burns. This rapid oxidization or burning occurs when a high-pressure stream of pure oxygen is directed on the metal after it has

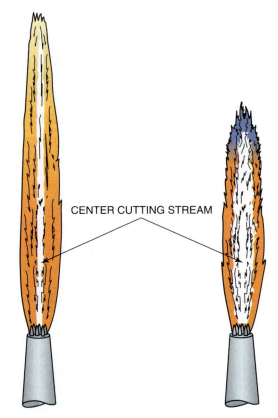

Figure 6.35 Center cutting stream
A clean cutting tip will have a long, well-defined oxygen stream

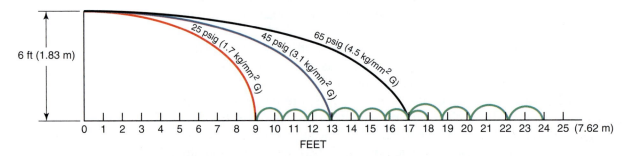

Figure 6.36 Limiting the distance sparks travel
The sparks from cutting a mild steel plate 3/8 in. (10 mm) thick, 6 ft (1.8 m) from the floor, will be thrown much farther if the cutting pressure is too high for the plate thickness. These cuts were made with a Victor cutting tip no. 0-1-101 using 25 psig (1.7 kg/mm²), as recommended by the manufacturer, and by excessive pressures of 45 psig (3.1 kg/mm²) and 65 psig (4.5 kg/mm²)

CAUTION

Some metals release harmful oxides when they are cut. Extreme caution must be taken when cutting used, oily, dirty, or painted metals. They often produce dangerous fumes when they are cut. You may need extra ventilation and a respirator to be safe. Check with the welding shop supervisor or shop safety officer before cutting any metal you are unfamiliar with.

been preheated to a temperature above its kindling point. **Kindling point** is the lowest temperature at which a material will burn. The kindling temperature of iron is 1600°F (870°C), at which temperature the metal is a dull red. Note that iron is the pure element and cast iron is an alloy made primarily of iron and carbon. The OFC process will work well on any metal that will rapidly oxidize, such as iron, low-carbon steel, magnesium, titanium, and zinc.

The OFC process is most often used to cut iron and low-carbon steels, because unlike with most other metals, little or no oxides are left on the metal and it can easily be welded.

Figure 6.37 Overheating during cutting
As a hole is cut, the center may be overheated
Source: Courtesy of Larry Jeffus. See Welding Principles and Practices on DVD. DVD 4—Oxyacetylene Welding

The burning away of the metal is a chemical reaction between iron (Fe) and oxygen (O). The oxygen forms an iron oxide, primarily Fe_3O_4, that is light gray. Heat is produced by the metal as it burns. This heat helps carry the cut along. On thick pieces of metal, once a small spot starts burning (being cut), the heat generated helps the cut continue quickly through the metal. With some cuts, the heat produced may overheat small strips of metal being cut from a larger piece. As an example, the center piece of a hole being cut will quickly become red hot and will start to oxidize with the surrounding air, Figure 6.37. This heat produced by the cut makes it difficult to cut out small or internal parts.

EXPERIMENT 6-1

Observing Heat Produced during a Cut

This experiment may require more skill than you have developed by this time. You may wish to observe your instructor performing the experiment or try it later.

Using a properly lit and adjusted cutting torch, welding gloves, appropriate eye protection and clothing, and one piece of clean mild steel plate 6 in. (152 mm) long × 1/4 in. (6 mm) to 1/2 in. (13 mm) thick, you will make an oxyfuel gas cut without the preheat flame.

Position the piece of metal so that the cutting sparks fall safely away from you. With the torch lit, pass the flame over the length of the plate until it is warm but not hot. Brace yourself and start a cut near the edge of the plate. When the cut has been established, have another student turn off the acetylene regulator. The cut should continue if you remain steady and the plate is warm enough. *Hint: Using a slightly larger tip size will make this easier.*

Complete a copy of the Student Welding Report in Appendix I or provided by your instructor. ■

THE PHYSICS OF A CUT

As a cut progresses along a plate, a record of what happened during the cut is preserved along both sides of the kerf. This record indicates to the welder what was correct or incorrect with the preheat flame, cutting speed, and oxygen pressure.

Preheat

The size and number of preheat holes in a tip have an effect on both the top and bottom edges of the metal. An excessive preheat flame results in the top edge of the plate being melted or rounded off and an excessive amount of hard-to-remove slag being deposited along the bottom edge. If the flame is too small, the travel speed must be slower. A reduction in speed may result in the cutting stream wandering from side to side. The torch tip can be raised slightly to eliminate some of the damage caused by too much preheat. However, raising the torch tip causes the cutting stream of oxygen to be less forceful and less accurate.

Speed

The cutting speed should be fast enough that the **drag lines** have a slight slant backward if the tip is held at a 90° angle to the plate, Figure 6.38. If the cutting speed is too fast, the oxygen stream may not have time to go completely through the metal, resulting in an incomplete cut, Figure 6.39. Too slow a cutting speed results in the cutting stream wandering, causing gouges in the side of the cut, Figures 6.40 and 6.41.

Pressure

A correct pressure setting results in the sides of the cut being flat and smooth. A pressure setting that is too high causes the cutting stream to expand as it leaves the tip, resulting in the sides of the kerf being slightly dished, Figure 6.42. When the pressure setting is too low, the cut may not go completely through the metal.

Module 8
Unit 2
Key Indicator 1, 3, 4, 5, 6

Module 9
Key Indicator 1

EXPERIMENT 6-2

Effect of Flame, Speed, and Pressure on a Machine Cut

Using a properly lit and adjusted automatic cutting machine, welding gloves, appropriate eye protection and clothing, a variety of tip sizes, and one piece of mild steel plate 6 in. (152 mm) long × 1/2 in. (13 mm) to

Figure 6.38 Correct cut
Source: Courtesy of Larry Jeffus

Figure 6.39 Poor cut
Too fast a travel speed, resulting in an incomplete cut; too much preheat and the tip is too close, causing the top edge to be melted and removed
Source: Courtesy of Larry Jeffus

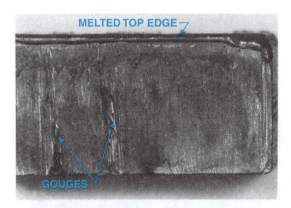

Figure 6.40 Poor cut
Too slow a travel speed results in the cutting stream wandering, causing gouges in the surface; preheat flame is too close, melting the top edge
Source: Courtesy of Larry Jeffus

Figure 6.41 Poor cut
Too slow a travel speed at the start; too much preheat
Source: Courtesy of Larry Jeffus

1 in. (25 mm) thick, you will observe the effect of the preheat flame, travel speed, and pressure on the metal being cut.

Using a variety of tips, speeds, and oxygen pressures, make a series of cuts on the plate. As each cut is made, listen to the sound it makes. Also look at the stream of sparks coming off the bottom. A good cut should have a smooth, even sound, and the sparks should come off the bottom of the metal more like a stream than a spray, Figure 6.43. When the cut is complete, look at the drag lines to determine what was correct or incorrect with the cut, Figure 6.44.

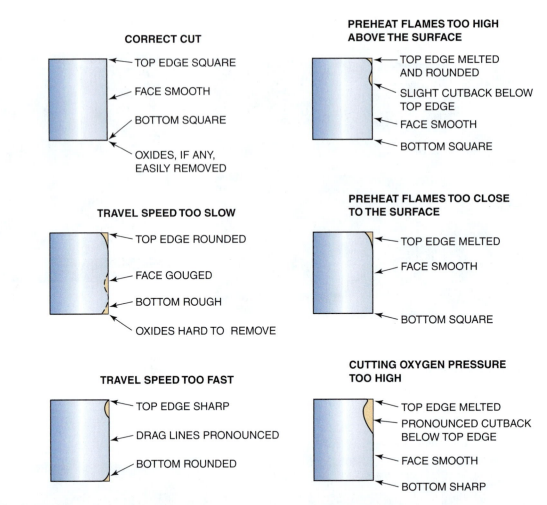

CORRECT CUT
TOP EDGE SQUARE
FACE SMOOTH
BOTTOM SQUARE
OXIDES, IF ANY, EASILY REMOVED

PREHEAT FLAMES TOO HIGH ABOVE THE SURFACE
TOP EDGE MELTED AND ROUNDED
SLIGHT CUTBACK BELOW TOP EDGE
FACE SMOOTH
BOTTOM SQUARE

TRAVEL SPEED TOO SLOW
TOP EDGE ROUNDED
FACE GOUGED
BOTTOM ROUGH
OXIDES HARD TO REMOVE

PREHEAT FLAMES TOO CLOSE TO THE SURFACE
TOP EDGE MELTED
FACE SMOOTH
BOTTOM SQUARE

TRAVEL SPEED TOO FAST
TOP EDGE SHARP
DRAG LINES PRONOUNCED
BOTTOM ROUNDED

CUTTING OXYGEN PRESSURE TOO HIGH
TOP EDGE MELTED
PRONOUNCED CUTBACK BELOW TOP EDGE
FACE SMOOTH
BOTTOM SHARP

Figure 6.42 Profile of oxyfuel-cut plates

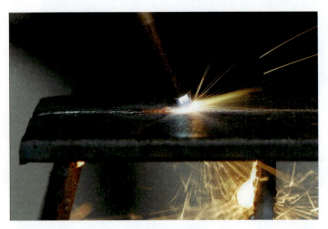

Figure 6.43 Good cut showing a steady stream of sparks flying out from the bottom of the cut
Source: Courtesy of Larry Jeffus. See Welding Principles and Practices on DVD. DVD 4 —Oxyacetylene Welding

Figure 6.44 Poor cut
The slag is backing up because the cut is not going through the plate
Source: Courtesy of Larry Jeffus

Repeat this experiment until you know a good cut by the sound it makes and the stream of sparks. A good cut has little or no slag left on the bottom of the plate.

Using a properly lit and adjusted automatic cutting machine, welding gloves, appropriate eye protection and clothing, and one piece of mild steel plate 6 in. (152 mm) long × 3/8 in. (10 mm) thick, you will make a 45° bevel down the length of the plate.

Mark the plate in strips 1/2 in. (13 mm) wide. Set the tip for beveling and cut a bevel. The bevel should be within ±3/32 in. (2 mm) of a straight line and ±5° of a 45° angle. There may be some soft slag, but no hard slag, on the beveled plate. Repeat this practice until you can make the cut within tolerance.

Complete a copy of the Student Welding Report in Appendix I or provided by your instructor for both the straight cut and the bevel cut. ▪

EXPERIMENT 6-3

Effect of Flame, Speed, and Pressure on a Hand Cut

Using a properly lit and adjusted hand torch, welding gloves, appropriate eye protection and clothing, and the same tip sizes and mild steel plate, repeat Experiment 6-2 to note the effects of the preheat flame, travel speed, and pressure on hand cutting.

Complete a copy of the Student Welding Report in Appendix I or provided by your instructor. ▪

Module 8
Unit 1
Key Indicator 4

Module 9
Key Indicator 1

Slag

The two types of slag produced during a cut are soft slag and hard slag. **Soft slag** is very porous, brittle, and easily removed from a cut. There is little or no unoxidized iron in it. It may be found on some good cuts. Hard slag may be mixed with soft slag. **Hard slag** is attached solidly to the bottom edge of a cut, and its removal requires a lot of chipping and grinding. There is 30% to 40% or more unoxidized iron in hard slag. The higher the unoxidized iron content, the more difficult the slag is to remove. **Slag** is found on bad cuts as a result of dirty tips, too much preheat, too slow a travel speed, too short a coupling distance, or incorrect oxygen pressure.

The slag from a cut may be kept off one side of the plate being cut by slightly angling the cut toward the scrap side of the cut, Figure 6.45. The angle needed to force the slag away from the good side of the plate may be as small as 2° or 3°. This technique works best on thin sections; on thicker sections the bevel may show.

PLATE CUTTING

Low-carbon steel plate can be cut quickly and accurately, whether thin-gauge sheet metal or sections more than 4 ft (1.2 m) thick are used. It is possible to achieve cutting speeds as fast as 32 in./min (13.5 mm/s) in

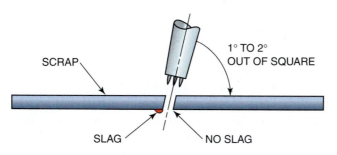

Figure 6.45 Keeping slag off one side of the plate
A slight angle on the torch will put the slag on the scrap side of the cut

Figure 6.46 Hand torches for thick sections
Source: Courtesy of Victor Equipment, a Thermadyne Company

1/8-in. (3-mm) plate, and accuracy on machine cuts of ±3/64 in. Some very large hand-cutting torches with an oxygen cutting volume of 600 cfh (2830 L/min) can cut metal that is 4 ft (1.2 m) thick, Figure 6.46. Most hand torches will not easily cut metal that is more than 7 in. (178 mm) to 10 in. (254 mm) thick.

The thicker the plate, the more difficult the cut is to make. Thin plate, 1/4 in. (6 mm) or less, can be cut and the pieces separated even if poor techniques and incorrect pressure settings are used. Thick plate, 1/2 in. (13 mm) or thicker, often cannot be separated if the cut is not correct. For very heavy cuts, on plate 12 in. (305 mm) or thicker, the equipment and operator technique must be nearly perfect or the cut will be faulty.

Plate that is properly cut can be assembled and welded with little or no postcut cleanup. Poor-quality cuts require more time to clean up than is needed for the required adjustments to make a good weld.

CUTTING TABLE

Because of the nature of the torch cutting process, special consideration is given to the flame cutting support. Any piece being cut should be supported so the torch flame will not cut through it into the table. Special cutting tables are used that expose only a small metal area to the torch flame. Some tables use parallel steel bars of metal and others use cast iron pyramids. All cutting should be set up so the flame and oxygen stream runs between the support bars or over the edge of the table.

If an ordinary welding table or another steel table is used, special care must be taken to avoid cutting through the tabletop. The piece being cut may be supported above the support table by firebrick. Another method is to cut the metal over the edge of the table.

TORCH GUIDES

In manual torch cutting, a guide or support is frequently used to allow for better control and more even cutting. It takes a very skilled welder to make a straight, clean cut even when following a marked line. It is more difficult still to make a radius cut to any accuracy. Guides and supports allow the height and angle of the torch head to remain constant. The speed of the cut, which is important to making a clean, even kerf, must be controlled by the welder.

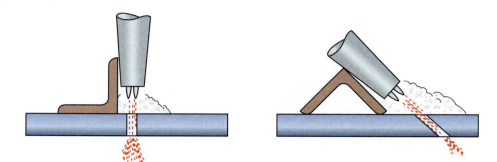

Figure 6.47 Using angle irons to aid in making cuts

Figure 6.48 Devices used to improve hand cutting
Source: Courtesy of Victor Equipment, a Thermadyne Company

Since the torch must be held in an exact position to make an accurate cut, the welder normally supports the torch weight with the hand. Supporting the torch weight this way not only allows for more accurate work but also cuts down on fatigue. A rest, such as a firebrick, can also be used to support the torch.

Various types of guides can be used to keep the torch in a straight line. Figure 6.47 shows an angle iron guide. The edge of the angle is followed to give the straight cut. Bevel cuts can be made freehand with the torch, but it is very difficult to keep them uniform. More accurate bevel cuts are made by resting the torch against the angle side of an angle iron.

Special roller guides, Figure 6.48, can also be attached to the torch head. The attachment holds the torch cutting tip at an exact height.

A circle-cutting attachment is used to cut circles. Figure 6.48 shows how the attachment fits on the torch head. The radius can be preset to any required distance. The cutter revolves around the center point when the cut is made. The roller controls the height of the torch tip above the plate surface.

PRACTICE 6-5

Flat, Straight Cut in Thin Plate

Using a properly lit and adjusted cutting torch and one piece of mild steel plate 6 in. (152 mm) long × 1/4 in. (6 mm) thick, you will cut off 1/2-in. (13-mm) strips.

Module 8
Unit 1
Key Indicator 5

Using a straightedge and soapstone, make several straight lines 1/2 in. (13 mm) apart. Starting at one end, make a cut along the entire length of plate. The strip must fall free, be slag-free, and be within ±3/32 in. (2 mm) of a straight line and ±5° of being square. Repeat this procedure until you can make the cut straight and slag-free. Turn off the cylinder valves, bleed the hoses, back out the pressure regulators, and clean up your work area when you are finished cutting.

Complete a copy of the Student Welding Report in Appendix I or provided by your instructor. ■

PRACTICE 6-6

Module 8
Unit 1
Key Indicator 5

Flat, Straight Cut in Thick Plate

Using a properly lit and adjusted cutting torch and one piece of mild steel plate 6 in. (152 mm) long × 1/2 in. (13 mm) thick or thicker, you will cut off 1/2-in. (13-mm) strips. *Remember that starting a cut in thick plate will take longer, and the cutting speed will be slower.*

Lay out, cut, and evaluate the cut as in Practice 6-5. Repeat this procedure until you can make the cut straight and slag-free. Turn off the cylinder valves, bleed the hoses, back out the pressure regulators, and clean up your work area when you are finished cutting.

Complete a copy of the Student Welding Report in Appendix I or provided by your instructor. ■

PRACTICE 6-7

Module 8
Unit 1
Key Indicator 5

Flat, Straight Cut in Sheet Metal

Using a properly lit and adjusted cutting torch and a piece of mild steel sheet 10 in. (254 mm) long and 18 gauge to 11 gauge thick, hold the torch at a very sharp leading angle, Figure 6.49, to cut the sheet along the line. The cut must be smooth and straight with as little slag as possible. Repeat this procedure until you can make the cut flat, straight, and slag-free. Turn off the cylinder valves, bleed the hoses, back out the pressure regulators, and clean up your work area when you are finished cutting.

Complete a copy of the Student Welding Report in Appendix I or provided by your instructor. ■

PRACTICE 6-8

Module 8
Unit 1
Key Indicator 6

Flame Cutting Holes

Using a properly lit and adjusted cutting torch, welding gloves, appropriate eye protection and clothing, and one piece of mild steel plate 1/4 in. (6 mm) thick, you will cut holes with diameters of 1/2 in. (13 mm) and 1 in. (25 mm).

Using the technique described for piercing a hole, start in the center and make an outward spiral until the hole is the desired size, Figure 6.50. The hole must be within ±3/32 in. (2 mm) of being round and ±5° of being square. The hole may have slag on the bottom. Repeat this procedure until you can make both small and large holes within tolerance. Turn off the cylinder valves, bleed the hoses, back out the pressure regulators, and clean up your work area when you are finished cutting.

Figure 6.49 Cutting sheet metal at a very sharp angle

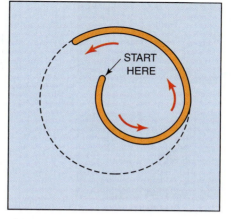

Figure 6.50 Starting a cut for a hole near the middle

Complete a copy of the Student Welding Report in Appendix I or provided by your instructor. ■

DISTORTION

Distortion occurs when the metal bends or twists out of shape as a result of being heated during the cutting process. This is a major problem when cutting a plate. If the distortion is not controlled, the end product might be worthless. There are two main methods of controlling distortion. One method involves making two parallel cuts on the same plate at the same speed and time, Figure 6.51. Because the plate is heated evenly, distortion is kept to a minimum, Figure 6.52.

The second method involves starting the cut a short distance from the edge of the plate, skipping over short tabs every 2 ft (0.6 m) to 3 ft (0.9 m) to keep the cut from separating. Once the plate cools, the remaining tabs are cut, Figure 6.53.

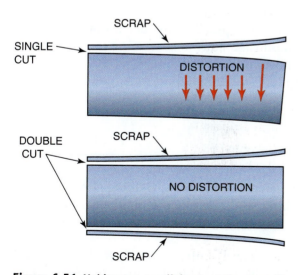

Figure 6.51 Making two parallel cuts at the same time to control distortion

Figure 6.52 Slitting adaptor for cutting machine
The adaptor can be used for parallel cuts from 1 in. (38 mm) to 12 in. (500 mm). Ideal for cutting test coupons
Source: Courtesy of Victor Equipment, a Thermadyne Company

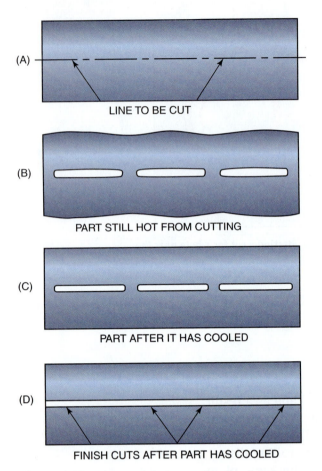

(A)

LINE TO BE CUT

(B)

PART STILL HOT FROM CUTTING

(C)

PART AFTER IT HAS COOLED

(D)

FINISH CUTS AFTER PART HAS COOLED

Figure 6.53 Steps used during cutting to minimize distortion

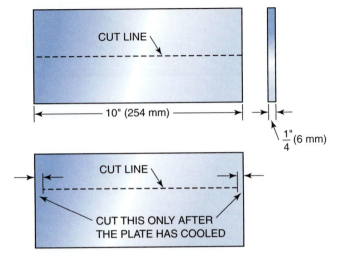

CUT LINE

10" (254 mm)

$\frac{1}{4}$" (6 mm)

CUT LINE

CUT THIS ONLY AFTER
THE PLATE HAS COOLED

Figure 6.54 Making two cuts with minimum distortion
The sizes of these and other cutting projects can be changed to fit available stock

EXPERIMENT 6-4

Minimizing Distortion

Using a properly lit and adjusted cutting torch, welding gloves, appropriate eye protection and clothing, and two pieces of mild steel 10 in. (254 mm) long × 1/4 in. (6 mm) thick, you will make two cuts and then compare the distortion. Lay out and cut out both pieces of metal as shown in Figure 6.54. Allow the metal to cool, and then cut the remaining tabs. Compare the four pieces of metal for distortion.

Complete a copy of the Student Welding Report in Appendix I or provided by your instructor. ∎

Module 8
Unit 1
Key Indicator 7

PRACTICE 6-9

Beveling a Plate

Using a properly lit and adjusted cutting torch, welding gloves, appropriate eye protection and clothing, and one piece of mild steel plate 6 in. (152 mm) long × 3/8 in. (10 mm) thick, you will make a 45° bevel down the length of the plate.

Mark the plate in strips 1/2 in. (13 mm) wide. Set the tip for beveling and cut a bevel. The bevel should be within ±3/32 in. (2 mm) of a straight line and ±5° of a 45° angle. There may be some soft slag, but no hard slag, on the beveled plate. Repeat this practice until you can make the cut within tolerance. Turn off the cylinder valves, bleed the hoses, back out the pressure regulators, and clean up your work area when you are finished cutting.

Complete a copy of the Student Welding Report in Appendix I or provided by your instructor. ■

PRACTICE 6-10

Vertical Straight Cut

Using a properly lit and adjusted cutting torch, welding gloves, appropriate eye protection and clothing, and one piece of mild steel plate 6 in. (152 mm) long × 1/4 in. (6 mm) to 3/8 in. (10 mm) thick, marked in strips 1/2 in. (13 mm) wide and held in the vertical position, you will make a straight-line cut. Make sure that the sparks do not present a safety hazard and that the metal being cut off will not fall on any person or object.

Starting at the top, make one cut downward. Then, starting at the bottom, make the next cut upward. The cut must be free of hard slag and within ±3/32 in. (2 mm) of a straight line and ±5° of being square. Repeat these cuts until you can make them within tolerance. Turn off the cylinder valves, bleed the hoses, back out the pressure regulators, and clean up your work area when you are finished cutting.

Complete a copy of the Student Welding Report in Appendix I or provided by your instructor. ■

Module 8
Unit 1
Key Indicator 5

PRACTICE 6-11

Overhead Straight Cut

Using a properly lit and adjusted cutting torch, welding gloves, appropriate eye protection and clothing, and one piece of mild steel plate 6 in. (152 mm) long × 1/4 in. (6 mm) to 3/8 in. (10 mm) thick, marked in strips 1/2 in. (13 mm) wide, you will make a cut in the overhead position. When you make overhead cuts, it is important to be completely protected from the hot sparks. In addition to the standard safety clothing, you should wear a leather jacket, leather apron, cap, ear protection, and a full face shield.

The torch can be angled so that most of the sparks will be blown away. The metal should fall free when the cut is completed. The cut must be within 1/8 in. (3 mm) of a straight line and ±5° of being square. Repeat this practice until you can make the cut within tolerance. Turn off the cylinder valves, bleed the hoses, back out the pressure regulators, and clean up your work area when you are finished cutting.

Complete a copy of the Student Welding Report in Appendix I or provided by your instructor. ■

Module 8
Unit 1
Key Indicator 5

CUTTING APPLICATIONS

Making practice cuts on a piece of metal that will become scrap is a good way to learn the proper torch techniques. If a bad cut is made, there is no loss. In a production shop, where each piece of metal is important, however, scrapping metal because of bad cuts decreases the shop's profits.

A number of factors that do not exist during practice can affect your ability to make a quality cut on a part. The following can become problems during cutting:

- *Changing positions*—Often, parts are larger than can be cut from one position, so you may have to move to complete the cut. Stopping and restarting a cut can result in a small flaw in the cut surface. If this flaw exceeds the acceptable limits, the cut surface must be repaired before the part can be used. To avoid this problem, always try to stop at corners if the cut cannot be completed without moving.
- *Sparks*—You will often be making cuts in large plates. Even an ideal cut can create sparks that bounce around the plate surface. These sparks often find their way into your glove, under your arm, or to any other place that will become uncomfortable. Experienced welders usually keep working if the sparks are not too large or too uncomfortable. With experience you will learn how to angle the torch, direct the cut, and position your body to minimize this problem.
- *Hot surfaces*—As you continue making cuts to complete the part, it will begin to heat up. Depending on the size of the part, the number of cuts per part, and the number of parts being cut, this heat can become uncomfortable. You may find it necessary to hold the torch farther back from the tip, but this will affect the quality of your cuts, Figure 6.55. Sometimes you might be able to rest your hand on a block to keep it off of the plate. Another problem with heat buildup is that it may become high enough to affect the cut quality. Heat becomes a problem when it causes the top edge of the plate to melt during a cut, as if the torch tip were too large. This is more of a problem when several cuts are being made in close proximity. Planning your cutting sequence and allowing cooling time will help control this potential problem.
- *Dirty tips*—In any cutting, the tip will catch small sparks and become dirty or clogged. You must decide how dirty or clogged you will let

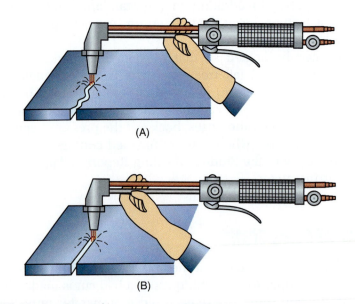

(A)

(B)

Figure 6.55 Bracing the torch
It is easier to make straight, smooth cuts if you can brace the torch closer to the tip, as in cut B

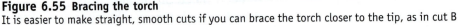

the tip get before you stop to clean it. Time spent cleaning the tip reduces productivity, unfortunately. On the other hand, if you do not stop occasionally to clean up, the quality of the cut will become so bad that postcut cleanup becomes excessive. It is your responsibility to decide when and how often to clean the tip.

- *Blowback*—As a cut progresses across the surface of a large plate, it may cross supports underneath the plate. During practice cuts this seldom if ever happens, but, depending on the design of the cutting table, it will occur even under the best of conditions in real work. If the support is small, the blowback may not cover you with sparks, plug the cutting tip, or cause a major flaw in the cut surface. If the support is large, then one or all of these events can occur. If you see that the blowback is not clearing quickly, it may be necessary to stop the cut. Stopping the cut halts the shower of sparks but leaves you with a problem restart.

Module 8
Unit 1
Key Indicator 2

Module 8
Unit 2
Key Indicator 2

PRACTICE 6-12

Cutting Out Internal and External Shapes

Using a properly lit and adjusted cutting torch, welding gloves, appropriate eye protection and clothing, and one piece of plate 1/4 in. (6 mm) to 3/8 in. (10 mm) thick, you will lay out and cut out one of the sample patterns shown in Figure 6.56, or any other design available.

Choose the pattern that best fits the piece of metal you have and mark it using a center punch. The exact size and shape of the layout are not as important as the accuracy of the cut. The cut must be made so that the center-punched line is left on the part and so that there is no more than 1/8 in. (3 mm) between the cut edge and the line, Figures 6.57 and 6.58. Repeat this practice until you can make the cut within tolerance. Turn off the cylinder valves, bleed the hoses, back out the pressure regulators, and clean up your work area when you are finished cutting.

Complete a copy of the Student Welding Report in Appendix I or provided by your instructor. ■

Module 8
Unit 1
Key Indicator 6

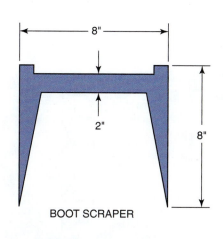

BOOT SCRAPER

Note: sizes of these and other cutting projects can be changed to fit available stock.

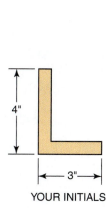

YOUR INITIALS

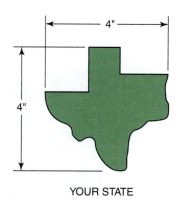

YOUR STATE

Figure 6.56 Suggested patterns for practice

Figure 6.57 Beginning a cut with the torch concentrating the flame on the top edge to speed starting
Source: Courtesy of Larry Jeffus

Figure 6.58 Rotating the torch to allow preheating of the plate ahead of the cut
This process speeds the cutting and provides better visibility of the line being cut
Source: Courtesy of Larry Jeffus

WASHING OR SCARFING

Oxyfuel torches and equipment are also used in metal-removal operations that are commonly called *washing* or *scarfing*. Specially shaped cutting tips are available that are designed to remove metal from a flat surface or from fillets or contoured joints, although it is also common to use a regular cutting tip for some washing or scarfing operations, Figure 6.59.

PRACTICE 6-13

Module 8
Unit 1
Key Indicator 8

Scarfing with the Washing Technique

Using a properly lit and adjusted cutting torch, welding gloves, appropriate eye protection and clothing, and a scarfing tip or a regular tip, you will use a washing technique to remove weld metal from a scrap piece of steel, a backing strip from a test specimen, or the vertical leg from a piece of angle iron, Figure 6.60.

Complete a copy of the Student Welding Report in Appendix I or provided by your instructor. ∎

Figure 6.59 Scarfing tips

Figure 6.60 Washing Operations

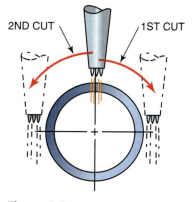

Figure 6.61 Cutting small-diameter pipe
Small-diameter pipe can be cut without changing the angle of the torch. After the top is cut, roll the pipe to cut the bottom

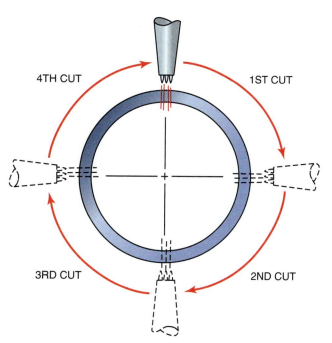

Figure 6.62 Cutting large-diameter pipe
On large-diameter pipe, the torch is turned to keep it at a right angle to the pipe. The pipe should be cut as far as possible before stopping and turning it

PIPE CUTTING

Freehand pipe cutting may be done in one of two ways. On small-diameter pipe, usually under 3 in. (76 mm), the torch tip is held straight up and down and moved from the center to each side, Figure 6.61. This technique can also be used successfully on larger pipe.

For large-diameter pipe, 3 in. (76 mm) and larger, the torch tip is always pointed toward the center of the pipe, Figure 6.62. This technique is also used on all sizes of heavy-walled pipe and can be used on some smaller pipes.

The torch body should be held so that it is parallel to the centerline of the pipe. Holding the torch parallel helps to keep the cut square.

CAUTION

When pipe is cut, hot sparks can come out of the end of the pipe nearest you, causing severe burns. For protection from hot sparks, plug up the open end of the pipe nearest you, put up a barrier to the sparks, or stand to one side of the material being cut.

PRACTICE 6-14

Square Cut on Pipe, 1G (Horizontal Rolled) Position

Using a properly lit and adjusted cutting torch, welding gloves, appropriate eye protection and clothing, and one piece of schedule 40 steel pipe with a diameter of 3 in. (76 mm), you will cut off 1/2-in.-long (13-mm-long) rings.

Using a template and a piece of soapstone, mark several rings, each 1/2 in. (13 mm) wide, around the pipe. Place the pipe horizontally on the cutting table. Start the cut at the top of the pipe using the proper piercing technique. Move the torch backward along the line and then forward; this will keep slag out of the cut. If the end of the cut closes in with slag, this will cause the oxygen to gouge the edge of the pipe when the cut is continued. Keep the tip pointed straight down. When you have gone as far with the cut as you can comfortably, quickly flip the flame away from the pipe. Restart the cut at the top of the pipe and cut as far as possible in the other direction. Stop and turn the pipe so that the end of the cut is on top and the cut can be continued around the pipe. When the cut is completed, the ring must fall free. When the pipe is placed upright on a flat plate, the pipe must stand within 5° of vertical and have no gaps higher than 1/8 in. (3 mm) under the cut. Repeat this procedure until you can make the cut within tolerance. Turn off the cylinder valves, bleed the hoses, back out the pressure regulators, and clean up your work area when you are finished cutting.

Complete a copy of the Student Welding Report in Appendix I or provided by your instructor. ■

PRACTICE 6-15

Square Cut on Pipe, 1G (Horizontal Rolled) Position

Using the same equipment, materials, and markings as described in Practice 6-14, you will cut off the 1/2-in.-long (13-mm-long) rings while keeping the tip pointed toward the center of the pipe.

Starting at the top, pierce the pipe. Move the torch backward to keep the slag out of the cut and then forward around the pipe, stopping when you have gone as far as you can comfortably. Restart the cut at the top and proceed with the cut in the other direction. Roll the pipe and continue the cut until the ring falls off freely. Stand the cut end of the pipe on a flat plate. The pipe must stand within 5° of vertical and have no gaps higher than 1/8 in. (3 mm). Repeat this practice until you can make the cut within tolerance. Turn off the cylinder valves, bleed the hoses, back out the pressure regulators, and clean up your work area when you are finished cutting.

Complete a copy of the Student Welding Report in Appendix I or provided by your instructor. ■

PRACTICE 6-16

Square Cut on Pipe, 5G (Horizontal Fixed) Position

Using the same equipment, materials, and markings as described in Practice 6-14, you will cut off 1/2-in.-long (13-mm-long) rings, using either technique, without rolling the pipe.

Start at the top and cut down both sides as far as you can comfortably. Reposition yourself and continue the cut under the pipe until the ring falls off freely. Stand the cut end of the pipe on a flat plate. The pipe must stand within 5° of vertical and have no gaps higher than 1/8 in. (3 mm). Repeat this practice until you can make the cut within tolerance. Turn off the cylinder valves, bleed the hoses, back out the pressure regulators, and clean up your work area when you are finished cutting.

Complete a copy of the Student Welding Report in Appendix I or provided by your instructor. ■

PRACTICE 6-17

Square Cut on Pipe, 2G (Vertical) Position

Using the same equipment, materials, and markings as described in Practice 6-14, you will cut off 1/2-in.-long (13-mm-long) rings, using either technique, from a pipe in the vertical position.

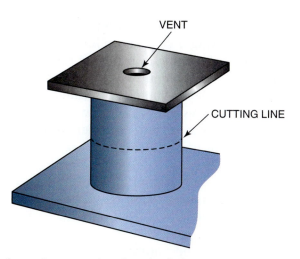

VENT

CUTTING LINE

Figure 6.63 Placing a plate on a short length of pipe to stop sparks flying around

Place a flat piece of plate over the open top end of the pipe to contain the sparks, Figure 6.63. Start on one side and proceed around the pipe until the cut is completed. Any buildup of slag may mean the ring has to be tapped free. Stand the cut end of the pipe on a flat plate. The pipe must stand within 5° of vertical and have no gaps higher than 1/8 in. (3 mm). Repeat this practice until you can make the cut within tolerance. Turn off the cylinder valves, bleed the hoses, back out the pressure regulators, and clean up your work area when you are finished cutting.

Complete a copy of the Student Welding Report in Appendix I or provided by your instructor. ■

SUMMARY

The oxyfuel cutting torch is one of the most commonly used (and misused) tools in the welding industry. Used properly it can produce almost machine-cut quality requiring no postcut cleanup. However, when it is misused, the oxyfuel torch can produce some of the most difficult problems for the welding fabricator to overcome. As you learn to use the cutting torch efficiently, you can dramatically reduce postcut cleanup time and increase productivity.

Proper equipment setup and torch tip cleaning are essential for the welder to produce quality oxyfuel cuts. Take your time when you are setting up and preparing to make a cut; this is not wasted time. A good setup will ensure the cut will meet the fabricator's quality needs.

A number of factors determine the travel speed for cutting, including plate thickness and the surface condition of the metal being cut. A good, clean, quality cut will progress at its own rate. Do not try to rush through too quickly. Learn to develop a sense for the cutting rate that produces the best-quality cut.

REVIEW

1. Using Table 6.1, list the six different fuel gases in rank order according to their temperature.

2. State one advantage of owning a combination welding–cutting torch as opposed to just having a cutting torch.
3. State one advantage of owning a dedicated cutting torch as compared to having a combination welding–cutting torch.
4. What is a mixing chamber? Where is it located?
5. Define the term *equal-pressure torch*. How does one work?
6. How does an injector-type mixing chamber work?
7. State the advantages of having two oxygen regulators on a machine-cutting torch.
8. Why are some copper alloy cutting tips chrome-plated?
9. What determines the preheat flame requirements of a torch?
10. What can happen if acetylene is used on a tip designed to be used with propane or another gas?
11. Why are some propane and natural gas tips made with a deep, recessed center?
12. What types of tip seals are used with cutting torch tips?
13. If a cutting tip sticks in the cutting head, how should it be removed?
14. How can cutting torch tip seals be repaired?
15. Why is the oxygen valve turned on before starting to clean a cutting tip?
16. Why does the preheat flame become slightly oxidizing when the cutting lever is released?
17. What causes the tiny ripples in a hand cut?
18. Why is a slight forward torch angle helpful for cutting?
19. Why should cans, drums, tanks, or other sealed containers not be opened with a cutting torch?
20. Why is the torch tip raised as the cutting lever is depressed when cutting a hole?
21. Why are the preheat holes not aligned in the kerf when making a bevel cut?
22. Sketch the proper end shape of a soapstone that is to be used for marking metal.
23. Using Table 6.4, answer the following:
    a. Oxygen pressure for cutting 1/4-in.-thick (6-mm-thick) metal
    b. Acetylene pressure for cutting 1-in.-thick (25-mm-thick) metal
    c. Tip cleaner size for a tip for 2-in.-thick (51-mm-thick) metal
    d. Drill size for a tip for 1/2-in.-thick (13-mm-thick) metal
24. What is the best way to set the oxygen pressure for cutting?
25. What metals can be cut using the oxyfuel gas process?
26. Why is it important to have extra ventilation and/or a respirator when cutting some used metals?
27. What factors of a cut can be read from the sides of the kerf after the cut is completed?
28. What is hard slag?
29. Why is it important to make good-quality cuts?
30. Describe the methods of controlling distortion when making cuts.
31. How does cutting small-diameter pipe differ from cutting large-diameter pipe?

CHAPTER

7

Plasma Arc Cutting

OBJECTIVES

After completing this chapter, the student should be able to

- describe plasma and describe a plasma torch
- explain how a plasma cutting torch works
- set up and use a plasma cutting torch
- name three advantages of PAC over OFC
- name two disadvantages of PAC compared with OFC
- describe heat input effects with PAC as they relate to distortion
- select the correct filter lens for a given job
- set up for a PAC gouge
- make straight cuts in thin plate
- make straight cuts in thick plate
- make a bevel on plate
- make a U-groove on plate
- pierce holes using PAC

KEY TERMS

arc cutting

arc plasma

cup

dross

electrode setback

electrode tip

heat-affected zone

high-frequency
    alternating current

ionized gas

joules

kerf

nozzle

nozzle insulator

nozzle tip

pilot arc

plasma

plasma arc

plasma arc gouging

stack cutting

standoff distance

water shroud

water table

AWS SENSE EG2.0

Key Indicators Addressed in this Chapter:

Module 8: Thermal Cutting Processes

Unit 3: Manual Plasma Arc Cutting (PAC)
Key Indicator 1: Performs safety inspections on manual PAC equipment
    and accessories

Key Indicator 2: Makes minor external repairs to manual PAC equipment and accessories

Key Indicator 3: Sets up for manual PAC operations on carbon steel, austenitic stainless steel and aluminum

Key Indicator 4: Operates manual PAC equipment on carbon steel, austenitic stainless steel and aluminum

Key Indicator 5: Performs manual PAC straight, square edge, cutting operations in all positions, on limited thickness range of carbon steel, austenitic stainless steel, and aluminum

Key Indicator 6: Performs manual PAC shape, square edge, cutting operations, in all positions, on limited thickness range of carbon steel, austenitic stainless steel and aluminum

Module 9: Welding Inspection and Testing Principles
Key Indicator 1: Examines cut surfaces and edges of prepared base metal parts

INTRODUCTION

The plasma process was developed in the mid-1950s as an attempt to create an arc using argon that would be as hot as the arc created when using helium gas. The early gas tungsten arc (GTA) welding process used helium gas and was called "heliarc." This early GTA welding process worked well with helium, but helium was expensive. The gas manufacturing companies had argon as a by-product from the production of oxygen. There was no good commercial market for this waste argon gas, but gas manufacturers believed there would be a good market if they could find a way to make argon perform similarly to helium for welding.

Early experiments found that when the arc was restricted in a fast-flowing column of argon, a plasma was formed. The plasma was hot enough to rapidly melt any metal. The problem was that the fast-moving gas blew the molten metal away. Researchers could not find a way to control this scattering of the molten metal, so they decided to introduce this as a cutting process, not a welding process, Figure 7.1.

Several years later, with the invention of the gas lens, plasma was successfully used for welding. Today the plasma arc can be used for plasma arc welding (PAW), plasma spraying (PSP), plasma arc cutting (PAC), and plasma arc gouging. Plasma arc cutting is the most often-used plasma process.

PLASMA

The word **plasma** has two meanings: it is the fluid portion of blood, and it is a state of matter that is found in the region of an electrical discharge (arc). The plasma created by an arc is an **ionized gas** that has electrons

Figure 7.1 Plasma arc cutting machine
This unit can have additional power modules added to the base of its control module to give it more power
Source: Courtesy of ESAB Welding & Cutting Products

and positive ions whose charges are nearly equal. For welding we use the electrical definition of plasma.

A plasma is present in any electrical discharge. A plasma consists of charged particles that conduct the electrons across the gap between the work and an electrode.

A plasma results when a gas is heated to a high enough temperature to convert into positive and negative ions, neutral atoms, and negative electrons. The temperature of an unrestricted arc is about 11,000°F, but the temperature created when the arc is concentrated to form a plasma is about 43,000°F, Figure 7.2. This is hot enough to rapidly melt any metal the arc comes in contact with.

ARC PLASMA

Arc plasma is gas that has been heated to at least a partially ionized state, enabling it to conduct an electric current (ANSI/AWS A3.0-89). **Plasma arc** is the term most often used in the welding industry to refer to the arc plasma used in welding and cutting processes. The plasma arc produces both the high temperature and intense light associated with all forms of arc welding and **arc cutting** processes.

PLASMA TORCH

A plasma torch is a device that allows the creation and control of the plasma for welding or cutting processes, depending on its design. The plasma is created in the cutting and welding torches in essentially the same way, and

Module 8
Unit 3
Key Indicator 2

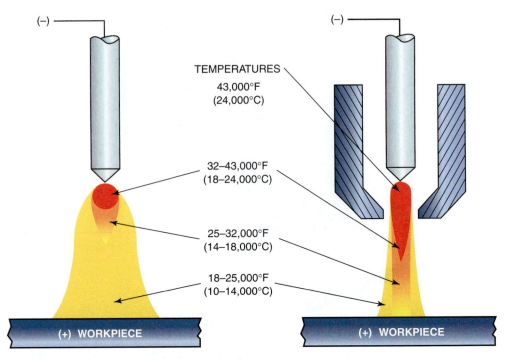

Figure 7.2 Approximate temperature differences between a standard arc and a plasma arc
Source: Courtesy of the American Welding Society

both torches have the same basic parts. A plasma torch supplies electrical energy to a gas to change it into the high-energy state of a plasma.

Torch Body

The torch body on a manual torch is made of a special plastic that is resistant to high temperatures, ultraviolet light, and impact, Figure 7.3. The torch body provides a good grip area and protects the cable and hose connections to the head. Torch bodies of a variety of lengths and sizes are available. Generally, the longer, larger torches are used for the higher-capacity machines; however, sometimes you might want a longer or larger torch to give yourself better control or a longer reach. On machine torches the body is often called a barrel and may come with a rack attached to its side. The rack is a flat gear that allows the torch to be raised and lowered manually to the correct height above the work.

Torch Head

The torch head is attached to the torch body where the cables and hoses attach to the electrode tip, nozzle tip, and nozzle. The torch and head may be connected at any angle (e.g., 90°, 75°, or 180° [straight]), or the head can be flexible. The 75° and 90° angles are popular for manual operations, and 180° straight torch heads are most often used for machine operations. Because of the heat produced in the head by the arc, some provisions for cooling the head and its internal parts must be made. For low-power torches this cooling may be either by air or by water. Higher-power torches must be liquid-cooled, Figure 7.4. It is possible to replace just the torch head on most torches if it becomes worn or damaged.

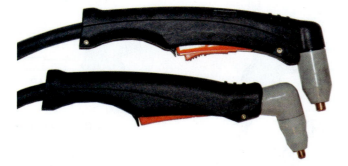

Figure 7.3 Hand-held torches
Hand-held plasma cutting torches come in a variety of sizes for different-capacity machines
Source: Courtesy of Thermadyne Holding Corporation

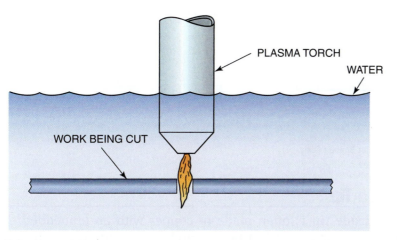

PLASMA TORCH

WATER

WORK BEING CUT

Figure 7.4 Water tables
Water tables can be used with plasma cutting to reduce the sound and sparks produced during cuts

Power Switch

Most hand-held torches have a manual power switch that is used to start and stop the power source, gas, and cooling water (if used). Most often, this is a thumb switch located on the torch body, but it may be a foot control or located on the panel for machine equipment. The thumb switch may be molded into the torch body or it may be attached to the torch body with a strap clamp. The foot control must be rugged enough to withstand the welding shop environment. Some equipment has an automatic system that starts the plasma when the torch is brought close to the work.

Common Torch Parts

The electrode tip, nozzle insulator, nozzle tip, nozzle guide, and nozzle are parts of the torch that must be replaced periodically as they wear out or become damaged from use, Figure 7.5.

The metal parts are usually made out of copper, and they may be plated. The plating of copper parts helps them stay spatter-free longer.

Caution

Improper use of the torch or assembly of torch parts may result in damage to the torch body and the need to replace these parts frequently.

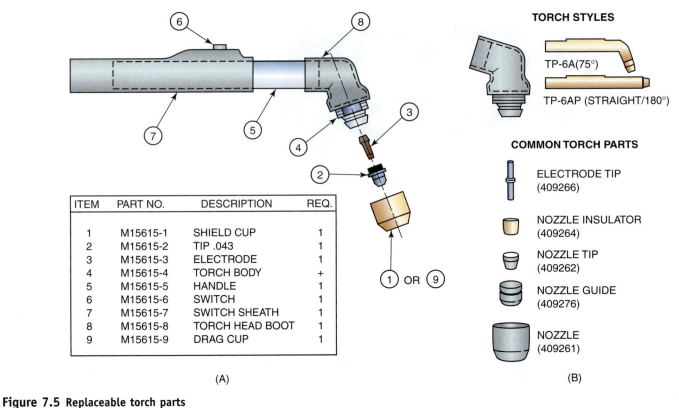

Figure 7.5 Replaceable torch parts
Source: (A) Courtesy of Lincoln Electric Company; (B) courtesy of Hobart Brothers Company

Electrode Tip

The **electrode tip** is often made of copper with an embedded tungsten tip. Copper/tungsten tips in newer torches have improved the quality of work they can be used to produce. Copper allows the heat generated at the tip to be conducted away faster. Keeping the tip as cool as possible lengthens its life and allows for better-quality cuts for a longer time. The newer torches are a major improvement over earlier torches, some of which required the welder to accurately grind the tungsten electrode into shape. If you are using a torch that requires the grinding of the electrode tip, you must have a guide to ensure that the tungsten is properly prepared.

Nozzle Insulator

The **nozzle insulator** is situated between the electrode tip and the nozzle tip. The nozzle insulator provides the critical gap spacing and the electrical separation of the parts. The spacing between the electrode tip and the nozzle tip, called **electrode setback**, is critical to the proper operation of the system.

Nozzle Tip

The **nozzle tip** has a small, cone-shaped, constricting orifice in the center. The electrode setback space, between the electrode tip and the nozzle tip, is where the electric current forms the plasma. The preset, close-fitting parts restrict the gas in the presence of the electric current

Figure 7.7 PAC torch replacement part kit
Kits contain the torch parts that are expected to wear out and need periodic replacement
Source: Courtesy of Thermadyne Holding Corporation

Figure 7.6 Nozzle tips
Different torches use different types of nozzle tips
Source: Courtesy of Larry Jeffus

so the plasma can be generated, Figure 7.6. The diameter of the constricting orifice and the electrode setback are major factors in the operation of the torch. Changes in the diameter of the orifice affect the action of the plasma jet. When the setback distance is changed, the arc voltage and current flow change.

Nozzle

The **nozzle**, sometimes called the **cup**, is made of ceramic or any other high-temperature-resistant substance. This helps prevent the internal electrical parts from accidental shorting and provides control of the shielding gas or water injection if they are used, Figure 7.7.

Water Shroud

A **water shroud** nozzle may be attached to some torches. The water surrounding the nozzle tip is used to control the potential hazards of light, fumes, noise, or other pollutants produced by the process.

POWER AND GAS CABLES

A number of power and control cables and gas and cooling-water hoses may be used to connect the power supply with the torch, Figure 7.8. The multipart cable is usually covered to provide some protection to the cables and hoses inside and to make handling the cable easier. The covering is heat resistant but will not prevent damage to the cables and hoses inside if it comes in contact with hot metal or is exposed directly to the cutting sparks.

Module 8
Unit 3
Key Indicator 2

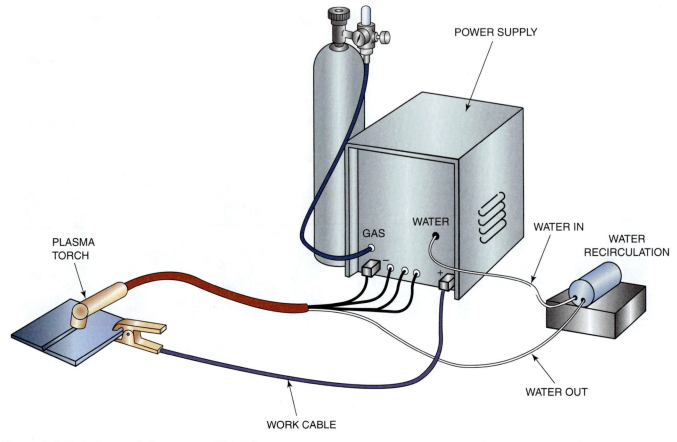

Figure 7.8 Typical manual plasma arc cutting setup

Power Cable

The power cable must have a high-voltage-rated insulation. It is made of finely stranded copper wire to allow for maximum flexibility of the torch, Figure 7.9. For all nontransfer-type torches and those that use a high-frequency pilot arc, there are two power conductors, one positive (+) and one negative (–). The size and current-carrying capacity of this cable are controlling factors for the power range of the torch. As the capacity of the equipment increases, the cable must be made larger to carry the increased current. Larger cables are less flexible and more difficult to manipulate. To make the cable smaller on water-cooled torches, the cable is run inside the cooling-water return line, which allows a smaller cable to carry more current. The water prevents the cable from overheating.

Gas Hoses

There may be two gas hoses running to the torch. One hose carries the gas used to produce the plasma, and the other provides a shielding gas. On some small-amperage cutting torches there is only one gas line. The gas line is made of a special heat-resistant, ultraviolet-light-resistant plastic. If it is necessary to replace the tubing because it is damaged, be sure to use tubing provided by the manufacturer or a welding supplier. The tubing must be sized to allow the required gas flow rate within the pressure range of the torch, and it must be free from solvents and oils that might contaminate the gas. If the pressure of the gas supplied is excessive, the tubing may leak at the fittings or rupture.

Figure 7.9 Portable plasma arc cutting machine with built-in air compressor
Source: Courtesy of Thermadyne Holding Corporation

Control Wire

The control wire is a two-conductor, low-voltage, stranded copper wire. This wire connects the power switch to the power supply. This allows the welder to start and stop the plasma power and gas as needed during the cut or weld.

Water Tubing

Medium- and high-amperage torches may be water-cooled. The water for cooling early torch models had to be deionized. Failure to use deionized water on these torches will result in the torch arcing out internally. Such arcing may destroy or damage the torch's electrode tip and the nozzle tip. To see whether your torch requires this special form of water, refer to the manufacturer's manual. If cooling water is required, it must be switched on and off at the same time as the plasma power. Allowing the water to circulate continuously might result in corrosion in the torch. When the power is reapplied, the contaminated water could cause internal arcing damage.

POWER REQUIREMENTS

Voltage

The production of the plasma requires a direct-current (DC), high-voltage, constant-current (drooping arc voltage) power supply. A constant-current machine allows for a rapid start of the plasma arc at the high open circuit voltage and a more controlled plasma arc as the voltage rapidly drops to the lower closed voltage level. The voltage required for most welding operations, such as shielded metal arc, gas metal arc, gas tungsten arc, and flux cored arc welding, ranges from 18 to 45 volts. The voltage for a plasma arc process ranges from 50 to 200 volts closed circuit and 150 to 400 volts open circuit. This higher electrical potential is required because the resistance of the gas increases as it is

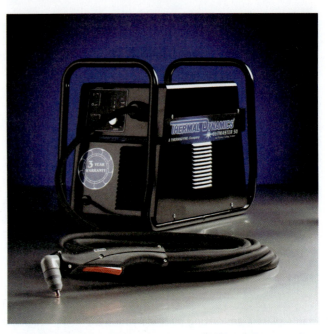

Figure 7.10 Inverter-type high-voltage plasma arc cutting power supply
Source: Courtesy of Thermal Dynamics®, a Thermadyne® Company

forced through a small orifice. The potential voltage of the power supplied must be high enough to overcome the resistance in the circuit in order for electrons to flow, Figure 7.10.

Amperage

Although the voltage is higher, the current (amperage) flow is much lower than for most other welding processes. Some low-powered PAC torches will operate with as low as 10 amps of current flow. High-powered plasma cutting machines can have amperages as high as 200 amps, and some very large automated cutting machines may have 1000-ampere capacities. The higher the amperage capacity, the faster the torch will cut and the thicker the material it will cut.

Wattage

The plasma process uses approximately the same amount of power, in watts, as a similar nonplasma process. Watts are the units of measure for electrical power. By determining the total wattage used for the non-plasma process and the plasma operation, you can make a comparison. Watts used in a circuit are determined by multiplying the voltage by the amperage, Figure 7.11. For example, a 1/8-in.-diameter E6011 electrode will operate at 18 volts and 90 amperes. The total watts used would be

$$W = V \times A$$
$$W = 18 \times 90$$
$$W = 1620 \text{ watts of power}$$

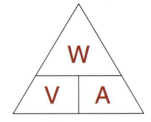

Figure 7.11 Ohm's Law

A low-power PAC torch operating with only 20 amperes and 85 volts would be using a total of

$$W = V \times A$$
$$W = 85 \times 20$$
$$W = 1700 \text{ watts of power}$$

HEAT INPUT

Although the total power used by plasma and nonplasma processes is similar, the actual energy input into the work per linear foot is less with plasma. The very high temperatures of the plasma process allow much higher traveling rates, so the same amount of heat input is spread over a much larger area. This has the effect of lowering the **joules** per inch of heat the weld or cut will receive. Figure 7.12 shows the cutting performance of a typical plasma torch. Note the relationship among amperage, cutting speed, and metal thickness. The lower the amperage, the slower the cutting speed or the thinner the metal that can be cut.

A high travel speed with plasma cutting will result in a heat input that is much lower than that of the oxyfuel cutting process. A steel plate cut using the plasma process may show only a slight increase in temperature after the cut. It is often possible to pick up a part only moments after it has been cut using plasma and find that it is cool to the touch. The same part cut with oxyfuel would be much hotter and require a longer time to cool off.

DISTORTION

Any time metal is heated in a localized zone or spot it expands in that area, and after the metal cools, it is no longer straight or flat, Figure 7.13. If a piece of metal is cut, there will be localized heating along the edge of the cut, and unless special care is taken, the part will not be usable as a result of its distortion, Figure 7.14. This distortion is a much greater problem with thin metals. By using a plasma cutter, an auto body

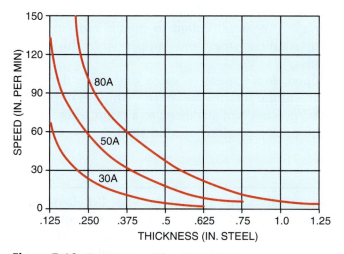

Figure 7.12 Plasma arc cutting parameters
Source: Courtesy of ESAB Welding & Cutting Products

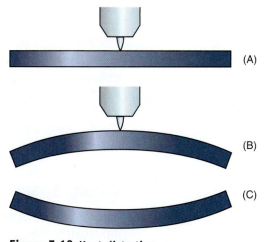

Figure 7.13 Heat distortion
(A) When metal is heated, (B) it bends up toward the heat. (C) As the metal cools, it bends away from the heated area

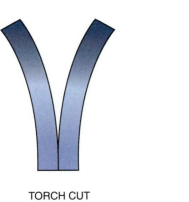

TORCH CUT   PLASMA CUT

Figure 7.14 Minimal bending with a plasma cut
The heat of a torch cut causes metal to bend, but the plasma cut is so fast little or no bending occurs

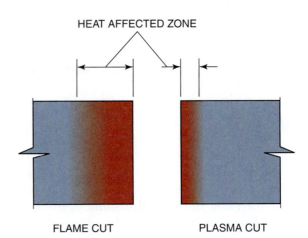

HEAT AFFECTED ZONE

FLAME CUT   PLASMA CUT

Figure 7.15 Heat-affected zone
A smaller heat-affected zone will result in less hardness or weakening along the cut edge

worker can cut the thin, low-alloy sheet metal of a damaged car with little problem from distortion.

On thicker sections, with the plasma process, the hardness zone along the edge of a cut will be so small that it is not a problem. In oxyfuel cutting of thick plate, especially higher-alloyed metals, this hardness zone can cause cracking and failure if the metal is shaped after cutting, Figure 7.15. Often the plates must be preheated before they are cut using oxyfuel to reduce the **heat-affected zone**. This preheating adds greatly to both the time and fuel costs of fabrication. By making it possible to perform most cuts without preheating, the plasma process greatly reduces fabrication cost.

APPLICATIONS

Early plasma arc cutting systems required that either helium or argon be used as the plasma and shielding gases. As the process improved, it was possible to start the PAC torch using argon or helium and then switch to less expensive nitrogen. The use of nitrogen as the plasma cutting gas greatly reduced the cost of operating a plasma system. Because of its operating expense, plasma cutting was limited to metals not easily cut using oxyfuel. Aluminum, stainless steel, and copper were the metals most often cut using plasma.

As the process developed, less expensive gases and even dry compressed air could be used, and the torches and power supplies improved. By the early 1980s, the PAC process had advanced to a point where it was used for cutting all but the thicker sections of mild steel.

Cutting Speed

High cutting speeds are possible, up to 300 in./min (25 ft/min, or approximately 1/4 mile/hour). The fastest oxyfuel cutting equipment can cut at only about one-fourth that speed. A problem with early high-speed machine cutting was that the cutting machines could not reliably make cuts as fast as the PAC torch. That problem has been resolved, and

Figure 7.16 Plasma arc machine cut in 2-in.-thick mild steel
Notice how smooth the cut edge is
Source: Courtesy of Larry Jeffus

the new machines and robots can operate at the upper limits of the plasma torch capacity. These machines and robots are capable of automatically maintaining the optimum torch standoff distance from the work. Some cutting systems even follow the irregular surfaces of preformed part blanks, Figure 7.16.

Metals

Any material that is electrically conductive can be cut using the PAC process. In a few applications nonconductive materials can be coated with conductive material so that they can be cut also. Although it is possible to make cuts in metal as thick as 7 in. (17.78 cm), it is not cost-effective to do so. The most popular materials cut are carbon steel up to 1 in., stainless steel up to 4 in., and aluminum up to 6 in. These are not the upper limits of the PAC process, but beyond these limits other cutting processes may be less expensive. A shop may use PAC for thicker material even if it is not cost-effective because it does not have ready access to the alternative process.

Other metals commonly cut using PAC are copper, nickel alloys, high-strength, low-alloy steels, and clad materials. PAC is also used to cut expanded metals, screens, and other items that would require frequent starts and stops if the oxyfuel process were used, Figure 7.17.

Standoff Distance

The **standoff distance** is the distance from the nozzle tip to the work, Figure 7.18. This distance is critical to producing quality plasma arc cuts. As the distance increases, the arc force is diminished and tends to spread out. This causes the kerf to be wider, the top edge of the plate to become rounded, and more dross to form on the bottom edge of the plate. However, if this distance becomes too close, the working life of the nozzle tip will be reduced. In some cases an arc can form between the nozzle tip and the metal that instantly destroys the tip.

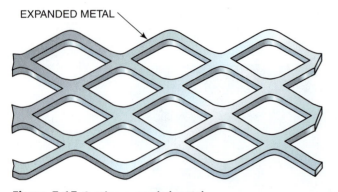

EXPANDED METAL

Figure 7.17 Cutting expanded metal
Expanded metal can easily be cut with the plasma arc process
but not very easily with an oxyfuel torch

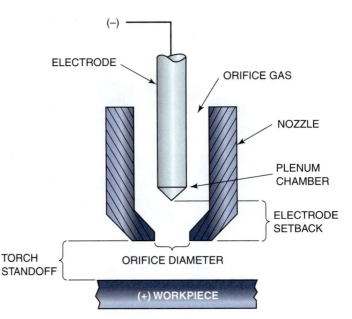

Figure 7.18 Conventional plasma arc terminology
Source: Courtesy of the American Welding Society

Figure 7.19 Castle nozzle tip
A castle nozzle tip can be used to
allow the torch to be dragged across
the surface

On some new torches, it is possible to drag the nozzle tip along the surface of the work without shorting it out. This is a great help for work on metal out of position or on thin sheet metal. Before you use your torch in this manner, you must check the owner's manual to see whether it will operate in contact with the work, Figure 7.19. This technique will allow the nozzle tip orifice to become contaminated more quickly.

Starting Methods

Because the electrode tip is located inside the nozzle tip and a high initial resistance to current flow exists in the gas flow before the plasma is generated, it is necessary to have a specific starting method. Two methods are used to establish a current path through the gas.

The most common method uses a **high-frequency alternating current** carried through the conductor, the electrode, and back from the nozzle tip. This high-frequency current ionizes the gas and allows it to carry the initial current to establish a pilot arc, Figure 7.20. After the pilot arc has been started, the high-frequency starting circuit can be stopped. A **pilot arc** is an arc between the electrode tip and the nozzle tip within the torch head. This is a nontransfer arc, so the workpiece is not part of the current path. Although the low current of the pilot arc is inside the torch, it does not create enough heat to damage the torch parts. When the torch is brought close enough to the work, the primary arc will follow the pilot arc across the gap, to the work, and the main plasma is started. Once the main plasma is started, the pilot arc power can be shut off.

The second method of starting requires the electrode tip and nozzle tip to be momentarily shorted together. This is accomplished by automatically moving them together and immediately separating them again. The momentary shorting allows the arc to be created without damaging the torch parts.

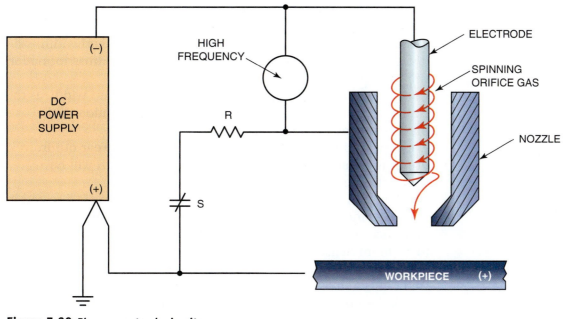

Figure 7.20 Plasma arc torch circuitry
Source: Courtesy of the American Welding Society

Kerf

The **kerf** is the space left in the workpiece as the metal is removed during a cut. A PAC kerf is often wider than an oxyfuel cut kerf. Several factors affect the width of the kerf, for example:

- standoff distance—The closer the torch nozzle tip is to the work, the narrower the kerf will be, Figure 7.21.
- orifice diameter—Keeping the diameter of the nozzle orifice as small as possible will keep the kerf smaller.

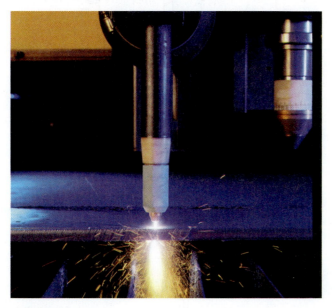

Figure 7.21 Torch height
When the torch is at the correct height, all of the metal and sparks are blown out the bottom side of the cut
Source: Courtesy of ESAB Welding & Cutting Products

- power setting—Too high or too low a power setting will cause an increase in the kerf width.
- travel speed—As the travel speed is increased, the kerf width will decrease; however, the bevel on the sides and the dross formation will increase if the speeds are excessive.
- gas—The type of gas or gas mixture will affect the kerf width, as the choice of gas affects travel speed, power, concentration of the plasma stream, and other factors.
- electrode and nozzle tip—As these parts begin to wear out from use or are damaged, the PAC quality and kerf width will be adversely affected.
- swirling of the plasma gas—On some torches, the gas is directed in a circular motion around the electrode before it enters the nozzle tip orifice. This swirling causes the plasma stream that is produced to be more dense with straighter sides. The result is an improved cut quality, including a narrow kerf, Figure 7.22.
- water injection—The injection of water into the plasma stream as it leaves the nozzle tip is not the same as the use of a water shroud. Water injection into the plasma stream will increase the swirl and further concentrate the plasma. This improves the cutting quality, lengthens the life of the nozzle tip, and makes a squarer, narrower kerf, Figure 7.23.

Table 7.1 lists some standard kerf widths for several metal thicknesses. These are to be used as a guide for nesting of parts on a plate to

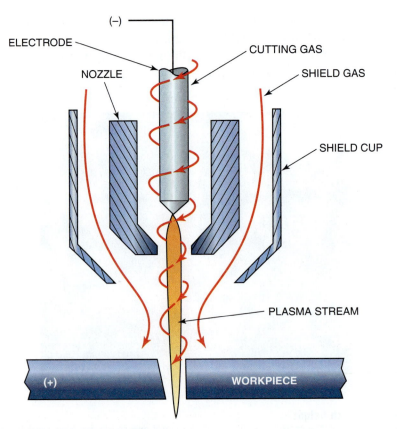

Figure 7.22 Swirling of the plasma gas
The cutting gas can swirl around the electrode to produce a tighter plasma column
Source: Courtesy of the American Welding Society

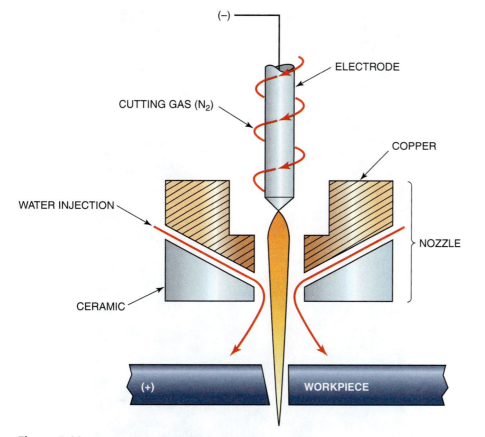

Figure 7.23 Water injection plasma arc cutting
Notice that the kerf is narrow, and one side is square
Source: Courtesy of the American Welding Society

Table 7.1 Standard Kerf Widths for Several Metal Thicknesses

Plate Thickness		Kerf Allowance	
in.	mm	in.	mm
1/8 to 1	3.2 to 25.4	+3/32	+2.4
1 to 2	25.4 to 51.0	+3/16	+4.8
2 to 5	51.0 to 127.0	+5/16	+8.0

maximize the material used and minimize scrap. The kerf size may vary from this depending on a number of variables with your PAC system. You should make test cuts to verify the size of the kerf before starting any large production cuts.

Because the sides of the plasma stream are not parallel as they leave the nozzle tip, a bevel is left on the sides of all plasma cuts. This bevel angle is from 1/2° to 3° depending on metal thickness, torch speed, type of gas, standoff distance, nozzle tip condition, and other factors affecting a quality cut. On thin metals, this bevel is undetectable and offers no problem in part fabrication or finishing.

The use of a plasma swirling torch and the direction in which the cut is made can cause one side of the cut to be square and the scrap side to have all of the bevel, Figure 7.24. This technique is only effective if one side of the cut is to be scrap.

Figure 7.24 Plasma cutting
Source: Courtesy of ESAB Welding & Cutting Products

GASES

Almost any gas or gas mixture can be used today for the PAC process. Changing the gas or gas mixture is one method of controlling the plasma cut. Although the type of gas or gases used will have a major effect on cutting performance, gas choice is only one of a number of changes that a technician can make to help produce a quality cut. The following are some of the cutting factors affected by changing the PAC gas or gases:

- force—The amount of mechanical impact on the material being cut; the density of the gas and its ability to disperse the molten metal.
- central concentration—Some gases will have a more compact plasma stream. This factor will greatly affect the kerf width and cutting speed.
- heat content—Changes in the electrical resistance of a gas or gas mixture will affect the heat content of the plasma produced. The higher the resistance, the higher will be the heat produced by the plasma.
- kerf width—A plasma that remains in a tightly compact stream will produce a deeper cut with less of a bevel on the sides.
- dross formation—The dross that may be attached along the bottom edge of the cut can be controlled or eliminated.

- top edge rounding—The rounding of the top edge of the plate can often be eliminated by selecting the correct gas or gas mixture.
- metal type—Because of the formation of undesirable compounds on the cut surface as the metal reacts to elements in the plasma, some metals should not be cut with a specific gas or gases.

Table 7.2 lists some of the popular gases and gas mixtures used for various metals in the PAC process. The selection of a gas or gas mixture for a specific operation to maximize the system performance must be tested with the equipment and setup being used. With constant developments and improvements in the PAC system, new gases and gas mixtures are continuously being added to the list. It is also important to have the correct gas flow rate for the tip size, metal type, and thickness. Too low a gas flow will result in a cut with excessive dross and sharply beveled sides, Figure 7.25. Too high a gas flow will produce a poor cut because of turbulence in the plasma stream and waste gas. A flow measuring kit can be used to test the flow at the plasma torch for more accurate adjustments.

Table 7.2 Gases for Plasma Arc Cutting and Gouging

Metal	Gas
Carbon and low alloy steel	Nitrogen Argon with 0 to 35% hydrogen air
Stainless steel	Nitrogen Argon with 0 to 35% hydrogen
Aluminum and aluminum alloys	Nitrogen Argon with 0 to 35% hydrogen
All plasma arc gouging	Argon with 35 to 40% hydrogen

Figure 7.25 Controlling pressure
Controlling the pressure is one way of controlling gas flow. Some portable plasma arc cutting machines have their own air pressure regulator and dryer. Air must be dried to provide a stable plasma arc
Source: Courtesy of Larry Jeffus

Stack Cutting

Because the PAC process does not rely on the thermal conductivity between stacked parts, as the oxyfuel process does, thin sheets can be stacked and cut efficiently. With the oxyfuel **stack cutting** of sheets, it is important that there be no air gaps between layers. Also, it is often necessary to make a weld along the side of the stack for the cut to start consistently.

The PAC process does not have these limitations. It is recommended that the sheets be held together for cutting, but this can be accomplished using standard C-clamps. The clamping needs to be tight because, if the space between layers is too great, the sheets may stick together. The only problem is that, because of the kerf bevel, the parts near the bottom might be slightly larger if the stack is very thick. This problem can be controlled using the same techniques as described for making the kerf square.

Dross

Dross is the metal compound that resolidifies and attaches itself to the bottom of a cut. This metal compound is made up mostly of unoxidized metal, metal oxides, and nitrides. It is possible to make cuts dross-free if the PAC equipment is in good operating condition and the metal is not too thick for the size of the torch being used. Because dross contains more unoxidized metal than most OFC slag, often it is much harder to remove if it sticks to the cut. The thickness that a dross-free cut can be made is dependent on a number of factors, including the gas or gas mixture used for the cut, travel speed, standoff distance, nozzle tip orifice diameter, wear condition of the electrode tip and nozzle tip, gas velocity, and plasma stream swirl.

Stainless steel and aluminum are easily cut dross-free. Carbon steel, copper, and nickel–copper alloys are much more difficult to cut dross-free.

MACHINE CUTTING

Almost any plasma torch can be attached to some type of semiautomatic or automatic device to allow it to make machine cuts. The simplest devices are oxyfuel portable flame-cutting machines that run on tracks, Figure 7.26. These portable machines are good for mostly straight or circular cuts. Complex shapes can be cut with a pattern cutter that uses a magnetic tracing system to follow the template's shape, Figure 7.27.

High-powered PAC machines may have amperages of up to 1000 amps. These machines must be used with some semiautomatic or automatic cutting system. The heat, light, and other potential hazards of these machines make them unsafe for manual operation.

Large, dedicated, computer-controlled cutting machines have been built specifically for PAC systems. These machines have the high travel speeds required to produce good-quality cuts and have a high volume of production. With these machines, the operator can input the specific cutting instructions such as speed, current, gas flow, location, and shape of the part to be cut, and the machine will make the cut with a high degree of accuracy once or any number of times.

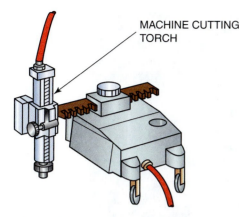

MACHINE CUTTING
TORCH

Figure 7.26 Machine cutting tool

Figure 7.27 Portable pattern cutter
A portable pattern cutter can cut shapes, circles, and straight lines
Source: Courtesy of ESAB Welding & Cutting Products

Robotic cutters are also available to perform high-quality, high-volume PAC, Figure 7.28. The advantage of using a robot is that, in most cases, the robot can be set up for multitasking. When a robot is programmed, it can cut the part out, change the tool itself and weld the parts together, change the tool and grind, drill, or paint the finished unit.

Water Tables

Machine cutting lends itself to the use of water cutting tables, although they can also be used with most hand torches. The **water table** is used to reduce the noise level, control the plasma light, trap the sparks, eliminate most of the fume hazard, and reduce distortion.

Water tables either support the metal just above the surface of the water or submerge the metal about 3 in. (7.62 cm) below the water's surface. Both types of water tables must have some method of removing the cut parts, scrap, and slag that build up in the bottom. Often the surface-type tables have the PAC torch connected to a water shroud nozzle, Figure 7.29. A water shroud nozzle offers the same advantages to the PAC process as the submerged table offers. In most cases, the manufacturers of this type of equipment have made provisions for a special dye to be added to the water. This dye helps control the harmful light produced by the PAC system. Check with the equipment's manufacturer for limitations and application of the use of dyes.

Figure 7.28 Automated cutting
Source: Courtesy of ESAB Welding & Cutting Products

MANUAL CUTTING

Manual plasma arc cutting is the most versatile of the PAC processes. It can be used in all positions, on almost any surface, and on most metals. This process is limited to low-power plasma machines; however, even these machines can cut up to 1 1/2-in. (3.81 cm)–thick metals. The limitation to low power, 100 amperes or less, is primarily for safety reasons. The higher-powered machines have extremely dangerous open circuit voltages that can kill a person if accidentally touched.

Setup

Module 8
Unit 3
Key Indicator 3

The setup is similar for most plasma equipment, but never attempt to set up a system without the owner's manual for the specific equipment.

Be sure all of the connections are tight and that there are no gaps in the insulation on any of the cables. Check the water and gas lines for leaks. Visually inspect the complete system for possible problems.

Before you touch the nozzle tip, be sure that the main power supply is off. The open circuit voltage on even low-powered plasma machines is high enough to kill a person. Replace all parts to the torch before the power is restored to the machine.

PLASMA ARC GOUGING

Plasma arc gouging is a recent addition to the PAC processes. The process is similar to air carbon arc gouging in that a U-groove can be cut into the metal's surface. The removal of metal along a joint before the

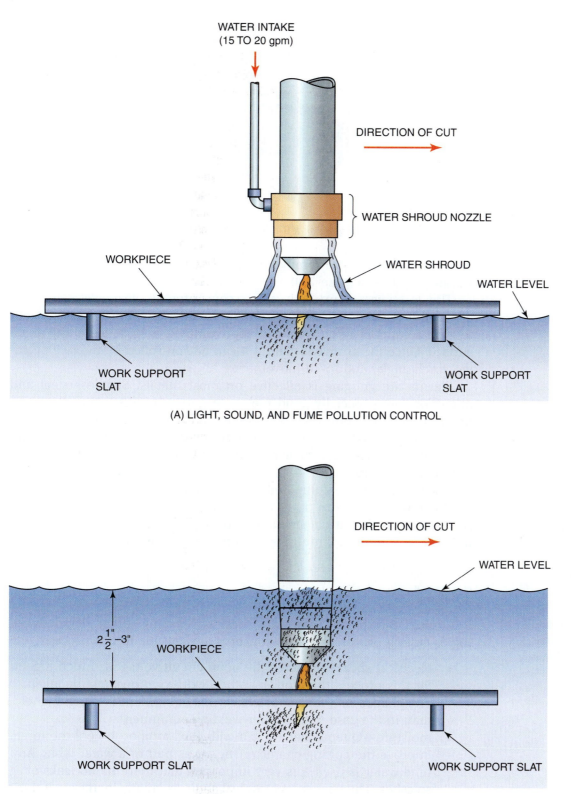

WATER INTAKE
(15 TO 20 gpm)

DIRECTION OF CUT

WATER SHROUD NOZZLE

WORKPIECE

WATER SHROUD

WATER LEVEL

WORK SUPPORT
SLAT

WORK SUPPORT
SLAT

(A) LIGHT, SOUND, AND FUME POLLUTION CONTROL

DIRECTION OF CUT

WATER LEVEL

$2\frac{1}{2}"$–3"

WORKPIECE

WORK SUPPORT SLAT

WORK SUPPORT SLAT

(B) UNDERWATER PLASMA CUTTING

Figure 7.29 Water shroud and underwater torch
A water table can be used either with (A) a water shroud or (B) underwater torches
Source: Courtesy of the American Welding Society

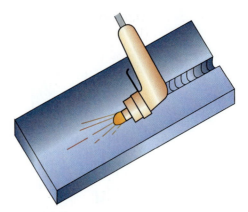

Figure 7.30 Plasma arc gouging a U-groove in a plate

metal is welded or the removal of a defect for repair is easy using this variation of PAC, Figure 7.30.

The torch is set up with a less-concentrated plasma stream. This allows the washing away of the molten metal instead of thrusting it out to form a cut. The torch is held at approximately a 30° angle to the metal surface. Once the groove is started it can be controlled by the rate of travel, torch angle, and torch movement.

Plasma arc gouging is effective on most metals. Stainless steel and aluminum are especially good metals to gouge because there is almost no cleanup. The groove is clean, bright, and ready to be welded. Plasma arc gouging is especially beneficial with these metals because there is no reasonable alternative available. The only other process that can leave the metal ready to weld is to have the groove machined, and machining is slow and expensive compared with plasma arc gouging.

It is important not to try to remove too much metal in one pass. The process will work better if small amounts are removed at a time. If a deeper groove is required, multiple gouging passes can be made.

SAFETY

Module 8
Unit 3
Key Indicator 1

PAC raises many of the same safety concerns as most other electric welding or cutting processes. Some concerns are specific to this process:

- electrical shock—Because the open circuit voltage is much higher for this process than for any other, extra caution must be taken. The chance of a fatal shock from this equipment is much higher than in the case of any other welding equipment.
- moisture—Often water is used with PAC torches to cool the torch, improve the cutting characteristic, or as part of a water table. Any time water is used it is very important that there be no leaks or splashes. The chance of electrical shock is greatly increased if there is moisture on the floor, cables, or equipment.
- noise—Because the plasma stream is passing through the nozzle orifice at a high speed, a loud sound is produced. The sound level increases as the power level increases. Even with low-power equipment the decibel (dB) level is above safety ranges. Some type of ear protection is required to prevent damage to the operator and other people in the area of the PAC equipment when it is in

operation. High levels of sound can have a cumulative effect on one's hearing. Over time, one's ability to hear will decrease unless proper precautions are taken. See the owner's manual for recommendations for the equipment in use.

- light—The PAC process produces light radiation in all three spectrums. The large quantity of visible light will cause night blindness if the eyes are unprotected. The most dangerous of the lights is ultraviolet. As in other arc processes, this light can cause burns to the skin and eyes. Infrared light can be felt as heat, and it is not as much of a hazard. Some type of eye protection must be worn when any PAC is in progress. Table 7.3 lists the recommended lens shade numbers for various power-level machines.

- fumes—The PAC process produces a large quantity of fumes that are potentially hazardous. A specific means for removing them from the work space should be in place. A downdraft table is ideal for manual work, but some special pickups may be required for larger applications. The use of a water table or a water shroud nozzle, or both, will greatly help to control fumes. Often the fumes cannot be exhausted into the open air without first being filtered or treated to remove dangerous levels of contaminants. Before installing an exhaust system, you must check with local, state, and federal officials to see whether specific safeguards are required.

- gases—Some plasma gas mixtures include hydrogen; because this is a flammable gas, extra care must be taken to ensure that the system is leak-proof.

- sparks—As with any process that produces sparks, the danger of an accidental fire is always present. This is a larger concern with PAC because the sparks are often thrown some distance from the work area and the operator's vision is restricted by a welding helmet. If there is any possibility that sparks will be thrown out of the immediate work area, a fire watch must be present. A fire watch is a person whose sole job is to watch for the possible starting of a fire. This person must know how to sound the alarm and have appropriate firefighting equipment handy. Never cut in the presence of combustible materials.

- operator checkout—Never operate any PAC equipment until you have read the manufacturer's owner's and operator's manual for the specific equipment to be used. It is a good idea to have some- one who is familiar with the equipment go through the operation after you have read the manual.

Table 7.3 Recommended Shade Densities for Filter Lenses

Current Range A	Minimum Shade	Comfortable Shade
Less than 300	8	9
300 to 400	9	12
400 plus	10	14

PRACTICE 7-1

Module 8
Unit 3
Key Indicator 4, 5

Flat, Straight Cuts in Thin Plate

Using a properly set up and adjusted PAC machine, proper safety protection, one or more pieces of mild steel, stainless steel, and aluminum 6 in. (152 mm) long and 16 gauge and 1/8 in. (3 mm) thick, you will cut off 1/2-in.-wide strips, Figure 7.31.

1. Starting at one end of the piece of metal that is 1/8 in. (3 mm) thick, hold the torch as close as possible to a 90° angle.
2. Lower your hood and establish a plasma cutting stream.
3. Move the torch in a straight line down the plate toward the other end, Figure 7.32.
4. If the width of the kerf changes, speed up or slow down the travel rate to keep the kerf the same size for the entire length of the plate.

Repeat the cut using both thicknesses of all three types of metals until you can make consistently smooth cuts that are within ±3/32 in. of a straight line and ±5° of being square. Turn off the PAC equipment and clean up your work area when you are finished cutting.

Module 9
Key Indicator 1

Complete a copy of the Student Welding Report in Appendix I or provided by your instructor. ∎

PRACTICE 7-2

Module 8
Unit 3
Key Indicator 4, 5

Flat, Straight Cuts in Thick Plate

Using a properly set up and adjusted PAC machine, proper safety protection, one or more pieces of mild steel, stainless steel, and aluminum 6 in. (152 mm) long and 1/4 in. and 1/2 in. thick, you will cut off 1/2-in.-wide strips. Follow the same procedure as outlined in Practice 7-1.

Repeat the cut using both thicknesses of all three types of metals until you can make consistently smooth cuts that are within ±3/32 in. of a

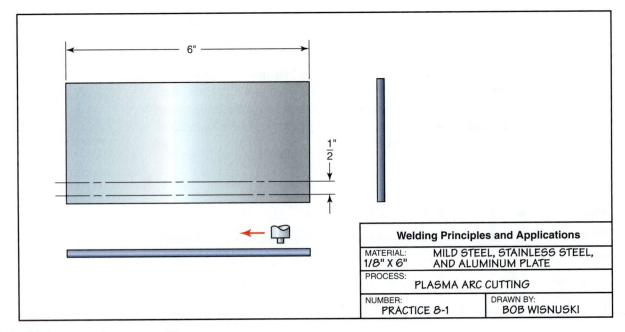

Welding Principles and Applications	
MATERIAL: 1/8" X 6"	MILD STEEL, STAINLESS STEEL, AND ALUMINUM PLATE
PROCESS: PLASMA ARC CUTTING	
NUMBER: PRACTICE 8-1	DRAWN BY: BOB WISNUSKI

Figure 7.31 Straight square plasma arc cutting

Figure 7.32 Flat, straight cut in thin plate
Starting at the edge of a plate, like starting an oxyfuel cut, move smoothly in a straight line toward the other end. Note the roughness left along the side of the plate from a previous cut
Source: Courtesy of Larry Jeffus

straight line and ±5° of being square. Turn off the PAC equipment and clean up your work area when you are finished cutting.

Complete a copy of the Student Welding Report in Appendix I or provided by your instructor. ■

**Module 9
Key Indicator 1**

PRACTICE 7-3

Flat Cutting Holes

Using a properly set up and adjusted PAC machine, proper safety protection, one or more pieces of mild steel, stainless steel, and aluminum 16 gauge, 1/8 in., 1/4 in., and 1/2 in. thick, you will cut 1/2-in. and 1-in. holes.

**Module 8
Unit 3
Key Indicator 4**

1. Starting with the piece of metal that is 1/8 in. (3 mm) thick, hold the torch as close as possible to a 90° angle.
2. Lower your hood and establish a plasma cutting stream.
3. Move the torch in an outward spiral until the hole is the desired size, Figure 7.33.

Repeat the hole cutting process until both sizes of holes are made using all the thicknesses of all three types of metals and you can make consistently smooth cuts that are within ±3/32 in. of being round and ±5° of being square. Turn off the PAC equipment and clean up your work area when you are finished cutting.

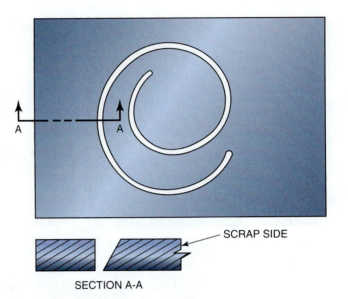

SCRAP SIDE

SECTION A-A

Figure 7.33 Flat cutting holes
When cutting a hole, make a test to see which direction to make the cut so that the beveled side is on the scrap piece

Module 9
Key Indicator 1

Complete a copy of the Student Welding Report in Appendix I or provided by your instructor. ■

PRACTICE 7-4

Beveling of a Plate

Module 8
Unit 3
Key Indicator 4, 6

Using a properly set up and adjusted PAC machine, proper safety protection, one or more pieces of mild steel, stainless steel, and aluminum 6 in. (152 mm) long and 1/4 in. and 1/2 in. thick, you will cut a 45° bevel down the length of the plate.

1. Starting at one end of the piece of metal that is 1/4 in. thick, hold the torch as close as possible to a 45° angle.
2. Lower your hood and establish a plasma cutting stream.
3. Move the torch in a straight line down the plate toward the other end, Figure 7.34.

Repeat the cut using both thicknesses of all three types of metals until you can make consistently smooth cuts that are within ±3/32 in. of a straight line and ±5° of a 45° angle. Turn off the PAC equipment and clean up your work area when you are finished cutting.

Complete a copy of the Student Welding Report in Appendix I or provided by your instructor. ■

Module 9
Key Indicator 1

PRACTICE 7-5

U-Grooving of a Plate

Module 8
Unit 3
Key Indicator 4, 6

Using a properly set up and adjusted PAC machine, proper safety protection, one or more pieces of mild steel, stainless steel, and aluminum 6 in. (152 mm) long and 1/4 in. or 1/2 in. thick, you will cut a U-groove down the length of the plate.

1. Starting at one end of the piece of metal, hold the torch as close as possible to a 30° angle, Figure 7.35.

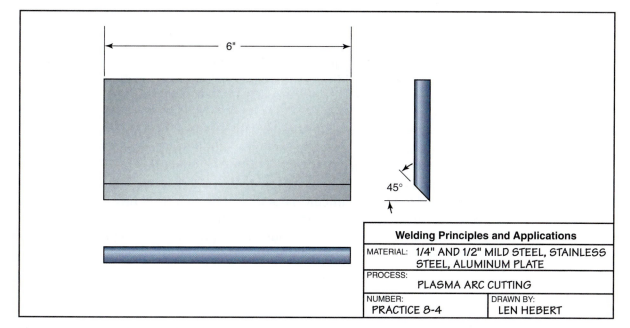

Figure 7.34 Beveled plasma arc cutting

2. Lower your hood and establish a plasma cutting stream.
3. Move the torch in a straight line down the plate toward the other end.
4. If the width of the U-groove changes, speed up or slow down the travel rate to keep the groove the same width and depth for the entire length of the plate.

Repeat the gouging of the U-groove using all three types of metals until you can make consistently smooth grooves that are within ±3/32 in. of a straight line and uniform in width and depth. Turn off the PAC equipment and clean up your work area when you are finished cutting.

Complete a copy of the Student Welding Report in Appendix I or provided by your instructor. ∎

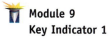
Module 9
Key Indicator 1

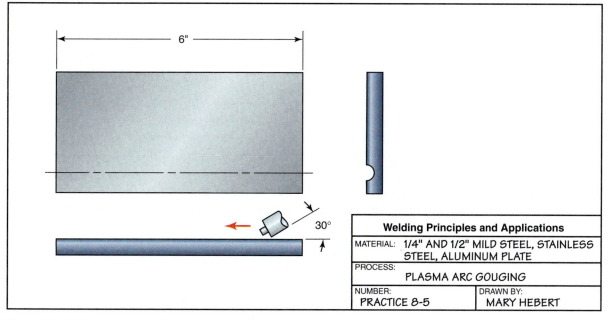

Figure 7.35 Plasma arc gouging

SUMMARY

Plasma arc cutting is quickly becoming one of the most popular cutting processes in the welding industry. Developments and changes in the equipment and torch design have extended the effective cutting life of the consumable torch parts. This has reduced the cutting cost and improved cut quality significantly. This process was once found only in large shops that cut stainless steel and aluminum; today it is being used by almost every segment of the industry, including small shops and home hobbyists.

One of the most difficult parts of a plasma cutting process to be mastered by the beginning student is the high cutting rate. Developing an eye and ear for the sights and sounds of the process as a high-quality cut is being produced will be a significant aid in your skill development.

REVIEW

1. What is electrical plasma?
2. Approximately how many times hotter is the plasma arc than an unrestricted arc?
3. Why is the body on a manual plasma torch made from a special type of plastic?
4. How are the torch heads of a plasma torch cooled?
5. What three types of power switches are used on plasma torches?
6. Why are some copper parts plated?
7. How have copper/tungsten tips helped the plasma torch?
8. What provides the gap between the electrode tip and the nozzle tip?
9. What are the advantages of a water shroud nozzle for plasma cutting?
10. How is water used to control power cable overheating?
11. What factors must be considered when selecting a material for plasma arc gas hoses?
12. Why must the cooling water be turned off when the plasma cutting torch is not being used?
13. What type of voltage is required for a plasma cutting torch?
14. A plasma arc torch that is operating with 90 volts (close circuit) and 25 amps is using how many watts of power?
15. Why do plasma cuts have little or no distortion?
16. What has limited the upper cutting speed of the plasma arc cutting process?
17. What metals can be easily cut using the PAC process?
18. What happens to torches without drag nozzle tips if the standoff distance is not maintained?
19. What are the two methods for establishing the plasma path to the metal being cut?
20. List eight things that will affect the quality of a PAC kerf.
21. List seven things that are affected by the choice of PAC gas or gases for a cut.

22. Describe stack cutting.
23. How does PAC dross compare with OFC slag?
24. Why must high-power PAC machines be used by a semiautomatic or automatic cutting system?
25. What is the advantage of using a water table for PAC?
26. How is plasma arc gouging performed?

CHAPTER

8

Related Cutting Processes

Key Indicator 3: Sets up for manual CAC-A scarfing and gouging operations, to remove base and weld metal on carbon steel

Key Indicator 4: Operates manual CAC-A equipment on carbon steel

Key Indicator 5: Performs manual CAC-A scarfing and gouging operations, to remove base and weld metal, in the 1G and 2G positions, on carbon steel

Module 9: Welding Inspection and Testing Principles

Key Indicator 1: Examines cut surfaces and edges of prepared base metal parts

INTRODUCTION

The number of specialized cutting processes being developed and improved increases every year. Many specialized cutting processes have been perfected and are currently in use. These new cutting methods are being used by a much wider group than just the welding industry. These processes can be used to cut a wide variety of materials, such as glass, plastic, printed circuit boards, cloth, and fiberglass insulation—and more materials are being added to this list almost daily. Material cutting and even hole drilling using a welding-related process have become common in the workplace.

Only a few of the more common cutting processes are covered in this chapter. These are the ones that a welder might be required to be able to perform or to have a working knowledge of.

LASERS

A laser produces a form of **monochromatic light** (light of a single-color wavelength) that travels in parallel waves, Figure 8.1. This form of light is generated when particular materials are excited either by intense light or with an electric current. The atoms or molecules of the lasing material

WHITE LIGHT

LASER LIGHT

Figure 8.1 White light and laser light
White light is made up of different frequencies (colors) of light waves. Laser light is made up of single-frequency light waves traveling parallel to each other

release their energy in the form of light. The laser light is produced as the atom or molecule falls from a high energy state to a lower energy state, Figure 8.2. The laser light is then bounced back and forth between one fully reflective and one partially reflective surface at the ends of the lasing material. As the light is reflected, it begins to form a **synchronized wave form**. The light that passes through the partially reflective mirror is the laser beam, Figure 8.3.

Once the laser beam emerges from the laser rod or chamber, it can be manipulated in the same way as any other type of light. It is possible to focus, reflect, absorb, or diffuse the light. Because the light waves travel in such a uniform manner, they exhibit special characteristics. The light beam tends to remain in a very tight column without spreading out, unlike ordinary light, Figure 8.4. This characteristic allows the laser beam to travel some distance without being significantly affected by the distance or the air it is traveling through.

Because the beam is not adversely affected by its travel, it can be used to carry information or transmit energy. The military uses lasers to aim weapons or identify targets for guided missiles or bombs.

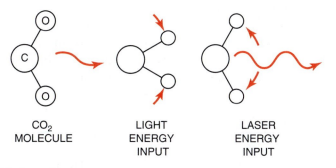

CO$_2$
MOLECULE

LIGHT
ENERGY
INPUT

LASER
ENERGY
INPUT

Figure 8.2 Priming a laser
Energy is stored in a carbon dioxide (CO$_2$) molecule until the molecule cannot hold any more. The energy is released suddenly, as when a balloon pops. This quick release results in a burst of light energy

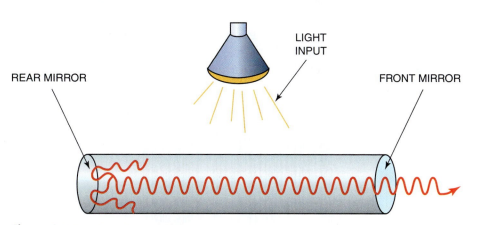

LIGHT
INPUT

REAR MIRROR

FRONT MIRROR

Figure 8.3 Creating the laser beam
Light energy reflects back and forth between the end mirrors until it forms a parallel laser light beam

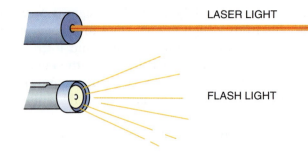

Figure 8.4 Laser light
Laser light stays in a very tight column unlike most other lights

Laser Equipment

Most lasers are in the power range 400 watts to 1500 watts. Some large machines have as much as 25 kW of power. Although the power of most lasers is relatively low compared with other welding processes, it is the laser's ability to concentrate the power into a small area that makes it work so well. The power density of a cutting laser can be equal to 65 million watts per square inch.

Laser equipment is larger than most other welding or cutting power supplies. A typical unit requires about as much floor space as a large desk, about 10 ft^2 (3 m^2).

Recent technological advances, increased competition, and a rapidly growing market have helped lower the high cost of the equipment. But even with the current cost of equipment, laser cutting and drilling have one of the lowest average hourly operating costs. The high reliability and long life of the equipment have also helped reduce operating costs.

Applications

Lasers have grown from the first bright-red spot generated by the ruby rod to a multibillion-dollar industry. We see lasers in use every day. They help speed our checking out at stores when they are used to read the uniform product code (UPC) on our purchases, Figure 8.5. Lasers are used by surveyors to measure distances and to verify that a building under construction is level. Doctors use lasers to make precise cuts during the most delicate operations. Lasers are even used to entertain us at shows and concerts.

Manufacturers use lasers for everything from marking products to joining parts, communicating between machines, guiding, punching, or cutting a variety of materials. The industrial laser can guide a machine

Figure 8.5 Laser applications
Bar codes are read by lasers to input information into computers

cut accurately or measure distances from a fraction of an inch to thousands of miles.

A laser can be used to drill holes through the hardest materials, such as synthetic diamonds, tungsten carbide, quartz, glass, or ceramics. Lasers are used to weld materials that are too thin or too hard to be welded with other heat sources. They can be used on materials that must be cut without overheating delicate parts located just a few thousandths of an inch away.

The benefits and capacities of the laser have made laser technology one of the most rapidly growing areas in manufacturing.

Laser light beams can be focused onto a very compact area. The photo energy of the light is converted into heat energy as it strikes the surface of a material. The highly concentrated energy can cause the instantaneous melting or vaporization of the target material. The equipment used for **laser beam welding (LBW)**, **laser beam cutting (LBC)**, and **laser beam drilling (LBD)** is similar in design and operation. Most laser welding and cutting operations use a gas laser, and most drilling operations use a solid state laser.

The ability of the surface of the material to either absorb or reflect the laser greatly affects the operation's efficiency. Some materials reflect more of the laser's light than others. The absorption rate for any material increases as the laser beam heats the surface. Once this surface-temperature threshold is reached, the process continues at a much higher level of efficiency.

Laser Types

Early lasers used a synthetic ruby rod to produce the laser light. Today, a large number of materials can be used, including such common items as glass and such exotic items as neodymium-doped yttrium aluminum garnet (Nd-YAG), often referred to as a **YAG laser**, Figure 8.6.

Lasers can be divided into two major types: those that use a solid material and those that use a gas. Each of these two types is divided into two groups on the basis of mode of operation: continuous or pulsed.

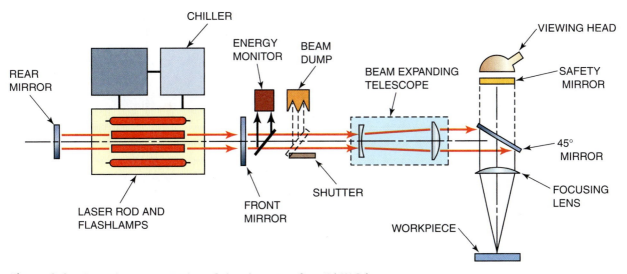

Figure 8.6 Schematic representation of the elements of an Nd-YAG laser
Source: Courtesy of the American Welding Society

Solid State Lasers

The first lasers were all **solid state lasers**. They used a ruby rod to produce the laser beam. Today, the most popular solid laser is the Nd-YAG. When this synthetic crystal is exposed to the intense light from flash tubes, it can produce high quantities of laser energy. High-powered solid state lasers have the disadvantage that the internal temperatures of the laser rod increase with operation time. These lasers are most often used for low-power continuous or high-power pulse operation. The solid state laser is capable of generating the highest-power laser pulses.

Gas Laser

Gas laser uses one or more gases to produce the laser beam. Popular gas lasers use nitrogen, helium, carbon dioxide (CO_2), or mixtures of these three gases. Gas lasers have either a gas-charged cylinder where the gas is static or a chamber that circulates the laser gas. The highest continuous power output is achieved using gas lasers with a blower to circulate the gas through a heat exchanger; such lasers can be rapidly pulsed to improve their cutting capacity, Figure 8.7.

Laser Beam Cutting

Both laser welding and laser cutting bring the material to a molten state. In the welding process the material is allowed to flow together and cool to form the weld metal. In the cutting process a jet of gas is directed into the molten material to expel it through the bottom of the cut. Although lasers are used primarily to cut very thin materials, they can be used to cut up to 1 in. of carbon steel.

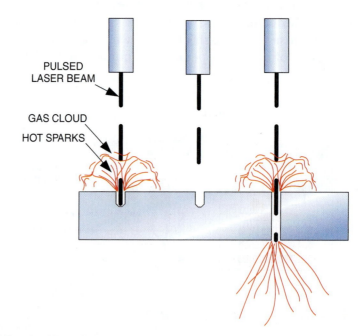

Figure 8.7 Pulsed laser beams
The gas cloud and hot sparks caused by the cutting laser beam dissipate between the pulses of the laser beam, which results in a better cut

The **cutting gas assist** can be either a nonreactive gas or an exothermic gas. Table 8.1 lists various gases and the materials they are used to cut. Nonreactive gases do not add any heat to the cutting process; they simply remove the molten material by blowing it out of the kerf. **Exothermic gases** react with the material being cut, as with an oxyfuel cutting torch. The additional heat produced as the exothermic cutting gas reacts with the metal being cut helps blow the molten material out of the kerf.

Among the advantages of laser cutting are

- a narrow heat-affected zone—Little or no heating of the surrounding material is observed, so it is possible to make very close parallel cuts without damaging the strip that is cut out, Figure 8.8.
- no electrical conductivity requirement—The part being cut does not have to be electrically conductive, so materials such as glass, quartz, and plastic can be cut. There is also no chance that a stray electrical charge might damage delicate computer chips while they are being cut using a laser.
- no contact with the part being cut except the laser beam—Small parts that have finished surfaces or small surface details can be cut without the danger of disrupting or damaging the surface. It is also not necessary to hold the parts securely, as it is when a cutting tool is used.

Table 8.1 Various Cutting Assist Gases and the Materials They Cut

Assist Gases	Material
Air	Aluminum
	Plastic
	Wood
	Composites Alumina
	Glass
	Quartz
Oxygen	Carbon steel
	Stainless steel
	Copper
Nitrogen	Stainless steel
	Aluminum
	Nickel alloys
Argon	Titanium

Figure 8.8 Detailed small part not much larger than a dime made with a laser
Source: Courtesy of Larry Jeffus

Figure 8.9 Nesting small parts
Because of the narrow kerf, small, detailed parts can be nested one inside the other
Source: Courtesy of Larry Jeffus

- a narrow kerf—Parts can be nested close together, which reduces waste of expensive materials, Figure 8.9.
- automation and robotics—The laser beam can easily be directed through an articulated guide to an automated machine or a robot.
- top edge finish—The top edge will be smooth and square without being rounded.

Laser Beam Drilling

The pulsed laser is the best choice for drilling operations. A short burst of high laser energy is concentrated on a small spot. The intense heat vaporizes the small spot with enough force to thrust the material out, leaving a small crater. Repeated blasting with the pulsed laser results in the crater becoming deeper, until the hole is drilled through the part or to a desired depth, Figure 8.10.

Holes with diameter as small as 0.0001 in. (0.0025 mm) can be drilled. The limitation on the hole's depth is the laser's focal length. Most holes are less than 1 in. deep.

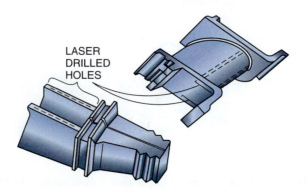

LASER
DRILLED
HOLES

Figure 8.10 Jet engine blades and a rotor component showing laser drilled holes
Source: Courtesy of the American Welding Society

AIR CARBON ARC CUTTING

The **air carbon arc cutting (CAC-A)** process was developed in the early 1940s and was originally named air arc cutting (AAC). Air carbon arc cutting was an improvement of the carbon arc process, which was used in the vertical and overhead positions and removed metal by melting a large enough spot that gravity would cause it to drip off the base plate, Figure 8.11. This process was slow and could not be accurately controlled. It was found that the molten metal could be blown away using a stream of air. This greatly improved the speed, quality, and control of the process.

In the late 1940s the first air carbon arc cutting torch was developed. Before this development, the process required two welders, one to control the carbon arc and the other to guide the air stream. The new torch housed the carbon electrode holder and the air stream in the same unit. This basic design is still in use today, Figure 8.12.

Unlike the oxyfuel process, the air carbon arc cutting process does not require that the base metal be reactive with the cutting stream. Oxyfuel cutting can be performed only on metals that can be rapidly oxidized by the cutting stream of oxygen. The air stream in the CAC-A process blows the molten metal away. This greatly increases the list of metals that can be cut, Table 8.2.

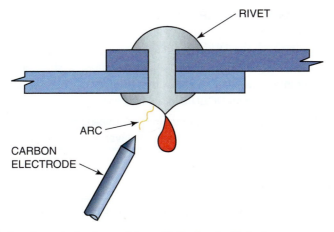

Figure 8.11 A carbon electrode used to melt the head off rivets

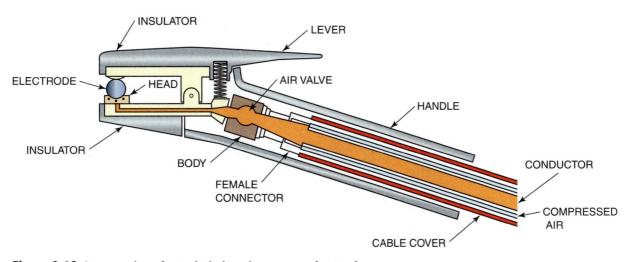

Figure 8.12 Cross section of a typical air carbon arc gouging torch
Source: Courtesy of the American Welding Society

Table 8.2 Recommended Procedures for Air Carbon Arc Cutting of Various Metals

Base Metals	Recommendations
Carbon steel and low alloy steel	Use DC electrodes with DCEP current. AC can be used but with a 50% loss in efficiency.
Stainless steel	Same as for carbon steel.
Cast iron, including malleable and ductile iron	Use of 1/2 inch or larger electrodes at the highest rated amperage is necessary. There are also special techniques that need to be used when gouging these metals. The push angle should be at least 70° and depth of cut should not exceed 1/2 inch per pass.
Copper alloys (copper content 60% and under)	Use DC electrodes with DCEN (electrode negative) at maximum amperage rating of the electrode.
Copper alloys (copper content over 60%, or size of workpiece is large)	Use DC electrodes with DCEN at maximum amperage rating of the electrode or use AC electrodes with AC.
Aluminum bronze and aluminum nickel bronze (special naval propeller alloy)	Use DC electrodes with DCEN.
Nickel alloys (nickel content is over 80%)	Use AC electrodes with AC.
Nickel alloys (nickel content less than 80%)	Use DC electrodes with DCEP.
Magnesium alloys	Use DC electrodes with DCEP. Before welding, surface of groove should be wire brushed.
Aluminum	Use DC electrodes with DCEP. Wire brushing with stainless wire brushes is mandatory prior to welding. Electrode extension (length of electrode between electrode torch and workpiece) should not exceed 3 inches for good-quality work. DC electrodes with DCEN can also be used.
Titanium, zirconium, hafnium, and their alloys	Should not be cut or gouged in preparation for welding or remelting without subsequent mechanical removal of surface layer from cut surface.

Module 8
Unit 1
Key Indicator 3

Manual Torch Design

The air carbon arc cutting torch is designed differently than the shielded metal arc electrode holder:

- The lower electrode jaw has a series of air holes.
- The jaw has only one electrode-locating groove.
- The electrode jaw can pivot.
- There is an air valve on the torch lead.

Having only one electrode-locating groove in the jaw and pivoting the jaw means that the air stream will always be aimed correctly. The air must be aimed just under and behind the electrode and always in the same direction, Figure 8.13. This ensures that the air stream is directed at the spot where the electrode arcs to base the metal.

Torches are available in a number of amperages. The larger torches have greater capacity but are less flexible to use on small parts.

The torch can be permanently attached to a welding cable and air hose or it can be attached to the welding power supply by gripping a tab at the end of the cable with the shielded metal arc electrode holder, Figure 8.14. The temporary attachment is easier if the air hose is

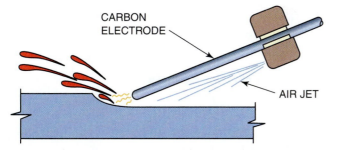

Figure 8.13 Air carbon arc gouging

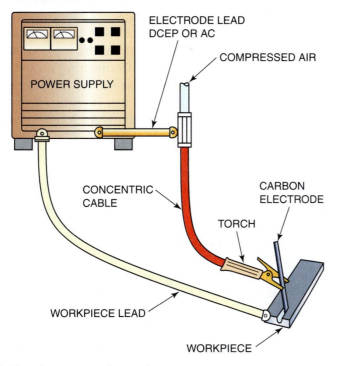

Figure 8.14 Air carbon arc gouging equipment setup
Source: Courtesy of the American Welding Society

equipped with a quick disconnect, which will allow it to be used for other air tools such as grinders or chippers. Greater flexibility for a work station can be achieved with this arrangement.

Electrodes

Air carbon arc cutting electrodes are copper-coated or plain (without a coating). The copper coating helps reduce **carbon electrode** overheating by allowing the electrode to carry higher currents and improving heat dissipation. The copper coating provides increased strength, which reduces accidental breakage.

Electrodes can be round, flat, or semiround, Figure 8.15. Round electrodes are used for most gouging operations, and flat electrodes are most often used to scarf off a surface. Round electrodes are available in sizes from 1/8 in. (3.17 mm) to 1 in. (25 mm) in diameter. Flat electrodes are available in 3/8-in. and 5/8-in. sizes.

ROUND   FLAT   SEMI-ROUND

Figure 8.15 Cross sections of carbon electrodes

Electrodes are available for both direct-current electrode positive and alternating current. DCEP electrodes, made of carbon in the form of **graphite**, are the most commonly used. AC electrodes are less common; they have some elements added to the carbon to stabilize the arc, which is required for the AC power.

To reduce waste, electrodes are made so they can be joined together. The joint consists of a female tapered socket at the top end and a matching tang on the bottom end, Figure 8.16. Connecting a new electrode to the remaining setup allows the stub to be consumed with little loss of electrode stock. This connecting of electrodes is required for most track-type air carbon arc cutting operations to allow for longer cuts.

Power Sources

Most shielded metal arc welding power supplies can be used for air carbon arc cutting. The operating voltage for air carbon arc cutting needs to be 28 volts or higher. This voltage is slightly higher than is required for most SMA welding, but most welding power supplies will meet this requirement. Check the owner's manual to see whether your welder is approved for air carbon arc cutting. If the voltage is lower than the minimum, the arc will tend to sputter out, and it will be hard to make clean cuts.

Because most carbon arc cutting requires a high amperage setting, it may be necessary to stop some cuts so that the duty cycle of the welder is not exceeded. On large industrial welders this is not normally a problem.

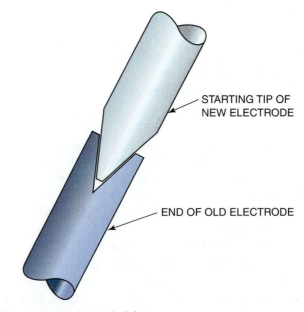

STARTING TIP OF
NEW ELECTRODE

END OF OLD ELECTRODE

Figure 8.16 Air carbon arc electrode joint

Air Supply

Air supplied to the torch should be between 80 psi and 100 psi (5.6 and 7 kg/cm). The minimum pressure is around 40 psi. The correct air pressure will result in cuts that are clean, smooth, and uniform. The air flow rate is also important. If the air line is too small or the compressor does not have the required capacity, there will be a loss in air pressure at the torch tip. This line loss will result in a lower than required flow at the tip. The resulting cut will be of lower quality.

Application

Air carbon arc cutting can be used on a variety of materials. It is a relatively low-cost way of cutting most metals, especially stainless steel, aluminum, nickel alloys, and copper. Air carbon arc cutting is most often used for repair work. Few cutting processes can match the speed, quality, and cost savings of this process for repair or rework. In repair or rework, the most difficult part is removing the old weld or cutting a groove so a new weld can be made. The air carbon arc can easily remove the worst welds, even if they contain slag inclusions or other defects. For repairs the arc can cut through thin layers of paint, oil, or rust and make a groove that needs little, if any, cleanup.

The highly localized heat results in only slight heating of the surrounding metal. As a result, usually there is no need to preheat hardenable metals to prevent hardness zones. Cast iron can be carbon arc gouged to prepare a crack for welding without causing further damage to the part by applying excessive heat.

Air carbon arc cutting can be used to remove a weld from a part. The removal of welds can be accomplished with such success that often the part needs no postcut cleanup. The root of a weld can be back-gouged so that a backing weld can be made to ensure 100% weld penetration, Figure 8.17.

> **CAUTION**
>
> Never cut on any material that might produce fumes that would be hazardous to your health without proper safety precautions, including adequate ventilation and/or a respirator.

Module 8
Unit 4
Key Indicator 1

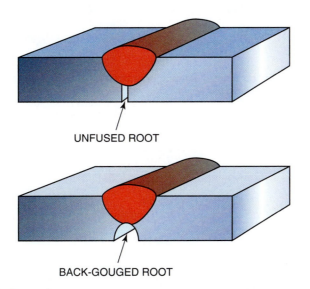

UNFUSED ROOT

BACK-GOUGED ROOT

Figure 8.17 Back-gouging
Back-gouging the root of a weld made in thick metal can ensure that a weld with 100% joint fusion can be made

The electrode should extend approximately 6 in. from the torch when starting a cut; as the cut progresses, the electrode is consumed. Stop the cut and readjust the electrode when its end is approximately 3 in. from the electrode holder. This will reduce the damage to the torch caused by the intense heat of the operation.

Gouging

Gouging is the most common application of air carbon arc cutting. Arc gouging is the removal of a quantity of metal to form a groove or bevel, Figure 8.18. The groove produced along an edge of a plate is usually a J-groove. The groove produced along a joint between plates is usually a U-groove, Figure 8.19. Both types of grooves are used to ensure that the weld applied to the joint has the required penetration into the metal.

Washing

Washing is the process sometimes used to remove large areas of metal so that hard surfacing can be applied, Figure 8.20. Washing can be used to remove large areas that contain defects, to reduce the transitional stresses of unequal-thickness plates, or to allow space for the capping of a surface with a wear-resistant material.

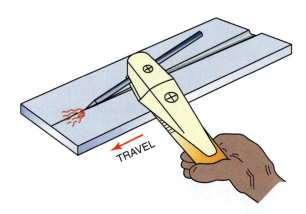

Figure 8.18 Manual air carbon arc gouging operation in the flat position
Source: Courtesy of the American Welding Society

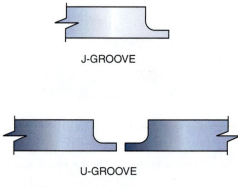

J-GROOVE

U-GROOVE

Figure 8.19 Air carbon arc gouging groove shapes

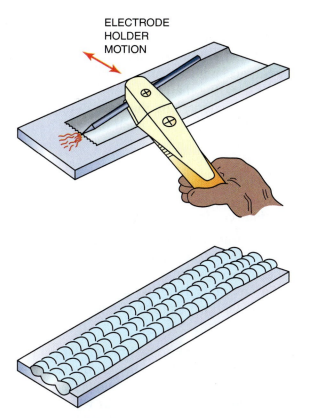

ELECTRODE
HOLDER
MOTION

Figure 8.20 Hardsurfacing weld applied in the carbon arc cut groove
Source: Courtesy of the American Welding Society

Safety

The air carbon arc requires several special precautions in addition to the safety requirements for the shielded metal arc:

Module 8
Unit 4
Key Indicator 1

- sparks—The quantity and volume of sparks and molten metal spatter generated during this process are a major safety hazard. Extra precautions must be taken to ensure that other workers, equipment, materials, or property in the area will not be affected by the spark stream.
- noise—The sound level is high enough to cause hearing damage if proper ear protection is not used.
- light—The arc light produced is the same as that produced by the shielded metal arc welding process, but because the arc has no smoke to diffuse the light and the amperages are usually much higher, the chances of receiving arc burns are much higher. Additional protection should be worn, such as thicker clothing, a leather jacket, and leather aprons.
- eyes—Because of the intense arc light, a darker welding filter lens should be used for the helmet.
- fumes—The combination of the air and the metal being removed results in a high volume of fumes. Special consideration must be given to the removal of these fumes from the work area. Before installing a ventilation system, check with local, state, and federal laws. Some of the fumes may have to be filtered before they can be released into the air.

- surface contamination—Often this process is used to prepare damaged parts for repair. If the used parts have paint, oils, or other contamination that might generate hazardous fumes, they must be removed in an acceptable manner before any cutting begins.
- equipment—Check the manufacturer's owner's manual for specific safety information concerning the power supply and the torch before you start any work with each piece of equipment for the first time.

PRACTICE 8-1

Air Carbon Arc Straight Cut in the Flat Position

Module 8
Unit 1
Key Indicator 3, 4, 5

Using an air carbon arc cutting torch and welding power supply that have been safely set up in accordance with the manufacturer's specific instructions in the owner's manual and wearing safety glasses, welding helmet, gloves, and any other required personal protection clothing, you will make a 6-in.-long (15 cm) straight U-groove gouge in a carbon steel plate.

1. Adjust the air pressure to approximately 80 psi (5.6 kg/cm).
2. Set the amperage within the range for the diameter electrode you are using by referring to the box the electrodes came in.
3. Check to see that the stream of sparks will not start a fire or cause injury or damage to anyone or anything in the area.
4. Make sure the area is safe, and turn on the welder.
5. Wearing a good dry leather glove to avoid electrical shock, insert the electrode into the torch jaws so that about 6 in. (15 cm) is extending outward. Be sure not to touch the electrode to any metal parts, because it may short out.
6. Turn on the air at the torch head.
7. Lower your arc welding helmet.
8. Slowly bring the electrode down at about a 30° angle so that it will make contact with the plate near the starting edge, Figure 8.21. Be prepared for a loud, sharp sound when the arc starts.
9. Once the arc is struck, move the electrode in a straight line down the plate toward the other end. Keep the speed and angle of the torch constant.
10. When you reach the other end, lift the torch so the arc will stop.
11. Raise your helmet and stop the air.
12. Remove the remaining electrode from the torch so it will not accidentally touch anything.

When the metal is cool, chip or brush any slag or dross off of the plate. This material should be easy to remove. The groove must be within ±1/8 in. (3.1 mm) of being straight and within ±3/32 in. (2.3 mm) of uniformity in width and depth. Repeat this practice until you can make the cut within these tolerances. Turn off the CAC equipment and clean up your work area when you are finished cutting.

Module 9
Key Indicator 1

Complete a copy of the Student Welding Report in Appendix I or provided by your instructor. ∎

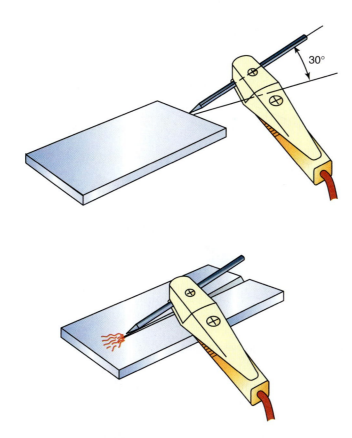

Figure 8.21 Air carbon arc U-groove gouging

PRACTICE 8-2

Air Carbon Arc Edge Cut in the Flat Position

Using the same equipment, adjustments, setup, and materials as described in Practice 8-1, you will make a J-groove along the edge of the plate.

1. Adjust the air pressure to approximately 80 psi (5.6 kg/cm).
2. Set the amperage within the range for the diameter electrode you are using.
3. Check to see that the stream of sparks will not start a fire or cause injury or damage to anyone or anything in the area.
4. Make sure the area is safe, and turn on the welder.
5. Wearing a good dry leather glove to avoid electrical shock, insert the electrode into the torch jaws so that about 6 in. (15 cm) is extending outward. Be sure not to touch the electrode to any metal parts, because it may short out.
6. Turn on the air at the torch head.
7. Lower your arc welding helmet.
8. Slowly bring the electrode down at about a 30° angle so that it will make contact with the plate near the starting edge, Figure 8.22.
9. Once the arc is struck, move the electrode in a straight line down the edge of the plate toward the other end. Keep the speed and angle of the torch constant.
10. When you reach the other end, lift the torch so the arc will stop.
11. Raise your helmet and stop the air.

Module 8
Unit 4
Key Indicator 3, 4, 5

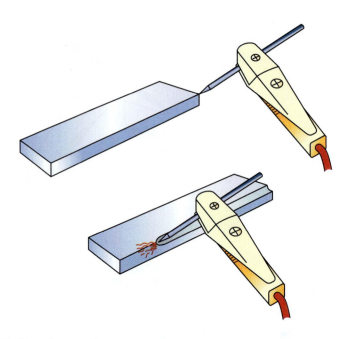

Figure 8.22 Air carbon arc J-groove gouging

12. Remove the remaining electrode from the torch so it will not accidentally touch anything.

When the metal is cool, chip or brush any slag or dross off of the plate. This material should be easy to remove. The groove must be within ±1/8 in. (3.1 mm) of being straight and within ±3/32 in. (2.3 mm) of uniformity in width and depth. Repeat this practice until you can make the cut within these tolerances. Turn off the CAC equipment and clean up your work area when you are finished cutting.

Module 9
Key Indicator 1

Complete a copy of the Student Welding Report in Appendix I or provided by your instructor. ■

PRACTICE 8-3

Module 8
Unit 4
Key Indicator 3, 4, 5

Air Carbon Arc Back-Gouging in the Flat Position

Using the same equipment, adjustments, setup, and materials as described in Practice 8-1, you will make a U-groove along the root face of a weld joint on a plate, Figure 8.23.

Follow the same starting procedure as you did in Practice 8-1.

1. Start the arc at the joint between the two plates.
2. Once the arc is struck, move the electrode in a straight line down the edge of the plate toward the other end. Watch the bottom of the cut to see that it is deep enough. If there is a line along the bottom of the groove, it needs to be deeper. Once the groove depth is determined, keep the speed and angle of the torch constant.
3. When you reach the other end, break the arc off.
4. Raise your helmet and stop the air.
5. Remove the remaining electrode from the torch so it will not accidentally touch anything.

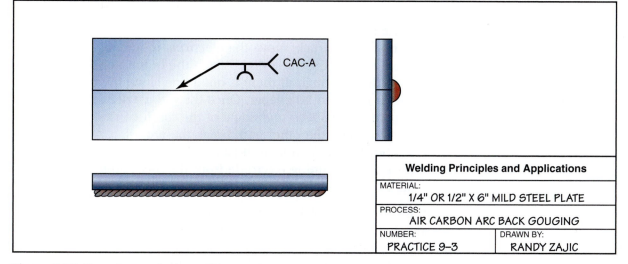

Figure 8.23 Air carbon arc back-gouging

When the metal is cool, chip or brush any slag or dross off of the plate. This material should be easy to remove. The groove must be within ±1/8 in. of being straight, but it may vary in depth so that all the unfused root of the weld has been removed. Repeat this practice until you can make the cut within these tolerances. Turn off the CAC equipment and clean up your work area when you are finished cutting.

Complete a copy of the Student Welding Report in Appendix I or provided by your instructor. ∎

Module 9
Key Indicator 1

PRACTICE 8-4

Air Carbon Arc Weld Removal in the Flat Position

Module 8
Unit 4
Key Indicator 3, 4, 5

Using the same equipment, adjustments, setup, and materials as described in Practice 8-1, you will make a U-groove to remove a weld from a plate, Figure 8.24.

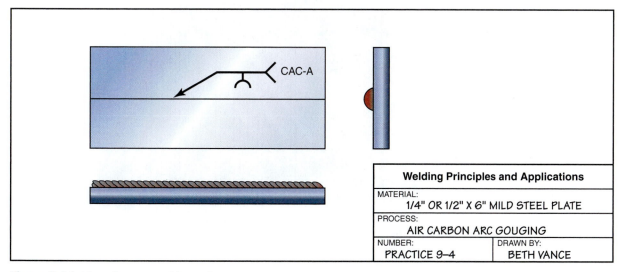

Figure 8.24 Air carbon arc weld gouging

Follow the same starting procedure as you did in Practice 8-1.

1. Start the arc on the weld.
2. Once the arc is struck, move the electrode in a straight line down the weld toward the other end. Watch the bottom of the cut to see that it is deep enough. If there is not a line along the bottom of the groove, it needs to be deeper. Once the groove depth is determined, keep the speed and angle of the torch constant.
3. When you reach the other end, break the arc off.
4. Raise your helmet and stop the air.
5. Remove the remaining electrode from the torch so it will not accidentally touch anything.

When the metal is cool, chip or brush any slag or dross off of the plate. This material should be easy to remove. The groove must be within ±1/8 in. (3.1 mm) of being straight, but it may vary in depth so that all the weld metal has been removed. Repeat this practice until you can make the cut within these tolerances. Turn off the CAC equipment and clean up your work area when you are finished cutting.

Module 9
Key Indicator 1

Complete a copy of the Student Welding Report in Appendix I or provided by your instructor. ∎

OXYGEN LANCE CUTTING

The **oxygen lance cutting** process uses a consumable alloy tube, Figure 8.25. The tip of the tube is heated to its kindling temperature. A high-pressure oxygen flow is started through the lance. The oxygen reacts with the hot lance tip, releasing sufficient heat to sustain the reaction, Figure 8.26.

Figure 8.25 Oxygen lance rods
Oxygen lance rods may be rolled out of flat strips of special alloys to form a tube
Source: Courtesy of Arcair®, a Thermodyne® Company

Figure 8.26 Oxygen lance cutting
Arcair® SLICE® portable cutting system that uses sparks created between the striker and tube to ignite the oxygen lance rod for cutting
Source: Courtesy of Arcair®, a Thermodyne® Company

The rod tip is heated to a red-hot temperature using an oxyfuel torch or electric resistance. Once the oxygen stream is started, it reacts with the lance material, which results in the creation of both a high-temperature and heat-releasing reaction.

The intense reaction allows the lance to cut through a variety of materials. The hot metal leaving the lance tip has not completed its exothermic reaction. As this reactive mass hits the surface of the material being cut, it releases a large quantity of energy into that surface. Thermal conductivity between the molten metal and the base material is a very efficient method of heat transfer. This transfer, along with the continued burning of the lance material on the surface, causes the base material to become molten.

Once the base material is molten, it may react with the burning lance material, forming fumes or slag, which is then blown from the cut. Any molten material that does not become reactive is carried out of the cut with the slag or blown out by the oxygen stream.

The addition of steel rods or other metals to the center of oxygen lance tubes has increased their productivity. The improved lances last longer and cut faster.

Application

The oxygen lance's unique method of cutting allows it to be used to cut material not normally cut using a thermal process. Films have portrayed the oxygen lance as a tool used by thieves to cut into safes. In reality, this would result in the valuables in the safe being destroyed.

Oxygen lances can be used to cut reinforced concrete in the demolition of buildings. Oxygen lance cutting allows the quick removal of thick sections of the building without the dangerous vibration caused by most conventional methods of demolition. This has been a life-saving factor in rescue work following earthquakes; for example, the oxygen lance saved thousands of hours and countless lives in Mexico City after the devastation from the city's worst earthquake in 1985. Local and national news agencies showed oxygen lances being used to cut large sections of concrete that fell from buildings into manageable pieces.

The oxygen lance can also be used to cut thick sections of cast iron, aluminum, and steel. Often in the production of these metals thick sections must be cut. Occasionally equipment failure will stop metal production. If the metal in production is allowed to cool, it may need to be cut into sections so that it can be removed from the machine. The oxygen lance cutting process is effective in this type of work.

Safety

It is important to follow all safety procedures when using this process. Manufacturers list specific safety precautions for the oxygen lances they produce. Read and follow those instructions carefully. The major safety concerns are

- fumes—The large volume of fumes generated is often a health hazard. An approved ventilation system must be provided if this work is to be done in a building or any other enclosed area.
- heat—This operation produces both high levels of radiant heat and plumes of molten sparks and slag. The operators must wear special heat-resistant clothing.
- noise—The sound produced is well above safety levels. Ear protection must be worn by anyone in the area.

WATER JET CUTTING

Water jet cutting is not a thermal cutting process. It does not put any heat into the material being cut. The cut is made by the rapid erosion of the material by a high-pressure jet of water, Figure 8.27. An abrasive powder may be added to the stream of water. Abrasives are added when hard materials such as metals are being cut.

The lack of heat input to the material being cut makes this process unique. Materials that heat might distort, make harder, or cause to delaminate are ideally suited to this process. The lack of heat distortion allows thin material to be cut with the edge quality of a laser cut and as distortion-free as a shear cut. Delamination is not a problem when cutting composite or laminated materials, such as carbon fibers, resins, or computer circuit boards, Figure 8.28.

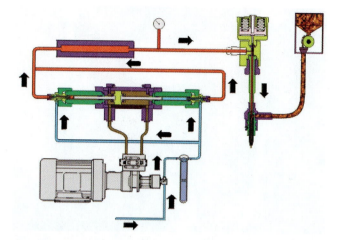

Figure 8.27 Schematic illustrating the elements of the water jet cutting system
Source: Courtesy of Ingersoll-Rand

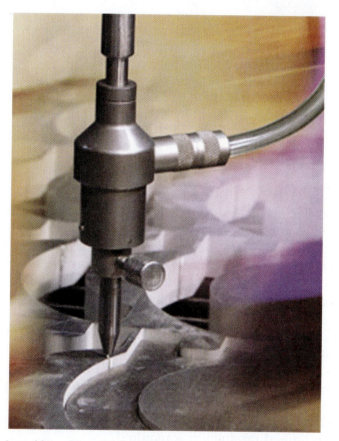

Figure 8.28 A machine with multiple cutting jets cutting out printed circuit boards
Water jet cutting is very versatile
Source: Courtesy of Ingersoll-Rand

Applications

The kerf width tends not to change unless too high a travel speed is being used. This results in a square, smooth finish on the cut surface. Postcut cleanup of the parts is totally eliminated for most materials, and only slight work is needed on a few others, Figure 8.29. The quality of the cut surface

Figure 8.29 Water jet cutting
Notice how narrow and clean this cut is
Source: Courtesy of Ingersoll-Rand

can be controlled so that even parts for the aerospace industry can often be assembled as cut.

The addition of an abrasive powder can speed up the cutting, allow harder materials to be cut, and improve the surface finish of a cut. The powder most often used is garnet. It is also commonly used as an abrasive on sandpaper. If an abrasive is used, the small water jet orifice will wear out faster.

Materials that often gum up a cutting blade, such as Plexiglas, ABS piping plastic, and rubber, can be cut easily. There is nothing for the material to adhere to that would disrupt the cut. The lack of heat also reduces the tendency of the material cut surface to become galled.

Most water jet cutting is performed by an automated or robotic system. A few band saw–type, hand-fed cutting machines are used for single cuts or when limited production is required.

SUMMARY

The welding field's ability to provide a wide variety of cutting processes has increased productivity in a variety of industries. One example of this diversity is the use of a laser beam to cut apart computer chips in the electronics field. Without such cutting processes, industry would have to rely on a much slower and more expensive mechanical or abrasive cutting process. No mechanical or abrasive cutting process can compete with the speed of these new cutting processes, and none can compete with their versatility.

Somewhat less attractive but nonetheless efficient processes such as carbon arc gouging and oxygen lance cutting fulfill specific industrial applications. Few cutting processes can rival the metal-removing capacity of the air carbon arc or oxygen lance processes. In addition, the low cost of the equipment and the flexibilities in application of these processes have lent themselves very successfully to the salvage and scrap industry. Air carbon arc gouging is extensively used for weld removal and repair work. A skilled technician can produce a groove that requires little or no postcut cleanup prior to rewelding.

REVIEW

1. Describe the use of the following processes: LBD, LBW, and LBC.
2. How is laser light formed?
3. Other than the lasing material, what is the difference between solid state lasers and gas lasers?
4. What effect does a material's surface have on the laser beam?
5. Using Table 8.1, list cutting assist gases and the material each can be used to cut.
6. What are reactive laser assist gases called? How do they work?
7. What type of laser beam is used for drilling? Why?
8. What is CAC-A?
9. Using Table 8.2, give the recommended procedure for air carbon arc gouging of carbon steel, magnesium alloys, and low-alloy copper.
10. Why are some carbon arc electrodes copper-coated?
11. What may happen if an SMA welding machine has below-minimum arc voltage for air carbon arc gouging?
12. How can carbon arc welding be used to ensure 100% weld penetration?
13. What is washing, and how can it be used?
14. How are oxygen lance cuts usually started?
15. What unusual material can be cut with an oxygen lance?
16. What are the advantages of abrasive powder in the water jet cutting stream?

CHAPTER

9

Testing and Inspection of Welds

INTRODUCTION

It is important to know that a weld will be fit for purpose and meet the requirements of codes or standards. It is also necessary to ensure the quality, reliability, and strength of a weldment. To this end, an active inspection program is required. The extent to which a welder and product are subjected to testing and inspection depends upon the intended use of the product. Items that are to be used in light, routine service, such as ornamental iron, fence posts, and gates, are not inspected as closely as products in critical use. Items in critical use include nuclear reactor containment vessels, oil refinery high-pressure vessels, aircraft airframes, and bridges. A weld that will be passed or acceptable for one welding application may not meet the requirements of another.

QUALITY CONTROL (QC)

Once a code or standard has been selected, a method is chosen for ensuring that the product meets the specifications. The two classifications of product **quality control (QC)** methods are destructive, or mechanical, testing and nondestructive testing. These methods can be used individually or in combination. **Mechanical testing (DT)** methods, except for hydrostatic testing, result in the product being destroyed. **Nondestructive testing (NDT)** does not destroy the part being tested.

Mechanical testing is commonly used to qualify welders and welding procedures. It can be used in a random-sample testing procedure in mass production. In many applications, a large number of identical parts are made, and a chosen number are destroyed by mechanical testing. The results of such tests are valid only for welds made under identical conditions: The only weld strengths known are the ones for the tested pieces, and it is assumed that the strengths of nontested pieces made under identical conditions are the same.

Nondestructive testing is used for welder qualification, welding procedure qualification, and product quality control. Since the weldment is not damaged, each weld can be tested and the part can still be used for its intended purpose. Because the parts are not destroyed, more than one testing method can be used on the same part. Frequently only a portion of the welds are tested to save time and money. The same random-sampling protocol applies to these tests as applies to mechanical testing. Critical parts or welds are usually 100% tested.

DISCONTINUITIES AND DEFECTS

Module 9
Key Indicator 1, 2

Discontinuities and flaws are interruptions in the typical structure of a weld. They may be a lack of uniformity in the mechanical, metallurgical, or physical characteristics of the material or weld. All welds have discontinuities and flaws, but not all of these are necessarily defects.

A **defect**, according to the AWS, is "a discontinuity or discontinuities that by nature or accumulated effect render a part or product unusable to meet minimum applicable acceptance standards or specifications. The

Table 9.1 Major Code Issuing Agencies*

American Bureau of Shipping

American Petroleum Institute

American Society of Mechanical Engineers

American Society for Testing and Materials

American Welding Society

British Welding Institute

United States government

*A more complete listing of agencies with addresses is included in the Appendix.

term designates rejectability (for example, total porosity or slag inclusion length that renders a part or product unable to meet minimum applicable acceptance standards or specifications)."

In other words, many acceptable products may have welds that contain discontinuities. But no products may have welds that contain defects. Nevertheless, a discontinuity may become so large, or there may be so many small discontinuities, that the weld is not acceptable under the standards for the code for that product. Some codes are stricter than others, so the same weld might be acceptable under one code but not under another.

Ideally, a weld should not have any discontinuities, but that is practically impossible. The difference between what is acceptable or fit for service and perfection is known as **tolerance**. In many industries, the tolerances for welds have been established and are available as codes or standards. Table 9.1 lists a few of the agencies that issue codes or standards. Each code or standard gives the tolerance that changes a discontinuity into a defect.

When a weld is evaluated, it is important to note the type of any discontinuity, the size of the discontinuity, and the location of the discontinuity. Any one or more of these factors, depending on the applicable code or standard, can change a discontinuity to a defect.

The 12 most common discontinuities are:

- porosity
- inclusions
- inadequate joint penetration
- incomplete fusion
- arc strikes
- overlap (cold lap)
- undercut
- cracks
- underfill
- laminations
- delaminations
- lamellar tears

Porosity

Porosity results from gas that was dissolved in the molten weld pool forming bubbles that are trapped as the metal cools to become solid. The bubbles that cause porosity form within the weld metal, so they cannot be seen as they form. These gas pockets form in the same way that bubbles form in a carbonated drink as it warms up or as air dissolved in water forms bubbles in the center of a cube of ice. Porosity forms either spherical (ball-shaped) or cylindrical (tube- or tunnel-shaped) bubbles. Cylindrical porosity is called *wormhole porosity*. The rounded edges tend to reduce the stresses around them; therefore, unless porosity is extensive, there is little or no loss in strength.

Porosity is most often caused by improper welding techniques, contamination, or an improper chemical balance between the filler and base metals.

Improper welding techniques may result in shielding gas not properly protecting the molten weld pool. For example, the E7018 electrode should not be weaved wider than two-and-one-half times the electrode diameter, because very little shielding gas is produced. If the electrode is weaved wider, parts of the weld are unprotected. Nitrogen from the air that dissolves in the weld pool and becomes trapped rather than escaping can produce porosity.

The intense heat of the weld can decompose paint, dirt, or oil from machining and rust or other oxides, producing hydrogen. Hydrogen, like nitrogen, can become trapped in the solidifying weld pool, producing porosity. When it causes porosity, hydrogen can also diffuse into the heat-affected zone and produce underbead cracking in some steels. The level of gas needed to crack welds is below that necessary to produce porosity.

Porosity can occur as a result of the following welding processes:

- plasma arc welding (PAW)
- submerged arc welding (SAW)
- gas tungsten arc welding (GTAW)
- gas metal arc welding (GMAW)
- flux cored arc welding (FCAW)
- shielded metal arc welding (SMAW)
- carbon arc welding (CAW)
- oxyacetylene welding (OAW)
- oxyhydrogen welding (OHW)
- pressure gas welding (PGW)
- electron beam welding (EBW)
- electroslag welding (ESW)
- laser beam welding (LBW)
- thermit welding (TW)

Porosity can be grouped into four major types:

- Uniformly scattered porosity is most frequently caused by poor welding techniques or faulty materials, Figure 9.1.
- Clustered porosity is most often caused by improper starting and stopping techniques, Figure 9.2.
- Linear porosity is most frequently caused by contamination within the joint, root, or interbead boundaries, Figure 9.3.
- Piping, or wormhole, porosity is most often caused by contamination at the root, Figure 9.4. This type of porosity forms when the gas escapes from the weld pool at the same rate as the pool is solidifying.

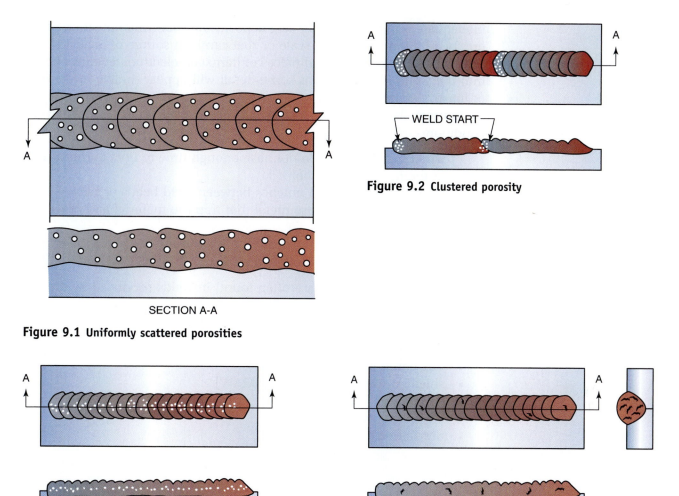

Figure 9.1 Uniformly scattered porosities

Figure 9.2 Clustered porosity

Figure 9.3 Linear porosity

Figure 9.4 Piping or wormhole porosity

Inclusions

Inclusions are nonmetallic materials, such as slag and oxides, that are trapped in the weld metal, between weld beads, or between the weld and the base metal. Inclusions may be jagged and irregularly shaped. They may also form in a continuous line. This causes stresses to concentrate and reduces the structural integrity (strength) of the weld.

Although not visible, their development can be expected if prior welds were improperly cleaned or had a poor contour. Unless care is taken in reading radiographs, the presence of slag inclusions can be misinterpreted as other defects.

Linear slag inclusions in radiographs generally contain shadow details; otherwise, they could be interpreted as lack-of-fusion defects. These inclusions result from poor manipulation that allows the slag to flow ahead of the arc, from not removing all the slag from previous welds, or from welding highly crowned, incompletely fused welds.

Scattered inclusions can resemble porosity; however, unlike porosity, they are generally not spherical. These inclusions can also result from inadequate removal of earlier slag deposits and poor manipulation of the arc. In addition, heavy mill scale or rust can be a source of inclusions, or they can result from unfused pieces of damaged electrode coatings falling into the weld. In radiographs some detail will appear, unlike in the case of linear slag inclusions.

Nonmetallic inclusions, Figure 9.5, are caused under the following conditions:

- when slag or oxides do not have enough time to float to the surface of the molten weld pool
- when there are sharp notches between weld beads or between the weld bead and the base metal that trap the material and prevent it floating out
- when the joint was designed with insufficient room for the correct manipulation of the molten weld pool

Nonmetallic inclusions are often found in welds produced by the following processes:

- submerged arc welding (SAW)
- gas metal arc welding (GMAW)
- flux cored arc welding (FCAW)
- shielded metal arc welding (SMAW)
- carbon arc welding (CAW)
- electroslag welding (ESW)
- thermit welding (TW)

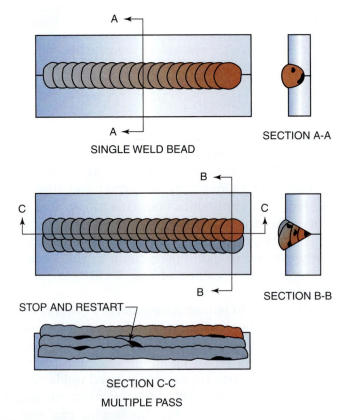

Figure 9.5 Nonmetallic inclusions

Inadequate Joint Penetration

Inadequate joint penetration occurs when the depth to which the weld penetrates the joint is less than that needed to fuse through the plate or into the preceding weld, Figure 9.6. A defect usually results that reduces the cross-sectional area of weld penetration in the joint to below the minimum required depth or could become a source of stress concentration that leads to fatigue failure. The criticality of such defects depends on the notch sensitivity of the metal and the factor of safety to which the weldment has been designed. Generally, if proper welding procedures are developed and followed, such defects do not occur.

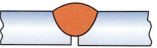
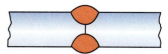

Figure 9.6 Inadequate joint penetration

The major causes of inadequate joint penetration are:

- improper welding technique—The most common cause is a misdirected arc. Also, the welding technique may require that both starting and run-out tabs be used so that the molten weld pool is well established before it reaches the joint. Sometimes a failure to back-gouge the root sufficiently provides a deeper root face than allowed for, Figure 9.7.
- insufficient welding current—Metals that are thick or have a high thermal conductivity are often preheated so that the weld heat is not drawn away so quickly by the metal that it cannot penetrate the joint.
- improper joint fitup—This problem results when the weld joints are not prepared or fitted accurately. Too small a root gap or too large a root face will keep the weld from penetrating adequately.
- improper joint design—When joints are accessible from both sides, back-gouging is often used to ensure 100% root fusion.

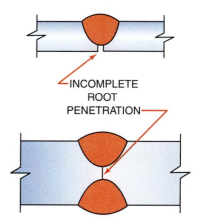

Figure 9.7 Incomplete root penetration

Inadequate joint penetration can occur as a result of the following welding processes:

- plasma arc welding (PAW)
- submerged arc welding (SAW)
- gas metal arc welding (GMAW)
- flux cored arc welding (FCAW)
- shielded metal arc welding (SMAW)
- carbon arc welding (CAW)
- electroslag welding (ESW)
- oxyacetylene welding (OAW)
- electron beam welding (EBW)

Incomplete Fusion

Incomplete fusion is the lack of coalescence between the molten filler metal and previously deposited filler metal or the base metal, Figure 9.8. A lack of fusion between the filler metal and previously deposited weld metal is called *interpass cold lap*. A lack of fusion between the weld metal and the joint face is called *lack of sidewall fusion*. Both problems usually travel along all or most of the weld's length.

Some major causes of lack of fusion are:

- inadequate agitation—Lack of weld agitation to break up oxide layers. The base metal or weld filler metal may melt, but a thin layer of oxide may prevent coalescence from occurring.

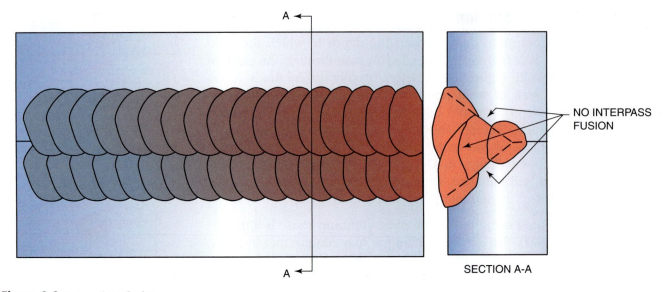

Figure 9.8 Incomplete fusion

- improper welding techniques—poor manipulation, such as moving too fast or using an improper electrode angle
- wrong welding process—An example is the use of short-circuiting transfer with GMAW to weld plate thicker than 1/4 in. (6 mm), which can cause lack of fusion because of the process's limited heat input to the weld.
- improper edge preparation—Not removing any notches or gouges in the edge of the weld joint. For example, if a flame-cut plate has notches along the cut, they could result in a lack of fusion in each notch, Figure 9.9.
- improper joint design—Incomplete fusion may also result from not applying enough heat to melt the base metal or the joint designer allowing too little space for correct molten weld pool manipulation.

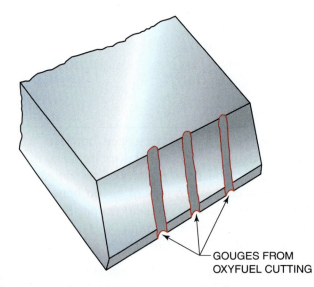

Figure 9.9 Gouge removal
Remove gouges along the surface of the joint before welding

- improper joint cleaning—failure to clean from the joint surfaces oxides resulting from the use of an oxyfuel torch to cut the plate or failure to remove slag from a previous weld

Incomplete fusion can be found in welds produced by all major welding processes.

Arc Strikes

Figure 9.10 shows arc strikes—small, localized points where surface melting occurred away from the joint. These spots may be caused by accidentally striking the arc in the wrong place or by faulty ground connections. Even though arc strikes can be ground smooth, they cannot be removed. These spots will always show if an acid etch is used. They also can be localized hardness zones or the starting point for cracking. Arc strikes, even when ground flush for a guided bend, will open up to form small cracks or holes.

Overlap

Overlap, also called *cold lap*, occurs in fusion welds when weld deposits are larger than the joint is conditioned to accept. The weld metal flows over the surface of the base metal without fusing to it, along the toe of the weld bead, Figure 9.11. Overlap generally occurs on the horizontal leg of a horizontal fillet weld under extreme conditions. It can also occur on both sides of flat-positioned capping passes. With GMA welding, overlap occurs when too much electrode extension is used to deposit metal at low power. Misdirecting the arc into the vertical leg and keeping the electrode nearly vertical will also cause overlap. To prevent overlap, the fillet weld must be correctly sized to less than 3/8 in. (9.5 mm), and the arc must be properly manipulated.

Undercut

Undercut is the result of the arc force removing from a joint face metal that is not replaced by weld metal, along the toe of the weld bead, Figure 9.12. It can result from excessive current. It is a common problem with GMA

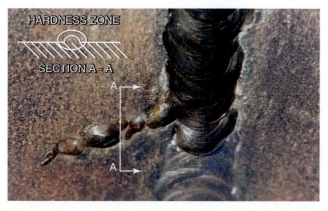

Figure 9.10 Arc strikes
Source: Courtesy of Larry Jeffus

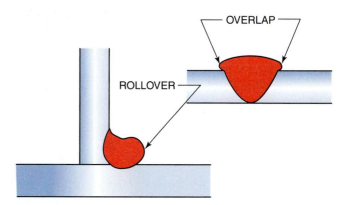

Figure 9.11 Rollover or overlap

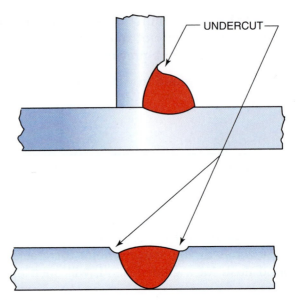

Figure 9.12 Undercut

welding when insufficient oxygen is used to stabilize the arc. Incorrect welding technique, such as incorrect electrode angle or excessive weave, can also cause undercut. To prevent undercutting, the welder can weld in the flat position using multiple passes instead of a single pass, change the shield gas, and improve manipulative techniques to fill the removed base metal along the toe of the weld bead.

Crater Cracks

Crater cracks are the tiny cracks that develop in weld craters as the weld pool shrinks and solidifies, Figure 9.13. Materials with a low melting temperature are rejected toward the crater center during freezing. Since these materials are the last to freeze, they are pulled apart or separated as a result of the weld metal's shrinking as it cools. The high shrinkage stresses aggravate crack formation. Crater cracks can be minimized, if not prevented, by not interrupting the arc quickly at the end of a weld. This allows the arc to lengthen, the current to drop gradually, and the crater to fill and cool more slowly. Some GMAW equipment has a crater-filling control that automatically gradually reduces the wire-feed speed at the

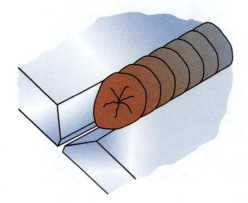

Figure 9.13 Crater or star cracks

end of a weld. For all other welding processes, the most effective way of preventing crater cracking is to slightly pull the weld back, allowing it to pool up on the weld bead before breaking the arc, Figure 9.14.

Underfill

Underfill on a groove weld occurs when the amount of weld metal deposited is insufficient to bring the weld's face or root surfaces to a level equal to that of the original plane or plate surface. For a fillet weld it occurs when the weld deposit has an insufficient effective throat, Figure 9.15. This problem can usually be corrected by slowing the travel rate or making more weld passes.

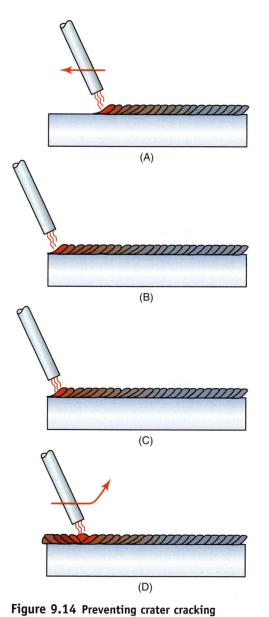

(A)

(B)

(C)

(D)

Figure 9.14 Preventing crater cracking
(A) Make a uniform weld; (B) Weld to the end of the plate; (C) Hold the arc for a second at the end of the weld; and (D) Quickly, so you deposit as little weld metal as possible, move the arc back up onto the weld and break the arc

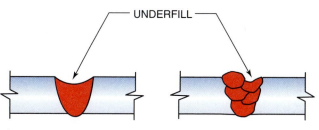

UNDERFILL

Figure 9.15 Underfill

Plate-Generated Problems

Not all welding problems are caused by the weld metal, the process, or the welder's lack of skill in depositing that metal. The material being fabricated can be at fault, too. Some problems result from internal plate defects that the welder cannot control. Others are the result of improper welding procedures that produce undesirable hard metallurgical structures in the heat-affected zone. The internal defects are the result of poor steelmaking practices. Steel producers try to keep their steels as sound as possible, but the mistakes that occur in steel production are blamed, too frequently, on the welding operation.

Lamination

Laminations differ from lamellar tearing (see below) in that they are more extensive and involve thicker layers of nonmetallic contaminants. Located toward the center of the plate, Figure 9.16, laminations are caused by insufficient cropping (removal of defects) of the pipe in ingots. The slag and oxidized steel in the pipe are rolled out with the steel, producing the lamination. Laminations can also be caused when the ingot is rolled at too low a temperature or pressure.

Delamination

When laminations intersect a joint being welded, the heat and stresses of the weld may cause some laminations to become delaminated. Contamination of the weld metal may occur if the lamination contained large amounts of slag, mill scale, dirt, or other undesirable materials. Such contamination can cause wormhole porosity or lack-of-fusion defects.

The problems associated with delaminations are not easily corrected. If a thick plate is installed in a compression load, an effective solution can be to weld over the delamination to seal it. A better solution is to replace the steel.

Lamellar Tears

Lamellar tears appear as cracks parallel to and under the steel surface. In general, they are not in the heat-affected zone, and they have a steplike configuration. They result from thin layers of nonmetallic inclusions that lie beneath the plate surface and have very poor ductility. Although barely noticeable, these inclusions separate when severely stressed, producing laminated cracks. These cracks are evident if the plate edges are exposed, Figure 9.17.

A solution to the problem of lamellar tearing is to redesign the joints to impose the lowest possible strain throughout the plate thickness. This can be accomplished by making smaller welds so that each subsequent weld pass heat-treats the previous pass to reduce the total stress in the finished weld, Figure 9.18. The joint design can be changed to reduce the stress on the weld through thickness of the plate, Figure 9.19.

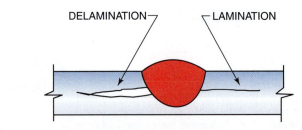

Figure 9.16 Lamination and delamination

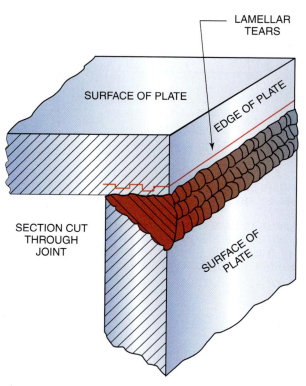

Figure 9.17 **Lamellar tearing**

Figure 9.18 **Using multiple welds to reduce weld stresses**

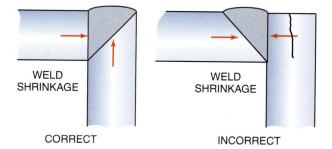

Figure 9.19 **Correct joint design to reduce lamellar tears**

DESTRUCTIVE TESTING (DT)

Destructive testing is used to determine actual values in weld metal and base metals in order to ensure that specific performance characteristics are obtained. These destructive tests are used to establish various mechanical properties, such as tensile strength, ductility, or hardness. Some destructive tests provide values for several properties, but most tests are designed to determine the value for a specific characteristic of a metal.

**Module 9
Key Indicator 2**

Tensile Testing

Tensile tests are performed with specimens prepared as round bars or flat strips. The simple round bars are often used for testing only the weld metal, sometimes called "all weld metal testing." This form of testing can be used on thick sections where base metal dilution into all of the weld metal is not possible. Round specimens are cut from the center of the weld metal. The flat bars are often used to test both the weld and the surrounding metal. Flat bars are usually cut at a 90° angle to the weld, Figure 9.20. Table 9.2 shows how a number of standard smaller-size bars can be used, depending on the thickness of the metal to be tested. Bar size also depends on the size of the tensile testing equipment available, Figure 9.21.

Two flat specimens are commonly used for testing thinner sections of metal. When welds are tested, the specimen should include the heat-affected zone and the base plate. If the weld metal is stronger than the plate, failure occurs in the plate; if the weld metal is weaker, failure occurs in the weld. This test, then, is open to interpretation.

After the weld section is machined to the specified dimensions, it is placed in the tensile testing machine and pulled apart. A specimen used

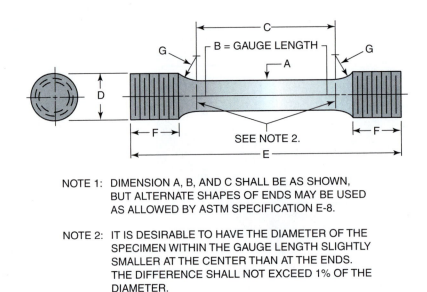

NOTE 1:   DIMENSION A, B, AND C SHALL BE AS SHOWN,
          BUT ALTERNATE SHAPES OF ENDS MAY BE USED
          AS ALLOWED BY ASTM SPECIFICATION E-8.

NOTE 2:   IT IS DESIRABLE TO HAVE THE DIAMETER OF THE
          SPECIMEN WITHIN THE GAUGE LENGTH SLIGHTLY
          SMALLER AT THE CENTER THAN AT THE ENDS.
          THE DIFFERENCE SHALL NOT EXCEED 1% OF THE
          DIAMETER.

Figure 9.20 Tensile testing specimen
Source: Courtesy of Hobart Brothers Company

to determine the strength of a welded butt joint for plate is shown in Figure 9.22.

The tensile strength, in pounds per square inch, is obtained by dividing the maximum load required to break the specimen by the original cross-sectional area of the specimen at the middle. The cross-sectional area is obtained as either $A = \pi r^2$ or $A = D^2 \times 0.785$ (where r^2 is the result of multiplying the radius by itself, and D^2 is the result of multiplying the diameter by itself). Using either formula will give you the same answer.

The elongation is found by fitting the fractured ends of the specimen together, measuring the distance between gauge marks, and subtracting the gauge length. The percentage of elongation is found by dividing the elongation by the gauge length and multiplying by 100:

$$E_1 = \frac{L_f - L_o}{L_o} \times 100,$$

where
E_1 = percentage of elongation,
L_f = final gauge length, and
L_o = original gauge length

Table 9.2 Dimensions of Tensile Testing Specimens

Specimen	Dimensions of Specimen						
	in./mm A	in./mm B	in./mm C	in./mm D	in./mm E	in./mm F	in./mm G
C-1	.500/12.7	2/50.8	2.25/57.1	.750/19.05	4.25/107.9	.750/19.05	.375/9.52
C-2	.437/11.09	1.750/44.4	2/50.8	.625/15.8	4/101.6	.750/19.05	.375/9.52
C-3	.357/9.06	1.4/35.5	1.750/44.4	.500/12.7	3.500/88.9	.625/15.8	.375/9.52
C-4	.252/6.40	1.0/25.4	1.250/31.7	.375/9.52	2.50/63.5	.500/12.7	.125/3.17
C-5	.126/3.2	.500/12.7	.750/19.05	.250/6.35	1.750/44.4	.375/9.52	.125/3.17

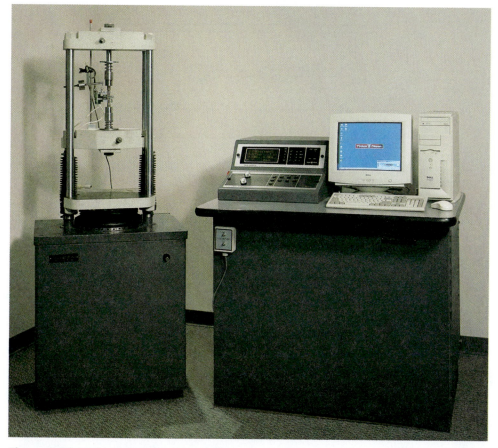

Figure 9.21 Typical tensile tester used for measuring the strength of welds (60,000-lb universal testing machines)
Source: Courtesy of Tinius Olsen Testing Machine Co., Inc.

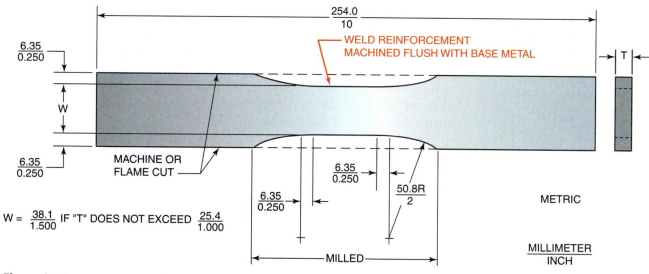

Figure 9.22 Tensile specimen for flat plate weld
Source: Courtesy of Hobart Brothers Company

Fatigue Testing

Fatigue testing is used to determine how well a weld can resist repeated fluctuating stresses or cyclic loading. The maximum value of the stresses is less than the tensile strength of the material. Fatigue strength can be

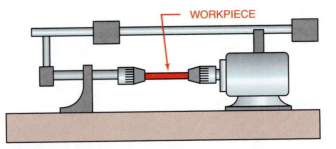

WORKPIECE

Figure 9.23 Fatigue testing
The specimen is placed in the chucks of the machine. As the machine rotates, the specimen is alternately bent twice for each revolution

lowered by improperly made weld deposits, which may be caused by porosity, slag inclusions, lack of penetration, or cracks. Any one of these discontinuities can act as a point of stress, eventually resulting in the failure of the weld.

In the fatigue test, the part is subjected to repeated changes in applied stress. Fatigue testing can be performed in several ways, depending upon the type of service the tested part must withstand. The results obtained are usually reported as the number of stress cycles that the part will resist without failure and the total stress used.

In one type of test, the specimen is bent back and forth. This test subjects the part to alternating compression and tension. A fatigue testing machine is used for this test, Figure 9.23. As the machine rotates, the specimen is alternately bent twice for each revolution. Failure is usually rapid.

Shearing Strength of Welds

The two forms of **shearing strength** of welds are transverse shearing strength and longitudinal shearing strength. To test transverse shearing strength, a specimen is prepared as shown in Figure 9.24. The width of the specimen is measured in inches or millimeters. A tensile load is applied, and the specimen is ruptured. The maximum load in pounds or kilograms is then determined.

The shearing strength of the weld, in pounds per linear inch or kilograms per linear millimeter, is obtained by dividing the maximum force by twice the width of the specimen:

$$\text{Shearing strength lb/in. (kg/mm)} = \frac{\text{maximum force}}{2(\text{width of specimen})}$$

To test longitudinal shearing strength, a specimen is prepared as shown in Figure 9.25. The length of each weld is measured in inches or millimeters. The specimen is then ruptured under a tensile load, and the maximum force in pounds or kilograms is determined.

The shearing strength of the weld, in pounds per linear inch or kilograms per linear millimeter, is obtained by dividing the maximum force by the sum of the length of welds that ruptured:

$$\text{Shearing strength lb/in. (kg/mm)} = \frac{\text{maximum force}}{\text{length of ruptured weld}}$$

CONVERSION TABLE – MILLIMETERS TO INCHES

DIM-mm	TOL	DIM-in.
9.52		0.375
9.52	± 1.58	0.375
12.70		0.500
19.05		0.750
50.60		2.000
63.50		2.500
228.60		9.000
114.30		4.500

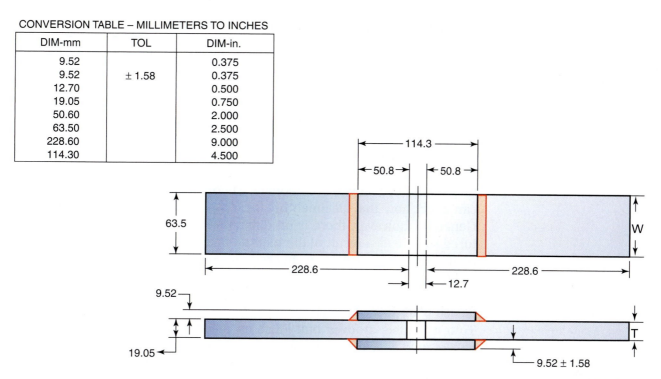

Figure 9.24 Transverse fillet weld shearing specimen after welding
Source: Courtesy of Hobart Brothers Company

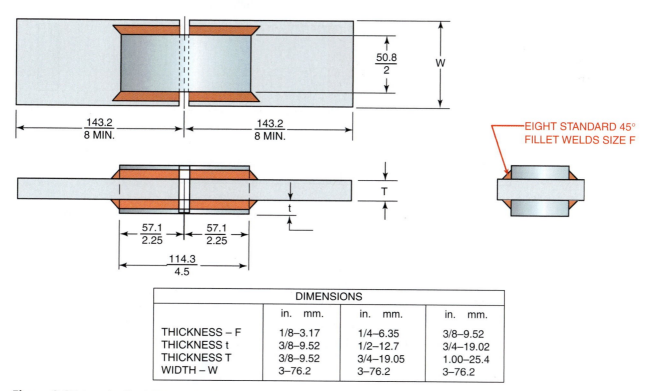

DIMENSIONS						
	in.	mm.	in.	mm.	in.	mm.
THICKNESS – F	1/8	3.17	1/4	6.35	3/8	9.52
THICKNESS t	3/8	9.52	1/2	12.7	3/4	19.02
THICKNESS T	3/8	9.52	3/4	19.05	1.00	25.4
WIDTH – W	3	76.2	3	76.2	3	76.2

Figure 9.25 Longitudinal fillet weld shear specimen

Welded Butt Joints

The three methods of testing welded butt joints are (1) the nick-break test, (2) the guided bend test, and (3) the free bend test. It is possible to use variations of these tests.

Nick-Break Test

A specimen for this test is prepared as shown in Figure 9.26A. The specimen is supported as shown in Figure 9.26B. A force is then applied, and the specimen is ruptured by one or more blows of a hammer. The force may be applied slowly or suddenly. Theoretically, the rate of application can affect how the specimen breaks, especially at a critical temperature. Generally, however, there is no difference in the appearance of the fractured surface as a result of the method of applying the force. The surfaces of the fracture should be checked to determine the soundness of the weld.

Guided Bend Test

To test welded, grooved butt joints on metal that is 3/8 in. (10 mm) thick or less, two specimens are prepared and tested—one face bend and one root bend, Figures 9.27A and 9.27B. If the welds pass this test, the welder is qualified to make groove welds on plate between 3/8 in. and 3/4 in. (10 mm and 19 mm) thick. These welds need to be machined as shown in Figure 9.28A. If these specimens pass, the welder will also be qualified to make fillet welds on materials of any (unlimited) thickness. For welded,

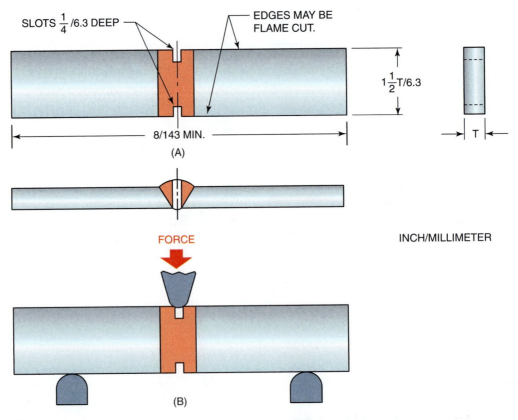

Figure 9.26 Nick-break specimens
(A) Nick-break specimen for butt joints in plate and (B) method of rupturing nick-break specimen
Source: Courtesy of Hobart Brothers Company

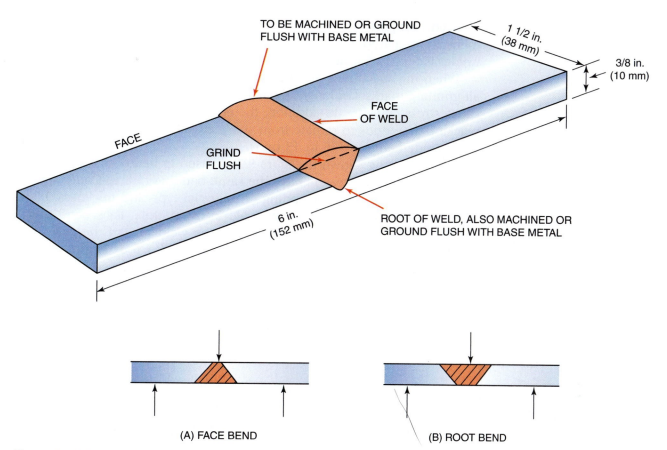

Figure 9.27 Root and face bend specimens for 3/8-in. (10-mm) plate

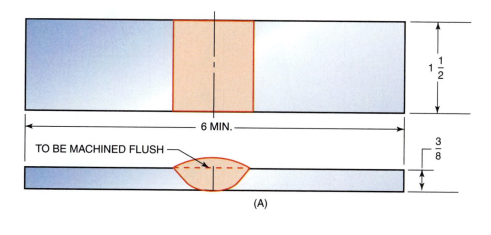

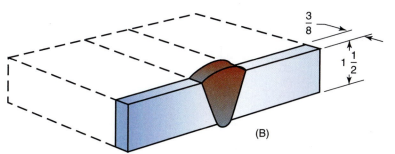

mm	CONVERSION in.
1.5	1/16
3.1	1/8
9.5	3/8
12.7	1/2
38.0	1 1/2
152.0	6

Figure 9.28 Root, face, and side bend specimens
(A) Root and face bend specimens and (B) side bend specimen

grooved butt joints on metal 1/2 in. (13 mm) thick, two side bend specimens are prepared and tested, Figure 9.28B. If the welds pass this test, the welder is qualified to weld on metals of unlimited thickness.

When the specimens are prepared, care must be taken to ensure that all grinding marks run longitudinally to the specimen so that they do not cause stress cracking. In addition, the edges must be rounded to reduce cracking that tends to radiate from sharp edges. The maximum radius of this rounded edge is 1/8 in. (3 mm).

The type of jig shown in Figure 9.29 is used to bend most specimens. Not all guided bend testers have the same bending radius. Codes specify different bending radii depending on material type and thickness. Place the specimens in the jig with the weld in the middle. Face bend specimens should be placed with the face of the weld toward the gap. Root bend specimens should be positioned so that the root of the weld is directed toward the gap. Side bend specimens are placed with either side facing up. The guided bend specimen must be pushed all the way

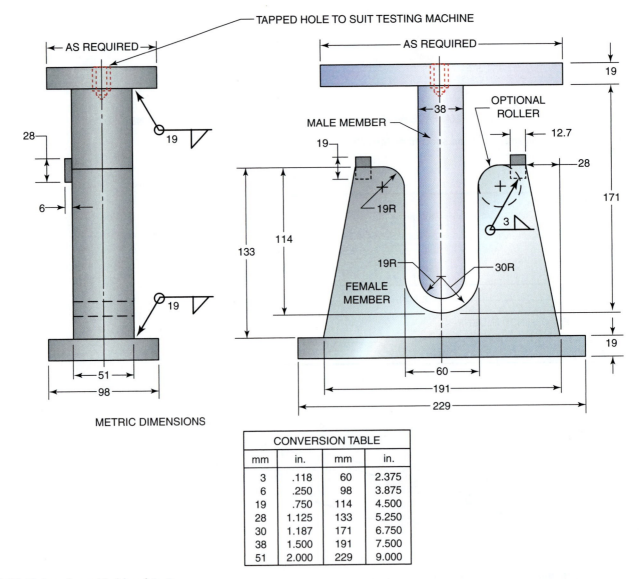

Figure 9.29 Fixture for guided bend test
Source: Courtesy of The Aluminum Association

through open (roller-type) bend testers and within 1/8 in. (3 mm) of the bottom on fixture-type bend testers.

Once the test is completed, the specimen is removed. The convex surface is then examined for cracks or other discontinuities and judged acceptable or unacceptable according to specified criteria. Some surface cracks and openings are allowable under codes.

Free Bend Test

The free bend test is used to test welded joints in plate. A specimen is prepared as shown in Figure 9.30. Note that the width of the specimen is 1.5 multiplied by the thickness of the specimen. Each corner lengthwise should be rounded in a radius not exceeding one-tenth the thickness of the specimen. Tool marks should run the length of the specimen.

Gauge lines are drawn on the face of the weld, Figure 9.31. The distance between the gauge lines is 1/8 in. (3.17 mm) less than the face of

CONVERSION TABLE							
in.	mm	in.	mm	in.	mm	in.	mm
1/4	6.35	9/16	14.2	1 1/8	28.5	2	50.8
3/8	9.52	3/4	19.05	1 1/4	31.7	6	152.4
1/2	12.7	15/16	23.8	1 1/2	38.1	8	203.2
5/8	15.8	1	25.4	1 7/8	47.6	9	228.6

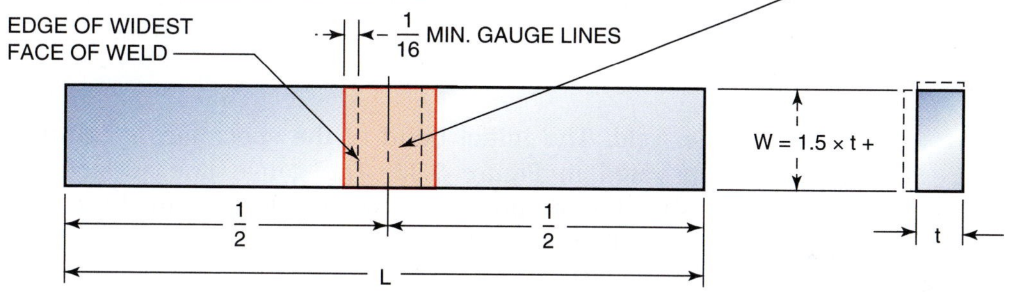

WELD REINFORCEMENT MACHINED FLUSH WITH BASE METAL →

EDGE OF WIDEST FACE OF WELD

$\frac{1}{16}$ MIN. GAUGE LINES

$\frac{1}{2}$ $\frac{1}{2}$

L

W = 1.5 × t +

t

DIMENSIONS							
T, inches –	1/4	3/8	1/2	5/8	3/4	1	1 1/4
W, inches –	3/8	9/16	3/4	15/16	1 1/8	1 1/2	1 7/8
L, min, inches –	6	8	9	10	12	13 1/2	15
B, min, inches –	1 1/4	1 1/4	1 1/4	2	2	2	2

Figure 9.30 Free bend test specimen
Source: Courtesy of Hobart Brothers Company

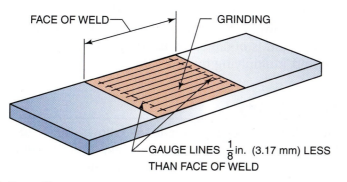

FACE OF WELD   GRINDING

GAUGE LINES $\frac{1}{8}$ in. (3.17 mm) LESS THAN FACE OF WELD

Figure 9.31 Gauge lines
Gauge lines are drawn on the weld face of a free bend specimen

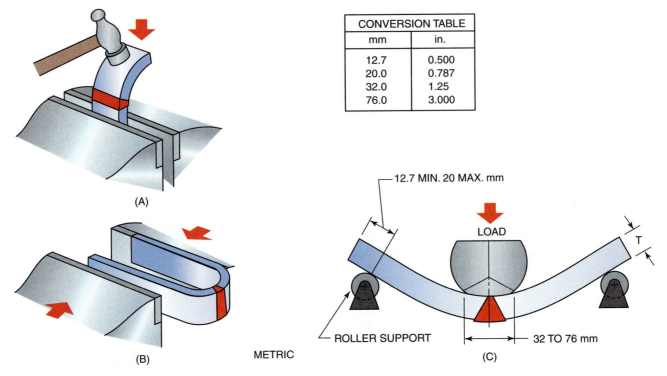

CONVERSION TABLE	
mm	in.
12.7	0.500
20.0	0.787
32.0	1.25
76.0	3.000

Figure 9.32 Free bend test
(A) The initial bend can be made in this manner; (B) a vise can be used to make the final bend; and (C) another method used to make the bend
Source: Courtesy of Hobart Brothers Company

the weld. The initial bend of the specimen is completed in the device illustrated in Figure 9.32. The gauge line surface should be directed toward the supports. The weld is located in the center of the supports and loading block.

Alternate Bend

The initial bend may be made by placing the specimen in the jaws of a vise with one-third the length projecting from the jaws. The specimen is bent away from the gauge lines through an angle of 30° to 45° using a hammer. The specimen is then inserted into the jaws of a vise, and pressure is applied by tightening the vise. The pressure is continued until a crack or depression appears on the convex face of the specimen. The load is then removed.

The elongation is determined by measuring the minimum distance between the gauge lines along the convex surface of the weld to the nearest 0.01 in. (0.254 mm) and subtracting the initial gauge length. The percentage of elongation is obtained by dividing the elongation by the initial gauge length and multiplying by 100.

Fillet Weld Break Test

The specimen for this test is made as shown in Figure 9.33A. In Figure 9.33B, a force is applied to the specimen until the rupture breaking of the specimen occurs. Any convenient means of applying the force may be used, such as an arbor press, a testing machine, or hammer blows. The

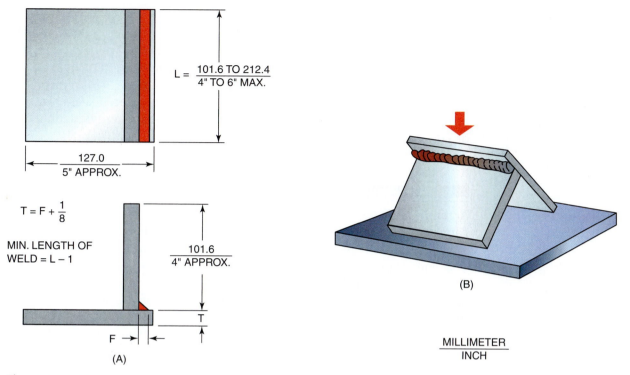

Figure 9.33 Fillet weld testing
(A) Fillet weld break test and (B) method of rupturing fillet weld break specimen
Source: Courtesy of Hobart Brothers Company

break surface should then be examined for soundness—i.e., slag inclusions, overlap, porosity, lack of fusion, or other discontinuities.

Testing by Etching

Testing by **etching** serves one of two purposes: (1) to determine the soundness of a weld or (2) to determine the location of a weld.

A test specimen is produced by cutting a portion from the welded joint so that a complete cross section is obtained. The face of the cut is filed and polished with fine abrasive cloth. The specimen can then be placed in the etching solution. The etching solution or reagent makes the boundary between the weld metal and base metal visible, if the boundary is not already distinctly visible.

The most commonly used etching solutions are hydrochloric acid, ammonium persulfate, and nitric acid.

Hydrochloric Acid

Equal parts by volume of concentrated hydrochloric (muriatic) acid and water are mixed by adding the acid into the water. The welds are immersed in the reagent at or near the boiling temperature. The acid usually enlarges gas pockets and dissolves slag inclusions, enlarging the resulting cavities.

Ammonium Persulfate

A solution is prepared consisting of one part of ammonium persulfate (solid) to nine parts of water by weight. The surface of the weld is rubbed with cotton saturated with this reagent at room temperature.

CAUTION

When diluting an acid, always pour the acid slowly into the water while continuously stirring the water. Carelessly handling this material or pouring water into the acid can result in burns, excessive fuming, or explosion. Be sure to wear safety glasses and gloves to prevent injuries.

Nitric Acid

One part of concentrated nitric acid is mixed with nine parts of water by volume. The reagent is applied to the surface of the weld with a glass stirring rod at room temperature. Nitric acid has the capacity to etch rapidly and should be used on polished surfaces only.

After etching, the weld is rinsed in clear, hot water. Excess water is removed, and the etched surface is then immersed in ethyl alcohol and dried.

Impact Testing

A number of tests can be used to determine the impact-withstanding capability of a weld. One common test is the Izod test, Figure 9.34A, in which a notched specimen is struck by an anvil mounted on a pendulum. The energy required to break the specimen (in footpounds, read from a scale mounted on the machine) is an indication of the impact resistance of the metal. This test compares the toughness of the weld metal with that of the base metal.

Another similar type of impact test is the Charpy test. Whereas the Izod test specimen is gripped on one end, held vertically, and usually tested at

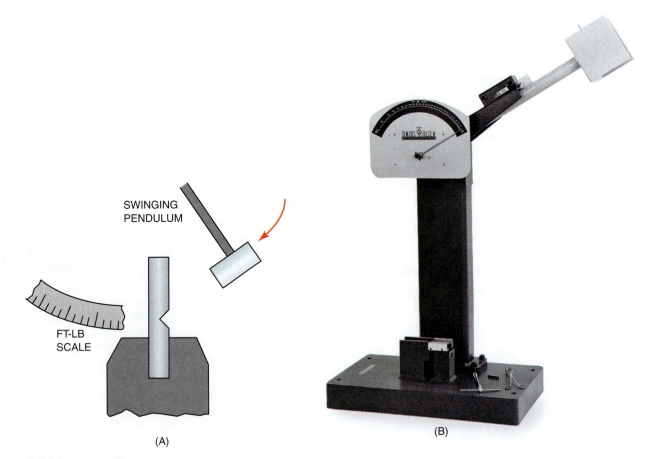

SWINGING
PENDULUM

FT-LB
SCALE

(A)

(B)

Figure 9.34 Impact testing
(A) Specimen mounted for Izod impact toughness testing and (B) a typical impact tester used for measuring the toughness of metals
Source: Courtesy of Tinius Olsen Testing Machine Co., Inc.

room temperature, the Charpy test specimen is held horizontally, supported on both ends, and usually tested at a specific temperature.

All impact test specimens must be produced according to ASTM specifications. A typical impact tester is shown in Figure 9.34B.

NONDESTRUCTIVE TESTING (NDT)

Module 9
Key Indicator 1, 2

Nondestructive testing (NDT) is a method used to test welds for surface defects such as cracks, arc strikes, undercuts, and lack of penetration. Internal or subsurface defects can include slag inclusions, porosity, and unfused metal in the interior of the weld.

Visual Inspection (VT)

Visual inspection (VT) is the most frequently used nondestructive testing method and is the first step in almost every other inspection process. The majority of welds receive only visual inspection. If the weld looks good, it passes; if it looks bad, it is rejected. This procedure is often mistakenly overlooked when more sophisticated nondestructive testing methods are used.

An active visual inspection schedule can reduce the finished weld rejection rate by more than 75%. Visual inspection can easily be used to check for fitup, interpass acceptance, welder technique, and other variables that will affect the weld quality. Minor problems can be identified and corrected before a weld is completed. This eliminates costly repairs or rejection.

Visual inspection should be used before any other nondestructive or mechanical tests are used to eliminate (reject) the obvious problem welds. Eliminating welds that have excessive surface discontinuities that will not pass the code or standard being used saves preparation time.

Penetrant Inspection (PT)

Penetrant inspection (PT) is used to locate minute surface cracks and porosity. Two types of penetrants are now in use: color-contrast and fluorescent versions. Color-contrast penetrants contain a colored (often red) dye that shows under ordinary white light. Fluorescent penetrants contain a more effective fluorescent dye that shows under black light.

The following steps should be followed when a penetrant is used:

1. The first step is precleaning. Suspected flaws are cleaned and dried so that they are free of oil, water, or other contaminants.
2. The test surface is covered with a film of penetrant by dipping, immersing, spraying, or brushing.
3. The test surface is then gently wiped, washed, or rinsed free of excess penetrant. It is dried with cloths or hot air.
4. A developing powder applied to the test surface acts as a blotter to speed the process by which the penetrant seeps out of any flaws open to the test surface.
5. Depending upon the type of penetrant applied, visual inspection is made under ordinary white light or near-ultraviolet black light, Figure 9.35. In the latter case, the penetrant fluoresces a yellow-green color, which clearly defines the defect.

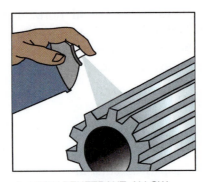

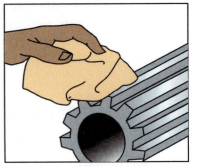

1. PRECLEAN INSPECTION AREA. SPRAY ON CLEANER/REMOVER –WIPE OFF WITH CLOTH.

2. APPLY PENETRANT, ALLOW SHORT PENETRATION PERIOD.

3. SPRAY CLEANER/REMOVER ON WIPING TOWEL AND WIPE SURFACE CLEAN.

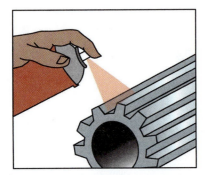

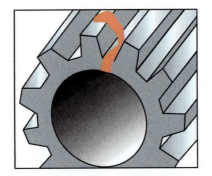

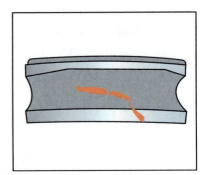

4. SHAKE DEVELOPER CAN AND SPRAY ON A THICK, UNIFORM FILM OF DEVELOPER.

5. INSPECT. DEFECTS WILL SHOW AS BRIGHT RED LINES IN WHITE DEVELOPER BACKGROUND.

Figure 9.35 Penetrant testing
Source: Adapted from Magnaflux Corporation

Magnetic Particle Inspection (MT)

Magnetic particle inspection (MT) uses fine ferromagnetic particles (powder) to indicate defects open to the surface or just below the surface on magnetic materials.

A magnetic field is induced in a part by passing an electric current through or around it. The magnetic field is always at right angles to the direction of current flow. Magnetic particle inspection registers an abrupt change in the resistance in the path of the magnetic field, such as would be caused by a crack lying at an angle to the direction of the magnetic poles at the crack: Finely divided ferromagnetic particles applied to the area will be attracted and outline the crack.

The flow or discontinuity interrupting the magnetic field in a test part can be either longitudinal or circumferential, Figure 9.36. A different type of magnetization is used to detect defects that run down the axis, as opposed to those occurring around the girth of a part. For some applications you may need to test in both directions.

In Figure 9.37A, longitudinal magnetization allows the detection of flaws running around the circumference of a part. The user places the test part inside an electrified coil to induce a magnetic field down the length of the test part. In Figure 9.37B, circumferential magnetization allows the detection of flaws running down the length of a test part by sending an electric current down the length of the part to be inspected.

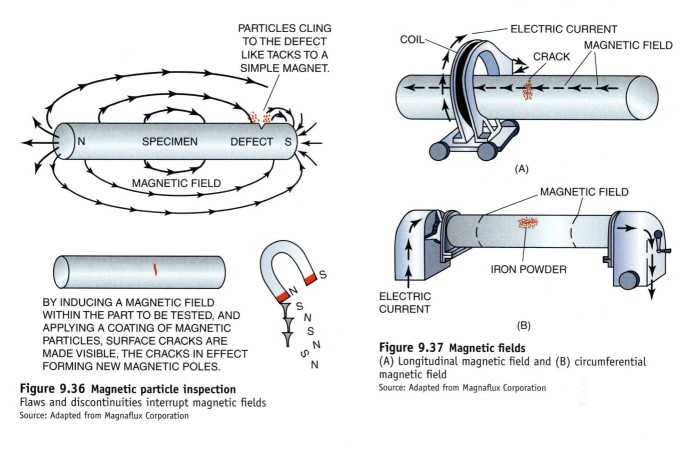

PARTICLES CLING TO THE DEFECT LIKE TACKS TO A SIMPLE MAGNET.

N SPECIMEN DEFECT S

MAGNETIC FIELD

BY INDUCING A MAGNETIC FIELD WITHIN THE PART TO BE TESTED, AND APPLYING A COATING OF MAGNETIC PARTICLES, SURFACE CRACKS ARE MADE VISIBLE, THE CRACKS IN EFFECT FORMING NEW MAGNETIC POLES.

Figure 9.36 Magnetic particle inspection
Flaws and discontinuities interrupt magnetic fields
Source: Adapted from Magnaflux Corporation

ELECTRIC CURRENT
COIL
CRACK
MAGNETIC FIELD

(A)

MAGNETIC FIELD
IRON POWDER
ELECTRIC CURRENT

(B)

Figure 9.37 Magnetic fields
(A) Longitudinal magnetic field and (B) circumferential magnetic field
Source: Adapted from Magnaflux Corporation

Radiographic Inspection (RT)

Radiographic inspection (RT) is a method for detecting flaws inside weldments. Radiography gives a picture of all discontinuities that are parallel (vertical) or nearly parallel to the source. Discontinuities that are perpendicular (flat) or nearly perpendicular to the source may not be seen on the X-ray film. Instead of using visible light rays, the operator uses invisible, short-wavelength rays developed by X-ray machines, radioactive isotopes (gamma rays), and variations on these methods. These rays are capable of penetrating solid materials and reveal most flaws in a weldment on an X-ray film or a fluorescent screen. Flaws are revealed on films as dark or light areas against a contrasting background after exposure and processing, Figure 9.38.

The defect images in radiographs measure differences in how the X rays are absorbed as they penetrate the weld. The weld itself absorbs most X rays. If something less dense than the weld is present, such as a pore or a lack-of-fusion defect, fewer X rays are absorbed, darkening the film. If something more dense is present, such as heavy ripples on the weld surface, more X rays will be absorbed, lightening the film. The X-ray image is a shadow of the flaw. The farther the flaw is from the X-ray film, the fuzzier the image appears.

Skilled readers of radiographs can interpret the significance of the light and dark regions by their shape and shading. Those skilled at interpreting weld defects in radiographs must also be very knowledgeable about welding.

Figure 9.39 shows samples of common weld defects and a representative radiograph for each.

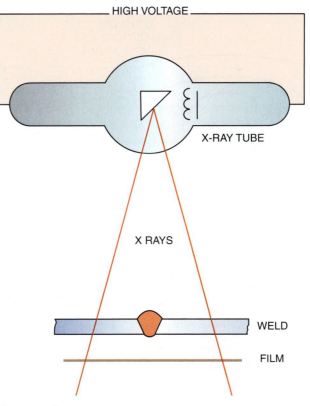

HIGH VOLTAGE

X-RAY TUBE

X RAYS

WELD

FILM

Figure 9.38 Schematic of an X-ray system

Four factors affect the selection of the radiation source:

- thickness and density of the material
- absorption characteristics
- time available for inspection
- location of the weld

Portable equipment is available for examining fixed or hard-to-move objects. The selection of the correct equipment for a particular application is determined by the specific voltage required, the equipment's degree of utility, the economics of inspection, and the production rates expected, Figures 9.40 (page 307) and 9.41 (page 307).

Ultrasonic Inspection (UT)

Ultrasonic inspection (UT) is fast and uses few consumable supplies, which makes it inexpensive for schools to use. However, because of the time required for most UT testing, it is not economic in the field as a non-destructive testing method. The ultrasonic inspection method employs electronically produced high-frequency sound waves (roughly 250,000 to 25 million cycles per second), which penetrate metals and many other materials at speeds of several thousand feet (meters) per second. A portable ultrasonic inspection unit is shown in Figure 9.42 (page 307).

The two types of ultrasonic equipment are pulse and resonance. The pulse-echo system, most often employed in the welding field, uses sound generated in short bursts or pulses. Since the high-frequency sound used is at a relatively low power, it has little ability to travel through air, so it must

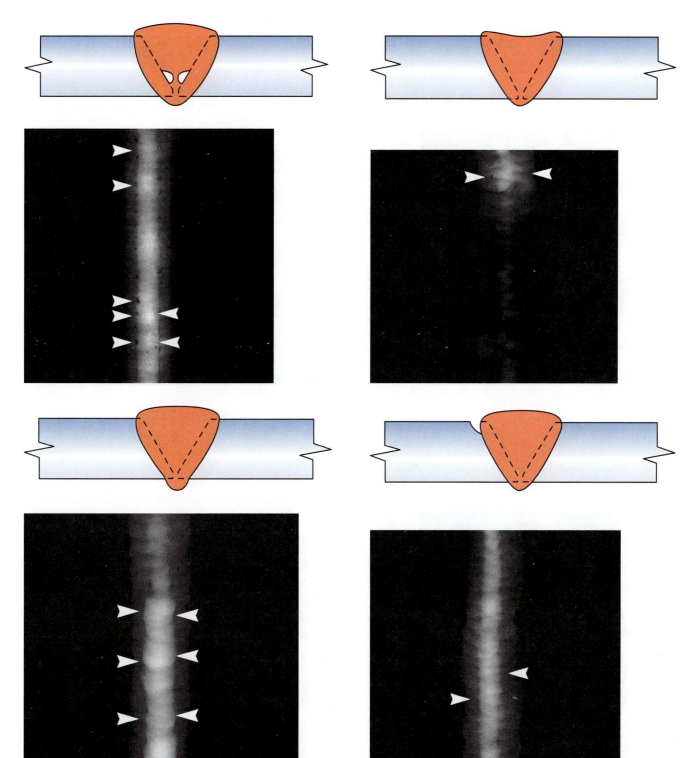

Figure 9.39 Welding defects and radiographic images
Source: Reprinted with permission of E. I. DuPont de Nemours & Co., Inc.

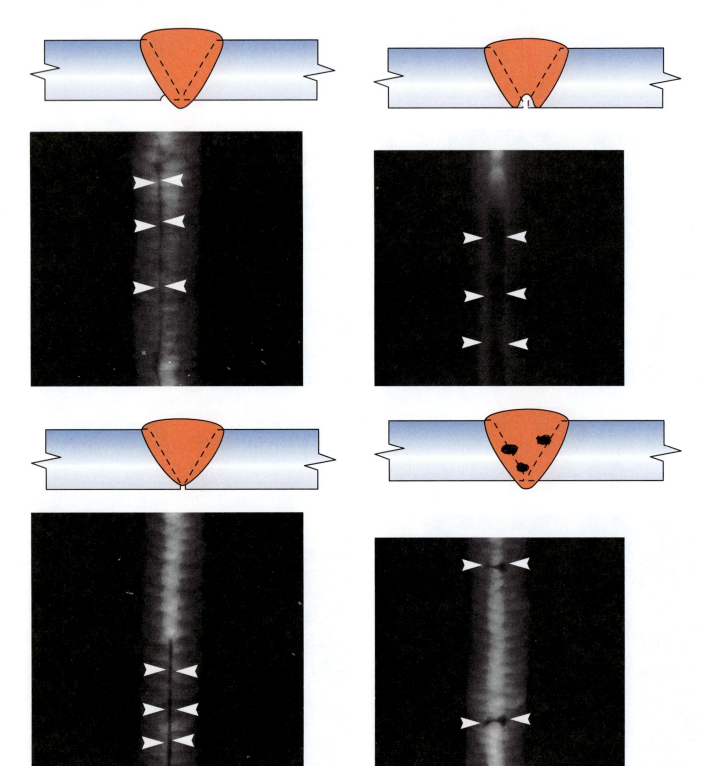

Figure 9.39 Continued

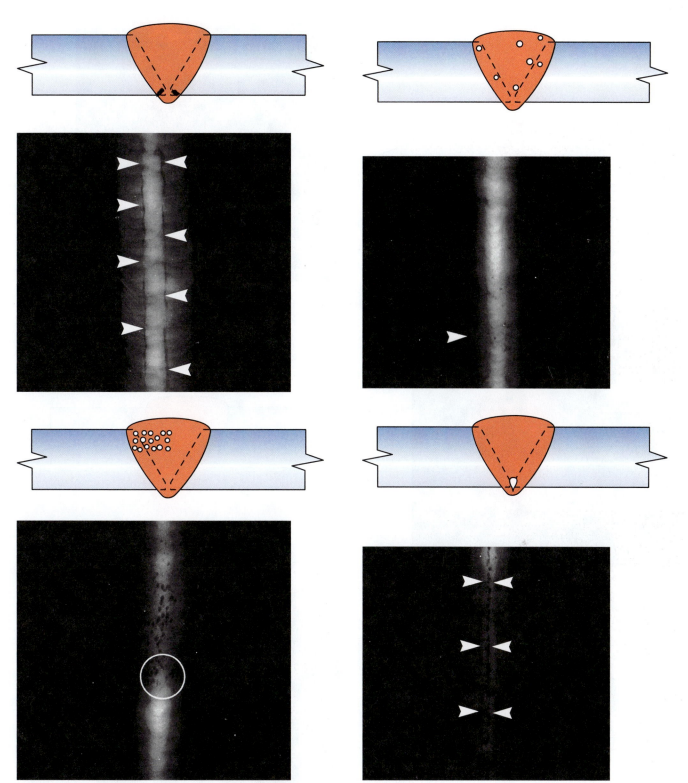

Figure 9.39 Continued

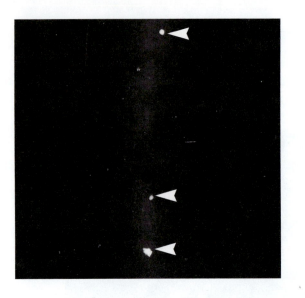

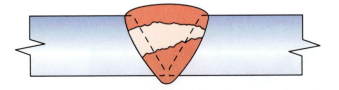

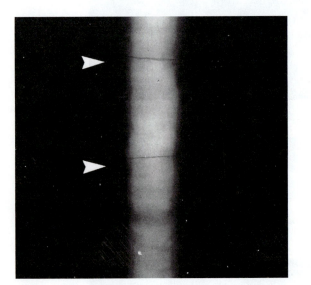

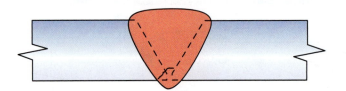

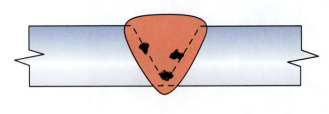

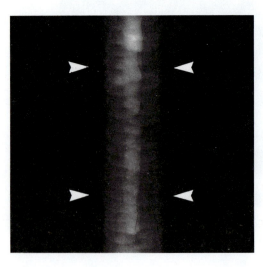

Figure 9.39 Continued

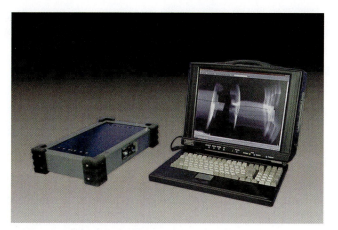

Figure 9.40 New mobile X-ray equipment
Source: Courtesy of CMOS X-ray

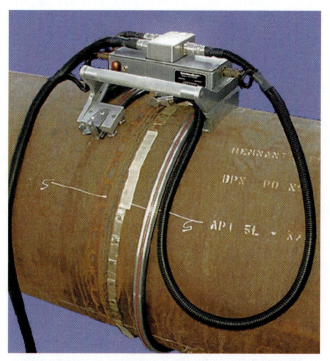

Figure 9.41 Preparing to test the quality of a weld on a pipe using X-ray equipment
Source: Courtesy of CMOS X-ray

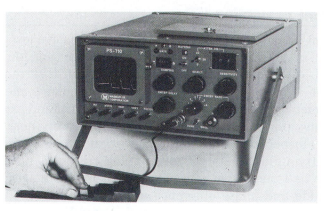

Figure 9.42 Portable ultrasonic inspection unit
Source: Courtesy of Magnaflux Corporation

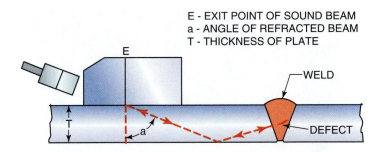

Figure 9.43 Ultrasonic testing

be conducted from the probe into the part through a medium such as oil or water.

Sound is directed into the part from a probe held at a preselected angle or in a preselected direction so that flaws will reflect some energy back to the probe. These ultrasonic devices operate much like depth sounders, or "fish finders." The speed of sound through a material is a known quantity. The equipment measures the time taken for a pulse to return from a reflective surface. Internal computers calculate the distance and present the information on a display screen so that an operator can interpret the results. The signals can be "monitored" electronically to operate alarms, print systems, or recording equipment. Sound not reflected by flaws continues into the part. If the angle is correct, the sound energy will be reflected back to the probe from the opposite side. Flaw size is determined by plotting the length, height, width, and shape using trigonometric rules.

Figure 9.43 shows the path of the sound beam in butt welding testing. The operator must know the exit point of the sound beam, the exact angle of the refracted beam, and the thickness of the plate when using shear wave and compression wave forms.

Leak Checking

In **Leak Checking**, leaks can be found by filling a welded container with either a gas or a liquid. Additional pressure may be applied to the material in the weldment. Water is the most frequently used liquid, although sometimes a liquid with a lower viscosity is used. If gas is used, it can be detected with an instrument when it escapes through a flaw in the weld or as bubbles from an air leak.

Eddy Current Inspection (ET)

Eddy current inspection (ET) is another nondestructive test. The method is based on a magnetic field that induces eddy currents within the material being tested. An eddy current is an induced electric current circulating wholly within a mass of metal. Eddy current inspection is effective in testing both nonferrous and ferrous materials for internal and external cracks, slag inclusions, porosity, and lack of fusion on or very near the surface. Eddy current inspection cannot locate flaws that are not near the surface.

A coil carrying high-frequency alternating current is brought close to the metal to be tested. A current is produced in the metal by induction. The magnitude and phase difference of these currents are indicated by the impedance value of the pick-up coil. Careful measurement of this impedance allows the detection of defects in the weld.

Hardness Testing

Hardness is the resistance of metal to penetration and is an index of the wear resistance and strength of the metal. Hardness tests can be used to determine the relative hardness of the weld and the base metal. The two **hardness testing** machines in common use are the Rockwell and the Brinell testers.

The **Rockwell hardness tester**, Figure 9.44, uses a 120° diamond cone for hard metals and a hardened steel ball of 1/16-in. (1.58-mm) or 1/8-in. (3.175-mm) diameter for softer metals. The method is based on measuring resistance to penetration. The depth of the impression is measured rather than the diameter. The hardness is read directly from a dial on the tester. The tester has two scales for reading hardness, known as the *B-scale* and the *C-scale*. The C-scale is used for harder metals, the B-scale for softer metals.

The **Brinell hardness tester** measures the resistance of material to the penetration of a steel ball under constant pressure (about 3000 kg) for a minimum of approximately 30 seconds, Figure 9.45. The diameter is measured microscopically, and the Brinell number is checked on a standard chart. Brinell hardness numbers are obtained by dividing the applied load by the area of the surface indentation.

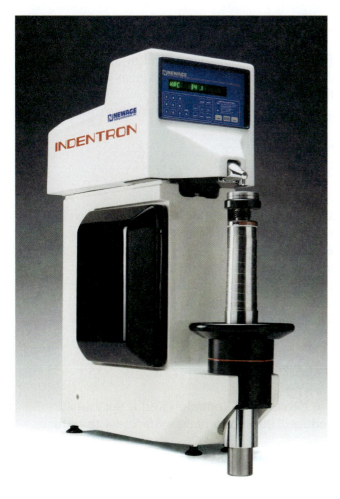

Figure 9.44 Rockwell hardness tester
Source: Courtesy of Newage Testing Instruments, Inc.

Figure 9.45 Brinell hardness tester
Source: Courtesy of Newage Testing Instruments, Inc.

SUMMARY

It is impossible to "inspect in" quality; no amount of inspection will produce a quality product. Quality is something that must be built into a product. The purpose of testing and inspection is to verify that the required level of quality is being maintained. The most important standard that a weld must meet is fitness for service. A weld must be able to meet the demands placed on the weldment without failure. No weld is perfect. It is important for both the welder and the welding inspector to know the appropriate level of weld discontinuities. Producing or inspecting welds to an excessively high standard will result in a product that is unnecessarily expensive.

Knowing the causes and effects of weld defects and discontinuities can aid you in producing higher-quality welds. Some welders believe that high-quality welds are a matter of luck, which is not true. Producing high-quality welds is a matter of skill and knowledge, which you must develop. Good welders often know whether welds they produce will or will not pass inspection. Inspecting your welds as part of your training program will help you develop this skill.

REVIEW

1. Why aren't all welds inspected to the same level or standard?
2. Why isn't the strength of all production parts known if a sample of parts is mechanically tested?
3. Why is it possible to do more than one nondestructive test on a weldment?
4. What is a discontinuity?
5. What is a defect?
6. What is tolerance?
7. What are the twelve most common discontinuities?
8. How can porosity form in a weld and not be seen by the welder?
9. What welding processes can cause porosity to form?
10. How is piping porosity formed?
11. What are inclusions, and how are they caused?
12. When does inadequate joint penetration usually become defective?
13. How can a notch cause incomplete fusion?
14. How can an arc strike appear on a guided bend test?
15. What is overlap?
16. What is undercut?
17. What causes crater cracks?
18. What is underfill?
19. What is the difference between a lamination and a delamination?
20. How can stress be reduced through a plate's thickness to reduce lamellar tearing?
21. What would be the tensile strength in pounds per square inch of a specimen measuring 0.375 in. thick and 1.0 in. wide if it failed at 27,000 pounds?

22. What would be the elongation for a specimen for which the original gauge length was 2 in. and final gauge length was 2.5 in.?
23. How are the results of a stress test reported?
24. What would be the transverse shear strength per inch of weld if a specimen that was 2.5 in. wide withstood 25,000 pounds?
25. What would be the longitudinal shearing strength per millimeter of a specimen that was 50.8 mm wide and 116 mm long and withstood 50.0 kg/mm?
26. What are the three methods of destructive testing of a welded butt joint?
27. How are the specimens bent for a guided bend root, face, and side bend test?
28. How wide should a specimen be if the material thickness is:
    a. 0.375 in.?
    b. 6.35 mm?
29. Why are guide lines drawn on the surface of a free bend specimen?
30. What part of a fillet weld break test is examined?
31. What can happen if acids are handled carelessly?
32. What information about the weld does an impact test provide?
33. Which is the most commonly used nondestructive test?
34. List the five steps to be followed for a penetrant test.
35. What properties must metal have before it can be tested using magnetic particle inspection?
36. Why will some flaws appear larger on an X ray than they are in the weld?
37. How is the size of a flaw determined using ultrasonic inspection?
38. What is the major limitation of eddy current inspection?
39. What information does a hardness test reveal?

CHAPTER 110

Welder Certification

OBJECTIVES

After completing this chapter, the student should be able to

- contrast practical and written weld examinations
- outline the steps to certify a weld and a welder
- integrate AWS SENSE workmanship standards into practical assignments
- contrast the minimum safety requirements for AWS SENSE qualification and the safety requirements to work in the school weld shop
- describe housekeeping responsibilities in a weld shop
- write a welder performance qualification test record (WPQR)
- make welds that comply with a standard

KEY TERMS

certified welders	weld test	welding procedure
entry-level welder	welder certification	specification (WPS)
housekeeping		workmanship standards

AWS SENSE EG2.0

Key Indicators Addressed in this Chapter:

Module 1: Occupational orientation

Key Indicator 1: Prepares time or job cards, reports and records
Key Indicator 2: Performs housekeeping duties
Key Indicator 4: Follows written instructions to complete work assignments

Module 9: Welding Inspection and Testing Principles

Key Indicator 1: Examines cut surfaces and edges of prepared base metal parts
Key Indicator 2: Examines tacks, root passes, intermediate layers and completed welds

INTRODUCTION

Welding is one of the few professions that require job applicants to demonstrate their skills even if they are already certified. Doctors, lawyers, and pilots do take a written test or require a license initially. But welders are often required to demonstrate their knowledge and their skills before being hired, since welding, unlike most other occupations, requires a high degree of eye–hand coordination.

A method commonly used to test welders' ability is the qualification or certification test. Welders who have passed such tests are referred to as *qualified welders;* if proper written records are kept of the test results, they are referred to as **certified welders**. Not all welding jobs require that the welder be certified: Some merely require that a basic **weld test** be passed before applicants are hired.

Welder certification can be divided into two major areas. The first area covers the traditional welder certification that has been used for years. This certification is used to demonstrate welding skills for a specific process on a specific weld, to qualify for a welding assignment or as a requirement for employment.

The American Welding Society has developed the second, newer area of certification. This certification has three levels. The first level is primarily designed for the new welder who needs to demonstrate **entry-level welder** skills. The other levels cover advanced welders and expert welders. This chapter covers the traditional form of certification and aspects of the AWS QC10 *Specification for Qualification and Certification for Entry Level Welder*.

Many employers now require applicants to pass a written examination before a practical weld test is given. AWS QC10 and AWS D9.1 *Sheet Metal Welding Code* both have provisions for written and oral examinations incorporated into them. This newer area of certification reflects a worldwide trend among employers and manufacturers, who are seeing the need for welders to have a basic understanding of process and procedures to go along with practical welding skills. ISO (International Organization for Standardization) and EN (European Norms) have also adopted many welding regulations that include written as well as practical examinations for welders.

PERFORMANCE TESTS

Some welders mistakenly believe that passing a welding certification test is a matter of luck. Passing a weld test is a function of welder skill and knowing detailed information regarding the acceptance level of the weld. Without both of these elements, passing the test would indeed be just luck.

You will need to practice taking a weld test to prepare yourself for passing a test for a job. As part of this practice, you will be required to lay out, cut out, fit up, and weld several different weldments following written and verbal instructions and using a variety of processes. It will also be your responsibility to keep the welding area clean and safe.

The instructions given in the experiments and practices in this text-book series provide the necessary instructions to perform quality welds, and AWS material provides the necessary codes and standards. Learning to read and interpret AWS materials is an important part of the certification process, since the AWS is the leading authority on the codes. Together, this chapter and the AWS material provide all of the knowledge and skill standards necessary to become a qualified and certified welder.

Layout

Most entry-level welders will be required to read and interpret simple drawings and sketches including welding symbols. Attention to detail is an important part of becoming a qualified welder, and the layout portion of a certification test or a production project can determine whether a weldment passes or fails inspection. Learning to pay attention to and self-inspect measurements, angles, tolerances, and fitups will be an important part of the practical welding exercises covered in this text series and throughout a welder's career.

(See Chapters 3 and 4 for more specific direction in reading drawings, layout, and fabrication.)

Safety

A score of 90% on safety is acceptable for the knowledge portion of the AWS SENSE certification; however, before doing any work in the welding shop, the student must pass a safety test with 100% accuracy. The test may be repeated as needed until 100% accuracy is obtained. The test must have questions relating to protection of people in the area and the welding shop in general, including precautionary information, shop and welding area ventilation, fire prevention and protection, and other general aspects of welding safety. See chapter 2 and ANSI Z49.1 for specific safety information.

Housekeeping

Cleaning the welder's work area is usually the responsibility of the welder. By its nature, the welding and cutting process produces large quantities of scrap, including electrode stubs, scrap metal, welding and cutting slag, and grinding dust. It is also necessary to keep welding leads, electrical cords, hoses, guns, torches or electrode holders, hand tools, and power tools up and out of harm's way, Figure 10.1.

Module 1
Key Indicator 2

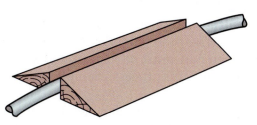

Figure 10.1 Cable safety
To prevent people from tripping if cables must be placed in walkways, lay two blocks of wood beside the cables

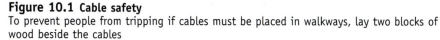

Figure 10.2 Housekeeping
Minor repairs of equipment may be the welder's responsibility
Source: Courtesy of Larry Jeffus

Some **housekeeping** chores may be performed by support staff, but the cleaning and picking up of the welding area and equipment is up to the welder. Cleanliness is as important to safety as many other responsibilities are. Keeping the area picked up can also improve welding productivity.

Routine repairs of equipment may also fall under housekeeping chores. Most welders are expected to change their own welding hood lenses as needed, and some are expected to make minor repairs and adjustments to guns, torches, and electrode holders, Figure 10.2. Before making any such repairs or adjustments, you must check with the manufacturer's literature for the equipment and your instructor.

Cutting Out Parts

Module 9
Key Indicator 1

Parts for the practices and SENSE workmanship samples will be cut out using the manual oxyfuel gas cutting (OFC), machine oxyfuel gas cutting (OFC–Track Burner), air arc cutting (CAC-A), and manual plasma arc cutting (PAC) processes.

To cut out the various parts, you will need to make straight, angled, and circular cuts. In some cases you will decide which method and process to use. On other cuts the method and process will be specified on the drawing. Students and instructors will inspect all cut surfaces for flaws and defects and repair them if necessary, Figure 10.3.

Fitup and Assembly of the Parts

Module 1
Key Indicator 4

Putting the parts of a weldment together in preparation for welding requires special skills. The more complex the weldment, the more difficult the assembly. Each part must be located and squared to other parts, Figure 10.4. To complicate this process, clamping may be necessary if not all parts are flat and straight.

The cutting process may distort the metal. Any such distortion and any bend already in the metal must be corrected during the fitup process. Sometimes grinding can be required to make a correct fitup, Figure 10.5.

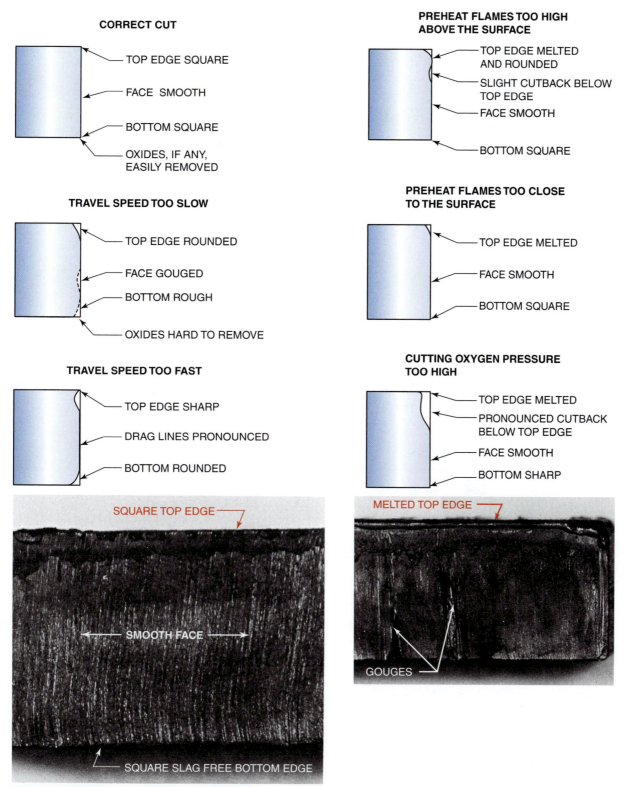

Figure 10.3 Flame-cut profiles and standards
Source: Photos courtesy of Larry Jeffus

Other times the parts can be fitted together correctly using C-clamps or pliers, Figure 10.6. In more difficult cases, tack welds and cleats or dogs must be used to achieve the proper fitup, Figure 10.7. Mismatch in groove welds may be eliminated by hammering, Figure 10.7B.

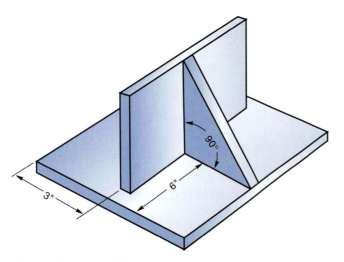

Figure 10.4 Assembly
Locate and square parts to be welded

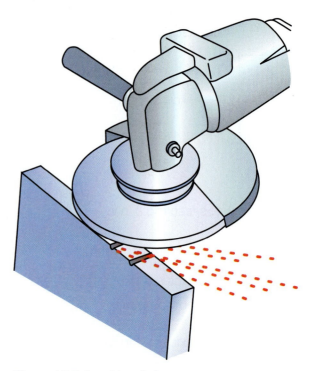

Figure 10.5 Portable grinder
A portable grinder can be used to correct a cutting or fitting problem

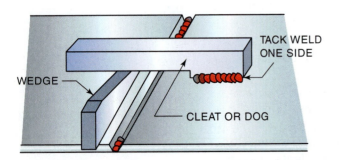

Figure 10.6 C-clamp being used to hold plates for tack welding

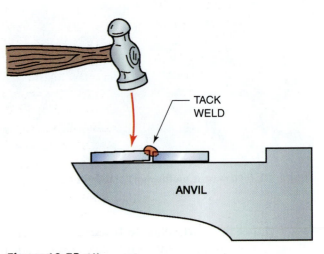

Figure 10.7A Alignment
Wedges and cleats can be used on heavier metal to pull the joint into alignment

Figure 10.7B Alignment
The plates can also be forced into alignment by striking them with a hammer

Workmanship Standards for Welding

Workmanship standards and definitions including tolerances, finish classes, and other general characteristics of fabricated and machined items are specified in welding codes and communicated to welders through the WPS, blueprints, and production drawings. Customer requirements (typically stated on prints) may override the welding code, but when this happens, the customer requirement must exceed the code criteria in the area that is overridden. The code or standard is to be used to advise designers and engineers and dictate requirements when not specified by the customer.

The weldment must be positioned so that the welds are being made within 15° of the specified position, Figure 10.8.

All arc strikes must be within the groove. Arc strikes outside the groove will be considered defects.

Tack welds must be small enough that they do not interfere with the finished weld. They must, however, be large enough to withstand the shrinkage forces from the welds as they are being made. Sometimes it is a good idea to use several small tacks on the same joint to ensure that the parts are held in place, Figure 10.9.

The weld bead size is important. Beads must be sized in accordance with the **welding procedure specification (WPS)** or drawing for each specific weld. All weld bead starts and stops must be smooth, Figure 10.10. For bend specimens the starts and stops should be made outside of the areas from which the weld test specimens will be taken.

The weld beads must be cleaned, either with a hand wire brush, a hand chipping hammer, a punch and hammer, or a needle-scaler. All weld cleaning must be performed with the test plate in the welding

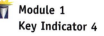

Module 1
Key Indicator 4

Module 9
Key Indicator 2

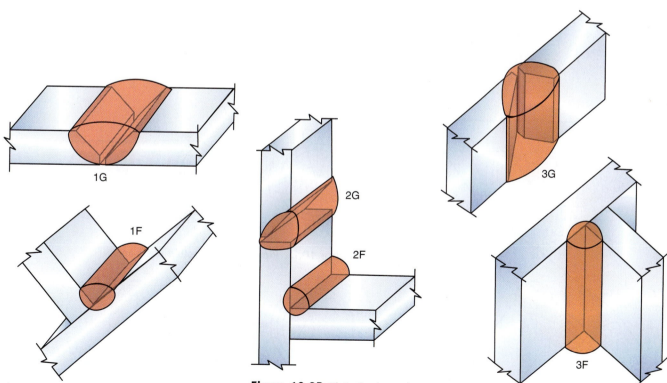

Figure 10.8A Plate flat position

Figure 10.8B Plate horizontal position

Figure 10.8C Plate vertical position

position. A grinder may not be used to remove weld control problems such as undercut, overlap, or trapped slag.

Because of the size of the test plate and the quantity of weld metal being deposited, the plate may have a tendency to become overheated. If overheating occurs, allow the test plate to air-cool, but do not quench it. A good practice is not to waste this time. You can start on another weld test position, fit up the next set of test plates, or find another productive use of your time while the plate air-cools.

Weld Inspection

In order to meet AWS SENSE inspection criteria, each weld and weld pass shall be inspected visually:

- There shall be no cracks or incomplete fusion.
- There shall be no incomplete joint penetration in groove welds, except as permitted for partial joint penetration groove welds.
- The test supervisor shall examine the weld for acceptable appearance and shall be satisfied that the welder is skilled in the process and procedure specified for the test.
- Undercut shall not exceed the lesser of 10% of the base metal thickness or 1/32 in. (0.8 mm).
- Where visual examination is the only criterion for acceptance, all weld passes are subject to visual examination at the direction of the test supervisor.
- The frequency of porosity shall not exceed one in each 4 in. (100 mm) of weld length, and the maximum diameter shall not exceed 3/32 in. (2.4 mm).
- Welds shall be free from overlap (American Welding Society AWS QC10-2006 *Specification for Qualification and Certification for Entry Level Welders*).

(Additional inspection and testing information can be found in Chapter 5.)

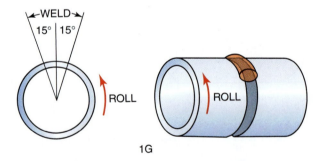

Figure 10.8E Pipe horizontal rolled position

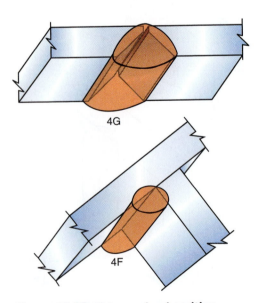

Figure 10.8D Plate overhead position

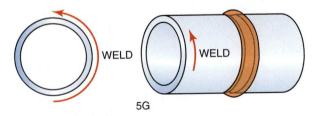

Figure 10.8F Pipe horizontal fixed position

Figure 10.9A Tack weld
Note the good fusion at the start and crater fill at the end
Source: Courtesy of Larry Jeffus

Figure 10.9B Tack weld size
The tack weld should be small and uniform to minimize its effect on the final weld
Source: Courtesy of Larry Jeffus

Figure 10.9C Tack weld number
Use enough tack welds to keep the joint in alignment during welding. Small tack welds are
easier to weld through without adversely affecting the weld
Source: Courtesy of Larry Jeffus

Figure 10.10 Smooth weld bead stops and starts
Taper down the weld when it is complete or when stopping a GTA weld to reposition. This will make the end smoother and make it easier to start the next weld bead if necessary
Source: Courtesy of Larry Jeffus

WELDING SKILL DEVELOPMENT

AWS EG2.0-2006 *Guide for the Training and Qualification of Welding Personnel Entry Level Welder* lists safety and related knowledge and welding skills that must be mastered as part of a training program that meets the requirements of AWS SENSE QC-10:2006 Specification for Qualification and Registration of Level 1—Entry Welders. The AWS SENSE requirements for welding and cutting skills are best obtained by following the skill development plan provided in this textbook series. For example, it is easier for a new welder to develop vertical up skills by starting welding on a plate that is at a 45° inclined angle. As you develop the skills to control the weld at this angle, slowly increase the angle until you are making welds in the vertical position.

Table 10.1 cross-references the list of cutting skills that the AWS states you should develop.

Module 1
Key Indicator 1

PRACTICE 10-1

Welder and Welder Operator Performance: Qualification Test Record (WPQR)

Using a completed weld and the following list of steps, you will complete the test record shown in Figure 10.11. This form is a composite of sample test recording forms provided in the AWS, ASME, and API codes. You may want to obtain a copy of one of the codes or standards and compare a weld you have made to the standard. This form is useful when you are testing one of the practice welds in this text.

The forms are designed to be used with a large variety of weld procedures, so they have spaces that will not be used every time.

1. Welder's name: the person who performed the weld
2. Identification no.: On a welding job, every person has an identification number that is used on the time card and paycheck. In this

Table 10.1 AWS Entry-Level Welder Cutting Process

Cutting Process	Straight Cutting	Beveling	Shape Cutting	Weld Removal (Washing)	Metal Removal (Goughing)
Oxyfuel gas cutting (OFC) plain carbon steel (manual)	X	X	X	X	
Oxyfuel gas cutting (OFC) plain carbon steel (machine [track burner])	X	X			
Air carbon arc cutting (CAC-A) plain carbon steel (manual)					X
Plasma arc cutting (PAC) plain carbon steel (manual)			X		
Plasma arc cutting (PAC) stainless steel (manual)			X		
Plasma arc cutting (PAC) aluminum (manual)			X		

space, you can write the class number or section number since you do not have a clock number.

3. Welding process(es): SMAW, GMAW, or GTAW
4. How the weld was accomplished: manually, semiautomatically, or automatically?
5. Test position: 1G, 2G, 3G, 4G, 1F, 2F, 3F, 4F, 5G, 6G, 6GR, Figure 10.12

WELDER AND WELDING OPERATOR QUALIFICATION TEST RECORD (WQR)

Welder or welding operator's name _____(1)_____ Identification no. _____(2)_____

Welding process ____(3)____ Manual ____(4)____ Semiautomatic ____(4)____
Machine ____(4)____

Position ____(5)____

(Flat, horizontal, overhead or vertical—if vertical, state whether up or down in accordance with welding procedure specification no.) ____(6)____

Material specification ____(7)____

Diameter and wall thickness (if pipe)—otherwise, joint thickness ____(8)____

Thickness range this qualifies ____(9)____

Filler Metal

Specification No. ____(10)____ Classification ____(11)____ F-number ____(12)____

Describe filler metal (if not covered by AWS specification) ____(13)____

Is backing strip used? ____(14)____

Filler metal diameter and trade name ____(15)____ Flux for submerged arc or gas metal arc or flux cored arc welding ____(16)____

Figure 10.11 Welder and welding operator qualification test record

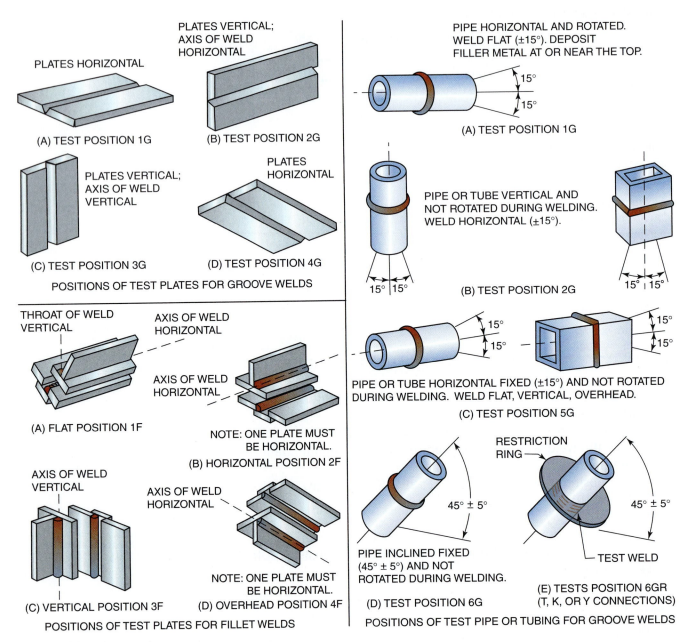

Figure 10.12 Welding positions

6. WPS used for this test
7. Base metal specification: the ASTM specification number, including the P-number, Table 10.2
8. Test material thickness (or) test pipe diameter (and) wall thickness: the thickness of the welded material or pipe diameter and wall thickness
9. Thickness range qualified (or) diameter range qualified: for both plate and pipe, a weld performed successfully on one thickness qualifies a welder to weld on material within that range. See Table 10.3 for a list of thickness range.
10. Filler metal specification number: the AWS has specifications for chemical composition and physical properties for electrodes. Some of these specifications are listed in Table 10.4.

Table 10.2 P-Numbers

	Type of Material
P-1	Carbon steel
P-3	Low alloy steel
P-4	Low alloy steel
P-5	Alloy steel
P-6	High alloy steel—predominantly martensitic
P-7	High alloy steel—predominantly ferritic
P-8	High alloy steel—austenitic
P-9	Nickel alloy steel
P-10	Specialty high alloy steels
P-21	Aluminum and aluminum-base alloys
P-31	Copper and copper alloy
P-41	Nickel

Table 10.3 Test Specimens and Range of Thickness Qualified

Plate Thickness (T) Tested in. (mm)	Plate Thickness (T) Qualified in. (mm)
$1/8 \leq T < 3/8$*	1/8 to 2T
$(3.1 \leq T < 9.5)$	(3.1 to 2T)
3/8 (9.5)	3/4 (19.0)
$3/8 < T < 1$	2T
$(9.5 < T < 25.4)$	2T
1 and over	Unlimited
(25.4 and over)	Unlimited

Pipe Size of Sample Weld	
Diameter in. (mm)	**Wall Thickness, T**
2 (50.8) or	Sch. 80
3 (76.2)	Sch. 40
6 (152.4) or	Sch. 120
8 (203.2)	Sch. 80

Diameter in. (mm)	Wall Thickness, in. (mm)	
3/4 (19.0) through 4 (101.6)	Minimum 0.063 (1.6)	Maximum 0.674 (17.1)
4 (101.6) and over	0.187 (4.7)	Any

*Thickness (T) is equal to or greater than 1/8 in. (≤) and thickness (T) is less than 3/8 in. (<).

11. Classification number: the standard number found on the electrode or electrode box, such as E6010, E7018, E316-15, ER1100
12. F-number: a specific grouping number for several classifications of electrodes with similar composition and welding characteristics. See Table 10.5 for the F-number corresponding to the electrode used.

Table 10.4 Specification Numbers

A5.10	Aluminum—bare electrodes and rods
A5.3	Aluminum—covered electrodes
A5.8	Brazing filler metal
A5.1	Steel, carbon, covered electrodes
A5.20	Steel, carbon, flux cored electrodes
A5.17	Steel-carbon, submerged arc wires and fluxes
A5.18	Steel-carbon, gas metal arc electrodes
A5.2	Steel—oxyfuel gas welding
A5.5	Steel—low alloy covered electrodes
A5.23	Steel—low alloy electrodes and fluxes—submerged arc
A5.28	Steel—low alloy filler metals for gas shielded arc welding
A5.29	Steel—low alloy, flux cored electrodes

13. Give the manufacturer's chemical composition and physical properties as provided.
14. Backing strip material specification: the ASTM specification number
15. Give the diameter of the electrode used and the manufacturer's identification name or number.
16. Flux for SAW or shielding gas(es) and flow rate for GMAW, FCAW, or GTAW

If the weld is a groove weld, follow steps 17 through 22 and then skip to step 27. If the weld test is a fillet weld, skip to space 22.

17. Visually inspect the weld and record any flaws.
18. Record the weld face, root face, and reinforcement dimensions.
19. Four (4) test specimens are used for 3/8-in. (10-mm) or thinner metal. Two (2) will be root bent and two (2) face bent. For thicker metal all four (4) will be side bent.
20. Visually inspect the specimens after testing and record any discontinuities.
21. Who witnessed the welding for verification that the WPS was followed?

Table 10.5 F-Numbers

Group Designation	Metal Types	AWS Electrode Classification
F1	Carbon steel	EXX20, EXX24, EXX27, EXX28
F2	Carbon steel	EXX12, EXX13, EXX14
F3	Carbon steel	EXX10, EXX11
F4	Carbon steel	EXX15, EXX16, EXX18
F5	Stainless steel	EXXX15, EXXX16
F6	Stainless steel	ERXXX
F22	Aluminum	ERXXXX

22. The identification number assigned by the testing agency
23. If the weld is a filler weld, fill spaces 23 through 28.
24. Visually inspect the weld and record any flaws.
25. Record the legs and reinforcement dimensions.
26. Measure and record the depth of the root penetration.
27. Polish the side of the specimens and apply an acid to show the complete outline of the weld.
28. Who witnessed the welding for verification that the WPS was followed?
29. The identification number assigned by the testing agency
30. If a radiographic test is used, fill spaces 29 through 33. If no radiographic test is used, go to space 34.
31. The number the lab placed on the X-ray film before it was exposed on the weld
32. Record the results of the reading of the film.
33. Whether the test passed or failed the specific code
34. Who witnessed the welding for verification that the WPS was followed?
35. The number assigned by the testing agency
36. The name of the company that requested the test
37. The name of the person who interpreted the results; usually a certified welding inspector (CWI) or other qualified person
38. Date the results of the test were completed ■

SUMMARY

Becoming a certified welder establishes your credentials in the industry. Not every industry requires certification; however, all of the welding fields recognize the importance of being certified. Advertisements in the newspaper, in the Yellow Pages, and outside of welding shops prominently display the words *Certified Welder*. Even people outside of the welding industry recognize the significance of someone having obtained the educational level and proficiency required to become a certified welder.

An important part of passing a certification test is your ability to follow all of the very specific details required by the certification process for which you are being tested. Read the qualification test procedures carefully. There may be specific requirements in this test you have not experienced before; do not assume you know what is required. There are many specific things that must be done in preparation for the certification test to ensure that the test results are valid and to ensure the test's successful completion. By diligently following the procedures you will certainly increase your chance of passing the certification. Often your first certification is the most difficult, but it is part of the learning process. Once you have learned the proper techniques and methods of performing certification welding, additional certification tests will become much easier. Experienced certified welders in the field have no difficulty routinely passing certification testing.

REVIEW

1. How does applying for a welding job differ from applying for most other types of jobs?
2. What processes can a welder be certified for?
3. List the variables that, if changed, would require that a new certification test be given.
4. What is required to become an AWS certified welder?
5. What is the tolerance allowed for laying out parts for the AWS entry-level welder certification test?
6. Why are written records about actual welding important to companies?
7. What score is needed on the safety test before welding in the shop, and what areas must be included on the test?
8. What cutting processes must you be able to use to pass the AWS entry-level welder certification test?
9. What are the visual inspection tolerances of the welds for the AWS entry-level welder certification test?
10. How smooth must the flame-cut surface of a bevel be before starting a welding test?
11. What shielding gases are required by the SMAW-S WPS?
12. Under what circumstances may a welding code be overridden?
13. How does a welder determine the proper bead size for a weld?

APPENDIX I—STUDENT WELDING REPORT

I. STUDENT WELDING REPORT

Student Name: _____ Date: _____

Instructor: _____ Class: _____

Experiment or Practice #: _____ Process: _____

Briefly describe task: _____

_____

_____

_____

INSPECTION REPORT

Inspection	Pass/Fail	Inspector's Name	Date
Safety:			
Equip. Setup:			
Equip. Operation:			
Welding	Pass/Fail	Inspector's Name	Date
Accuracy:			
Appearance:			
Overall Rating:			

Comments:

Student Grade: _____ Instructor Initials: _____ Date: _____

Glossary

acetone A fragrant liquid chemical used in acetylene cylinders. The cylinder is filled with a porous material and acetone is then added to fill. Acetylene is then added and absorbed by the acetone, which can absorb up to twenty eight times its own volume of the gas.

acetona Un liquido fragante químico que se usa en los cilindros del acetileno. El cilindro se llena de un material poroso y luego se le agrega la acetona hasta que se llene. El acetileno es absorbido por la acetona, la cual puede absorber 28 veces el propio volumen del gas.

acetylene A fuel gas used for welding and cutting. It is produced as a result of the chemical reaction between calcium carbide and water. The chemical formula for acetylene is C_2H_2. It is colorless, is lighter than air, and has a strong garlic-like smell. Acetylene is unstable above pressures of 15 psig (1.05 kg/cm^2 g). When burned in the presence of oxygen, acetylene produces one of the highest flame temperatures available.

acetileno Un gas combustible que se usa para soldar y cortar. Es producido a consecuencia de una reacción química de agua y calcio y carburo. La fórmula química para el acetileno es C_2H_2. No tiene color, es más ligero que el aire, y tiene un olor fuerte como a ajo. El acetileno es inestable en presiones más altas de 15 psig (1.05 kg/cm^2 g). Cuando se quema en presencia del oxígeno, el acetileno produce una de las llamás con una temperatura más alta que la que se utiliza.

air carbon arc cutting (CAC-A). A carbon arc cutting process variation that removes molten metal with a jet of air.

arco de carbón con aire Un proceso de cortar con arco de carbón variante que quita el metal derretido con un chorro de aire.

American Welding Society (AWS) A multifaceted, non-profit organization with a goal to advance the science, technology and application of welding and related joining disciplines.

Sociedad Americana de Soldadura (AWS) Una organización de múltiples facetas, sin fines de lucro, cuya meta es el avance de la ciencia, la tecnología y la aplicación de la soldadura y las disciplinas relacionadas a la soldadura.

arc cutting (AC) A group of thermal cutting processes that severs or removes metal by melting with the heat of an arc between an electrode and the workpiece.

corte con arco Un grupo de procesos termales para cortar que desúne o quita el metal derretido con el calor del arco en medio del electrodo y la pieza de trabajo.

arc plasma A state of matter found in the region of an electrical discharge (arc). See also plasma.

arco de plasma Un estado de la materia encontrado en la región de una descarga eléctrica (arco). Vea también plasma.

automated operation Welding operations are performed repeatedly by a robot or another machine that is programmed to perform a variety of processes.

operación automatizada Operaciones de soldaduras que se ejecutan repetidamente por un robot u otra maquina que está programada para hacer una variedad de procesos.

automatic operation Welding operations are performed repeatedly by a machine that has been programmed to do an entire operation without the interaction of the operator.

operación automática Operaciones de soldadura que se ejecutan repetidamente por una maquina que ha sido programada para hacer una operación entera sin influencia del operador.

AWS D1.1 Structural Welding Code-Steel code covers the welding requirements for any type of welded structure made from the commonly used carbon and low-alloy constructional steels.

AWS D1.1 Código de soldadura de estructural - Acero. Este código cubre los requistos de soldadura para cualquier tipo de estructura soldada hecha a partir de los aceros más usados en la construcción de carbón y baja aleación.

AWS welder certification A widely respected document certifying that a welder has passed a performance qualification test at an accredited test facility.

Certificación de soldador ASW Un documento ampliamente respetado que certifica que un soldador ha pasado la prueba de calificación de desempeño en una instalación de pruebas acreditada.

Brinell hardness tester A tool that characterizes the indentation hardness of materials through the scale of penetration of an indenter, loaded on a material testpiece.

Probador de dureza Brinell Una herramienta que caracteriza la dureza de indentación en materiales a través de la escala de penetración de un buril, cargado en una pieza para probar material.

carbon electrode A nonfiller metal electrode used in arc welding and cutting, consisting of a carbon or graphite rod, which may be coated with copper or other materials.

electrodo de carbón Un electrodo de metal que no se rellena usado en soldaduras de arco y para cortes consistiendo de varillas de carbón o grafito que pueden ser cubiertas de cobre u otros materiales.

certified welders Individuals who have demonstrated their welding skills for a process by passing a specific welding test.

soldador certificado Personas que han demostrado, mediante una prueba específica de soldadura, su habilidad para soldar en un proceso.

coalescence The growing together or growth into one body of the materials being welded.

coelescencia El crecimiento o desarrollo de un cuerpo de los materiales los cuales se están soldando.

combination welding symbol A welding symbol which indicates multiple operations.

símbolo de combinación de soldadura Un símbolo de soldadura que indica múltiples operaciones.

confined spaces A confined space has limited or restricted means for entry or exit, and it is not designed for continuous employee occupancy. Confined spaces include, but are not limited to, underground vaults, tanks, storage bins, manholes, pits, silos, process vessels, and pipelines.

espacios confinados Un espacio confinado tiene medios de entrada y salida limitados o restringidos, y no está diseñado para ocupación prolongada por empleados. Los espacios confinados incluyen, pero no se limitan a bóvedas subterráneas, tanques, cajas de almacenamiento, pozos de inspección, fosas, graneros, vehículos de proceso y tuberías.

conversion The operation of taking terms of measurement in one representation and converting to another, such as fractions converted into decimals.

conversión La operación de tomar términos de medidas en una representación y convertirlos a otra, como fracciones convertidas a decimales.

coupling distance The distance to be maintained between the inner cones of the cutting flame and the surface of the metal being cut, in the range of 1/8 in. (3 mm) to 3/8 in. (10 mm).

distancia de acoplamiento La distancia que debe de mantenerse entre los conos internos de la llama y la superficie del metal que se está cortando, varía de 1/8 pulgadas (3 mm) a 3/8 pulgadas (10 mm).

cup A nonstandard term for gas nozzle.

tazón Un término que no es la norma para de boquilla de gas.

cutting gas assist Supplimental gas added to enhance an oxyfuel cut.

asistir en el corte de gas Gas suplementario agregado para resaltar un corte de gasolina oxigenada.

cutting tip The part of an oxygen cutting torch from which the gases issue.

punta para cortar Esa parte de la antorcha para cortar con oxígeno por donde salen los gases.

defect A discontinuity that is unable to meet minimum acceptance standards or specifications.

defecto Una discontinuidad que no puede satisfacer las especificaciones o estándares mínimos de aceptación.

delamination A discontinuity that develops inside of the material, without being obvious on the surface.

delaminación Una discontinuidad que se desarrolla dentro del material sin ser obvia en la superficie.

discontinuity An interruption of the typical structure of a material, such as a lack of homogeneity in its mechanical, metallurgical, or physical characteristics. A discontinuity is not necessarily a defect.

discontinuidad Una interrupción de la estructura típica de un material, el que falta de homogenidad en sus caracteristicas mecánicas, metalúrgicas, o fisica.

drag (thermal cutting) The offset distance between the actual and straight line exit points of the gas stream or cutting beam measured on the exit surface of the base metal.

tiro (corte termal) La distancia desalineada entre la actual y la linea recta del punto de salida del chorro de gas o el rayo de cortar medido a la salida de la superficie del metal base.

drag lines High-pressure oxygen flow during cutting forms lines on the cut faces. A correctly made cut has up and down drag lines (zero drag); any deviation from the pattern indicates a change in one of the variables affecting the cutting process; with experience the welder can interpret the drag lines to determine how to correct the cut by adjusting one or more variables.

lineas del tiro La salida del oxígeno a presión elevada durante el corte forma lineas en las caras del corte. Un corte hecho correctamente tiene lineas hacia arriba y hacia abajo (zero tiro); cualquier desviación de la norma indica un cambio en uno de los variables que afectan el proceso de cortar; con experiencia el soldador puede interpretar las lineas de tiro y determinar como corregir el corte ajustando uno o más variables.

dross A mass of solid impurities floating on a molten metal.

escoria Una masa solida de impurezas flotando en un metal fundido.

earmuffs A type of hearing protection that covers the entire ear.

orejeras Un tipo de protección auditivo que cubre la oreja completa.

earplugs A type of hearing protection that is fitted into the ear.

tapones auditivos Un tipo de protección auditiva que coloca en el oído.

eddy current inspection (ET) Uses electromagnetic induction to detect flaws in conductive materials.

inspección de corriente parásita (ET) Usa inducción electromagnética para detectar fallas en los materiales conductores.

electrical ground   A common return path for electric current (earth return or ground return), or a direct physical connection to the Earth.

retorno de tierra eléctrico   Un retorno común para la corriente eléctrica (retorno de tierra o retorno a masa), o una conexión física directa a tierra.

electrical resistance   A ratio of the degree to which an object opposes an electric current through it, measured in Ohms.

resistencia eléctrica   Una proporción del grado en el que un objeto se opone a una corriente eléctrica a través de él, medida en Ohms.

electric shock   An electric shock can occur upon contact of a human's body with any source of voltage high enough to cause sufficient current through the muscles or hair.

descarga eléctrica   La descarga eléctrica puede ocurrir al contacto del cuerpo humano con cualquier superficie de voltaje lo suficientemente alta para provocar suficiente corriente a través de los músculos o el vello.

electrode setback   The distance the electrode is recessed behind the constricting orifice of the plasma arc torch or thermal spraying gun, measured from the outer face of the nozzle.

retroceso del electrodo   La distancia del hueco del electrodo que está detrás del orificio constringente de la antorcha de arco plasma o pistola de rocio termal, se mide de la cara de afuera a la boquilla.

electrode tip   The end of a welding electrode that is closest to the work.

punta del electrodo   El extremo del electrodo que está más cerca del trabajo.

entry-level welder   A person just entering the welding profession.

soldador principiante   Una persona que acaba de comenzar en la profesión de la soldadura.

equal-pressure torches   Medium-pressure torches are often called balanced-pressure or equal-pressure torches because the fuel gas and the oxygen pressure are kept equal. Operating pressures vary, depending on the type of tip used.

sopletes de presión equitativa   Los sopletes de presión media a menudo se llaman sopletes de presión balanceada o de presión equitativa debido a que la presión del gas de combustible y la del oxígeno se mantienen iguales.Las presiones de operación varían, dependiendo del tipo de punta usado.

etching   The process of using acids, bases or other chemicals to dissolve unwanted materials; used in welding inspection to identify grain boundries in cross sectional samples.

decapado   El proceso de usar ácidos, bases u otros químicos para disolver materiales indeseables; se usan en la inspección de soldadura para identificar límites de vetas en muestras de cruce seccional.

exhaust pickup   A component of a forced ventilation system that has sufficient suction to pick up fumes,

ozone, and smoke from the welding area and carry the fumes, etc., outside of the area.

recogedor de extracción   Un componente de un sistema de ventilación forzada que tiene suficiente succión para recoger vaho, ozono, y humo de la área de soldadura y lleva al vaho, etc. a fuera de la área.

exothermic gases   Combustable gasses providing heat as a byproduct.

gases exotérmicos   Gases combustibles que proporcionan calor como un producto derivado.

fillet weld   A weld of approximately triangular cross section joining two surfaces approximately at right angles to each other in a lap joint, tee joint, or corner joint.

soldadura de filete   Una soldadura de filete de sección transversa aproximadamente triangular que une dos superficies aproximademente en ángulos rectos de uno al otro en junta de traslape, junta en- T- o junta de esquina.

fitting   A mechanical connection for electrical current, gas flow or fluid.

ajuste   Una conexión mecánica para la corriente eléctrica, flujo de gas o de líquido.

flash burn   A burn caused by Ultra Violet (UV) light produced by a welding arc.

quemadura por radiación   Una quemadura causada por luz ultra violeta (UV) producida por un arco de soldadura.

flash glasses   Eye protection specifically designed to filter out UV light.

lentes de seguridad   Protección visual específicamente diseñada para filtrar la luz de los rayos UV.

flux cored arc welding (FCAW)   An arc welding process that uses an arc between a continuous filler metal electrode and the weld pool. The process is used with shielding gas from a flux contained within the tubular electrode, with or without additional shielding from an externally supplied gas, and without the application of pressure.

soldadura de arco con núcleo de fundente   Un proceso de soldadura de arco que usa un arco entre medio de un electrodo de metal rellenado continuo y el charco de la soldadura. El proceso es usado con gas de protección del flujo contenido dentro del electrodo tubular, y sin usarse protección adicional de abastecimiento de gas externo, y sin aplicarse presión.

forced ventilation   To remove excessive fumes, ozone, or smoke from a welding area, a ventilation system may be required to supplement natural ventilation. Where forced ventilation of the welding area is required, the rate of 200 cu ft (56 m^3) or more per welder is needed.

ventilación forzada   Para quitar excesivo vaho, ozono y humo de la área donde se solda, un sistema de ventilación puede ser requerido para suplementar la ventilación natural. Donde la ventilación forzada de la área de la soldadura es requerida, la rázon de 200 pies cúbicos (56 m³) o más es requerido por cada soldador.

forge welding (FOW) A solid state welding process that produces a weld by heating the workpieces to welding temperature and applying blows sufficient to cause permanent deformation at the faying surfaces.

soldadura por forjado Un proceso de soldadura de estado sólido que produce una soldadura calentando las piezas de trabajo a una temperatura de soldadura y aplicando golpes suficientes para causar una deformación permanente en las superficies del empalme.

full face shield Protective equipment designed to cover the entire face.

protector facial completo Equipo de protección diseñado para cubrir toda la cara.

fusion welding Any welding process or method that uses fusion to complete the weld.

soldadura de fusión Cualquier proceso de soldadura o método que usa fusión para completar la soldadura.

gas laser A laser in which an electric current is discharged through a gas to produce light.

láser de gas Un láser en el cual se descarga una corriente eléctrica a través de un gas para producir luz.

gas metal arc welding (GMAW) An arc welding process that uses an arc between a continuous filler metal electrode and the weld pool. The process is used with shielding from an externally supplied gas and without the application of pressure.

soldadura de arco metálico con gas Un proceso de soldar con arco que usa un arco en medio de un electrodo de metal para rellenar continuo y el charco de soldadura. El proceso usa protección de un abastecedor externo de gas y sin la aplicación de presión.

gas tungsten arc welding (GTAW) An arc welding process that uses an arc between a tungsten electrode (nonconsumable) and the weld pool. The process is used with shielding gas and without the application of pressure.

soldadura de arco de tungsteno con gas Un proceso de soldadura de arco que usa un arco en medio del electrodo tungsteno (no consumible) y el charco de la soldadura. El proceso es usado con gas de protección y sin aplicación de presión.

GFCI An electrical wiring device that disconnects a circuit whenever it detects that the electric current is not balanced between the phase ("hot") conductor and the neutral conductor.

GFCI Un dispositivo de cableado eléctrico que desconecta un circuito cada vez que detecta que la corriente eléctrica no está equilibrada entre el conductor de fase ("caliente") y el conductor neutral.

goggles Special eye protection designed to seal around each eye.

gafas de seguridad Protección visual especial diseñada para quedar ajustada firmemente alrededor de cada ojo.

gouging Removal of metal by the Air Carbon Arc or the Plasma Arc processes.

ranurado Quitar metal mediante el proceso de arco de carbón de aire o de arco de plasma.

graphite A mineral form of carbon.

grafito Una formación mineral del carbón.

groove An opening or a channel in the surface of a part or between two components, that provides space to contain a weld.

ranura Una abertura o un canal en la superficie de una parte o en medio de dos componentes, la cual provee espacio para contener una soldadura.

hard slag A form of slag caused by Oxyfuel cutting that is difficult to remove.

heat-affected zone The area of base material, either a metal or a thermoplastic, which has had its microstructure and properties altered by welding or heat intensive cutting operations.

high-frequency alternating current A low amperage current that is superimposed over the welding current to assist in arc initiation and stabilize AC Gas Tungsten Arc welding operations.

corriente alterna de alta frecuencia Una corriente de amperaje bajo que se superpone a la corriente de soldadura para ayudar a iniciar el arco y estabilizar la CA de las operaciones de arco de gas tungsteno.

high-speed cutting tip A special cutting tip usually constructed in two pieces that is designed for mechanized Oxyfuel cutting operations.

punta para corte de alta velocidad Una punta especial para corte generalmente hecha en dos partes que está diseñada para las operaciones de corte de gasolina oxigenada.

housekeeping Performance of duties to keep a welding shop or job site clean and free of hazards.

mantenimiento Llevar a cabo las tareas de mantenimiento de un taller de soldadura o lugar de trabajo limpio y libre de riesgos.

hot work permit A document required to be completed before beginning hot work operations in areas not specifically designated for welding or cutting.

permiso para trabajo en caliente Un documento que se debe completar antes de empezar el trabajo en caliente en áreas no diseñadas específicamente para soldadura o corte.

infrared light A form electromagnetic radiation whose wavelength is longer than that of visible light, but shorter than that of microwaves. Infrared radiation is heat that can be felt at a distance.

luz infrarroja Una forma de radiación electromagnética cuya longitud de onda es más larga que la de la luz visible pero más corta que la de microondas. La radiación infrarroja es calor que puede sentirse a distancia.

ionized gas See Plasma

gas ionizado Ver Plasma

joint dimensions The details set out in prints or plans to note the shape required.

dimensiones de juntas Los detalles establecidos en planos o dibujos para anotar la forma requerida.

joint type A weld joint classification based on the five basic arrangements of the component parts such as a butt joint, corner joint, edge joint, lap joint, and tee-joint.

tipo de junta Una clasificación de una junta de soldadura basada en los cinco arreglos del componente de partes como junta a tope, junta en esquina, junta de orilla, junta de solape, y junta en T.

joules The joule the SI unit of energy measuring heat, electricity and mechanical work.

julios El joule es la unidad de energía cin ética que mide el calor, la electricidad y el trabajo mecánico.

kerf The width of a saw, oxyfuel, or plasma cut.

corte El ancho del serrucho, la gasolina oxigenada o corte de plasma.

kindling point The temperature at which combustibles will ignite.

punto de ignición La temperatura a la cual los combustibles se encenderán.

laser beam cuts (LBC) A thermal cutting process that severs metal by locally melting or vaporizing with the heat from a laser beam. The process is used with or without assist gas to aid the removal of molten and vaporized metal. Laser in a acronym for Light Amplification by Stimulated Emission of Radiation.

cortes de rayo laser (LBC) Un proceso termal de corte que separa el metal al derretir localmente o vaporizar con el calor de un rayo láser. El proceso se utiliza con o sin asistencia de gas para ayudar a remover el metal derretido o vaporizado. Laser es un acrónimo para Luz – Amplificación – Simulando- Emisión – de Radiación.

laser beam drilling (LBD) A thermal drilling or piercing process that removes metal by local melting or vaporizing with the heat from a laser beam.

Perforación con rayo láser (LBD) Un proceso termal de perforación que quita material al derretir o vaporizar con el calor del rayo láser.

laser beam welds (LBW) A welding process that produces coalescence with the heat from a laser beam impinging on the joint.

soldadura con rayo láser (LBW) Un proceso que produce coalescencia con el calor de un rayo láser que choca con la junta.

layout Putting work pieces in place in preparation for joining, with or without the assistance of clamps or fixtures.

disposición Ordenar las piezas de trabajo en su lugar en preparación para soldarlas, con o sin la ayuda de abrazaderas o herramientas.

machine cutting torch Cutting with equipment that requires manual adjustment of the equipment controls in response to visual observation of the cutting, with the torch held by a mechanical device such as a track burner or a radiograph.

soldadura con máquina cortadora Cortar con equipo que requiere ajuste manual de los controles del equipo en respuesta a la observación visual del corte, con un soplete sostenido por un dispositivo mecánico como un porta quemador o radiógrafo.

machine operation Welding operations are performed automatically under the observation and correction of the operator.

operación de máquina Operaciones de soldadura son ejecutadas automáticamente bajo la observación y corrección del operador.

manual operation The entire welding process is manipulated by the welding operator.

operación manual Todo el proceso de soldadura es manipulado por un operador de soldadura.

material safety data sheet (MSDS) A form containing data regarding the properties of a particular substance.

hoja de datos de seguridad del material (MSDS) Un formulario que contiene la información sobre las propiedades de una sustancia en particular.

measuring The process of estimating the magnitude of some attribute of an object, such as its length or weight, relative to some standard (unit of measurement), such as a meter or a kilogram.

medir El proceso de calcular la magnitud de algún atributo de un objeto, como su longitud o peso, relativo a alguna unidad de medida estándar, como un metro o un kilogramo.

mechanical testing (DT) In mechanical, or destructive, testing, tests are carried out to the specimen's failure in order to understand a specimen's structural performance or material behavior under different loads. These tests are generally much easier to carry out, yield more information, and are easier to interpret than nondestructive testing.

prueba mecánica (DT) En pruebas mecánicas, o destructivas, las pruebas se llevan a cabo para comprobar la falla de la muestra, para poder entender el desempeño estructural o comportamiento del material bajo diferentes cargas Estas pruebas generalmente son mucho más sencillas de llevar a cabo, rinden mas información y son más fáciles de interpretar que las pruebas no destructivas.

monochromatic Monochromatic light is light of a single wavelength, though in practice it can refer to light of a narrow wavelength range. A monochromatic object or image is one whose range of colors consists of shades of a single color or hue.

monocromática La luz monocromática es luz de una sola longitud de onda, aunque en la práctica puede referirse a luz de un rango angosto de longitud de onda Un objeto o imagen monocromática es aquél cuyos colores consisten en matices de un solo color o tono.

MPS gases A group of flammable gasses containing Methylacetylene Propadiene, Stabilized.

Gases MPS   Gases pertenecientes a un grupo de gases inflamables que contienen metilacetileno propadieno, estabilizado.

natural ventilation   The process of supplying and removing air through an indoor space by natural means.

ventilación natural   El proceso de proporcionar y circular aire en un espacio interior por medios naturales.

nondestructive testing (NDT)   Testing that does not destroy the test object; also called nondestructive examination (NDE) and nondestructive inspection (NDI).

pruebas no destructivas (NDT)   Pruebas que no destruyen el objeto probado; también llamadas exámenes no destructivos (NDE) e inspección no destructiva (NDI).

nozzle   The end piece that directs the flow of gas of a GMAW, GTAW, or FCAWG welding torch or gun.

boquilla   La pieza del extremo que dirige el flujo de gas de un soplete de soldadura o pistola GMAW, GTAW, o FCAWG.

nozzle insulator   A non-conductive piece that separates the nozzle Mig gun from the contact tube.

aislante de boquilla   Una pieza no conductora que separa la boquilla de la pistola MIG del tubo de contacto.

nozzle tip   The end of a nozzle, sometimes replicable on heavy duty GMAW and FCAW equipment.

punta de la boquilla   El extremo de la boquilla, a veces replicable en equipo GMAW y FCAW para tareas pesadas.

orifice   A term used to describe the hole at the end of a tube or nozzle.

orificio   Término usado para describir el hoyo en el extremo de un tubo o una boquilla.

oxyacetylene hand torch   Handheld apparatus used for making oxyfuel cuts or welds oxygen lance cutting a tube that carries oxygen to a heated tool; used to cut metal; the tip is continuously consumed.

soplete de mano de oxiacetiléno   Aparato manual usado para hacer cortes o soldaduras de gasolina oxigenada lanza de oxigeno cortar un tubo que lleva oxígeno a una herramienta calentada usada para cortar metal; la punta se consume continuamente.

oxyfuel gas cutting (OFC)   A group of oxygen cutting processes that uses heat from an oxyfuel gas flame. See also oxygen cutting, oxyacetylene cutting, oxyhydrogen cutting, and oxypropane cutting.

gas para cortar oxicombustible   Un grupo de procesos para cortar con oxígeno que usa calor de una llama de gas oxicombustible. Vea también cortes con oxigeno, cortes con oxiacetileno, cortes con oxihidrógeno, y cortes con oxipropano.

oxyfuel gas welding (OFW)   A group of welding processes that produces coalescence of workpieces by heating them with an oxyfuel gas flame. The processes are used with or without the application of pressure and with or without filler metal.

soldadura con gas oxicombustible   Un grupo de procesos de soldadura que produce coalescencia de las piezas de trabajo calentándolas con una llama de gas oxicombustible. Los procesos son usados sin la aplicación de presión y con o sin el metal para rellenar.

part dimensions   The size, shape and contour of a part represented on a drawing, sketch or blueprint.

dimensiones de las piezas   El tamaño, la forma y el contorno de la pieza representada en un dibujo, bocetoeb o plano.

pilot arc   A low-current arc between the electrode and the constricting nozzle of the plasma arc torch to ionize the gas and facilitate the start of the welding arc.

piloto del arco   Un arco de corriente baja en medio del electrodo y la boquilla constreñida de la antorcha de arco de plasma para ionizar el gas y facilitar el arranque del arco para soldar.

plasma   A gas that has been heated to an at least partially ionized condition, enabling it to conduct an electric current.

plasma   Un gas que ha sido calentado a lo menos parcialmente a una condicón ionizada permitiendo que conduzca una corriente eléctrica.

plasma arc cutting (PAC)   An arc cutting process that uses a constricted arc and removes the molten metal with a high-velocity jet of ionized gas issuing from the constricting orifice.

cortes con arco de plasma   Un proceso de cortar con el arco que usa un arco constreñido y quita el metal derretido con un chorro de alta velocidad de gas ionizado que sale de la orifice constringente.

plasma arc gouging   A thermal gouging process that removes metal by melting with the heat of plasma arc torch.

ranurado con arco de plasma   Un proceso de ranurado termal que quita el metal al derretirlo con el calor del arco del soplete de plasma.

preheat flame   Brings the temperature of the metal to be cut above its kindling point, after which the high-pressure oxygen stream causes rapid oxidation of the metal to perform the cutting.

llama para precalentamiento   Sube la temperatura del metal que está para cortarse a una temperatura de encendimiento, después que la corriente del oxígeno de alta presión cause una oxidación rápida del metal para hacer el corte.

preheat holes   The cutting tip has a central hole through which the oxygen flows. Surrounding this central hole are a number of other holes called preheat holes. The differences in the type or number of preheat holes determine the type of fuel gas to be used in the tip.

agujeros para precalentamiento   La boquilla para cortar tiene un agujero central por donde corre el oxígeno. Rodeando este agujero central hay un numero de otros agujeros que se llaman agujeros para precalentar. Las diferencias en el tipo o número de agujeros percalentados determina el tipo de gas combustible que se usará en la boquilla.

projection drawings Reprehensive sketches, drawings or computer models presented in one of several view types, such as isometric or orthographic projections.

bocetos de proyección Dibujos, bocetos o modelos por computadora reprensibles presentados en una o varias formas de verse como isométricos o proyecciones ortográficas.

qualification See preferred terms welder performance qualification and procedure qualification.

calificación Vea términos preferidos calificación de ejecución del soldador y calificación de procedimiento.

quality control In engineering and manufacturing, quality control is a system that ensures products or services are designed and produced to meet or exceed customer requirements.

control de calidad En ingeniería y en manufactura, el control de calidad es una forma de asegurar que los productos o servicios están diseñados y son producidos para satisfacer o exceder los requisitos de los clientes.

radiographic inspection (RT) A nondestructive testing (NDT) method of inspecting materials for hidden flaws by using the ability of short wavelength electromagnetic radiation (high energy photons) to penetrate various materials.

inspección radiográfica (RT) Un método de prueba no destructiva (NDT) para inspeccionar materiales con defectos escondidos usando la habilidad de radiación electromagnética de onda corta (fotones de alta energía para penetrar diferentes materiales).

resistance welding (RW) A group of welding processes that produces coalescence of the faying surfaces with the heat obtained from resistance of the workpieces to the flow of the welding current in a circuit of which the workpieces are a part and by the application of pressure.

soldadura por resistencia Un grupo de procesos para soldar que producen coalescencia de las superficies empalmadas con el calor obtenido de la resistencia de las piezas de trabajo al correr la corriente de soldadura en un circuito en las cuales las piezas de trabajo forman parte, y por la aplicación de presión.

Rockwell hardness The Rockwell scale is a hardness scale based on the indentation hardness of a material. The Rockwell test determines the hardness by measuring the depth of penetration of an indenter under a large load compared to the penetration made by a preload.

dureza Rockwell La escala Rockwell es una escala de dureza basada en la dureza a indentación de un material La prueba Rockwell determina la dureza al medir la profundidad de penetración de un buril bajo una carga grande comparada a la penetración hecha por una precarga.

safety glasses Eye protection worn on the face to protect the eyes from impacts, ultra violet and infrared radiation.

gafas de seguridad Protección para los ojos utilizada en la cara para proteger los ojos de impactos, radiación ultra violeta e infrarroja.

semiautomatic operation During the welding process, the filler metal is added automatically, and all other manipulation is performed manually by the operator.

operación semiautomática Durante el proceso de la soldadura, el metal de relleno es añadido automáticamente, y todas las otras manipulaciones son ejecutadas manualmente por el operador.

shear strength As applied to a soldered or brazed joint, it is the ability of the joint to withstand a force applied parallel to the joint.

fuerza cizallada Asi como es aplicada a una junta de soldadura fuerte o soldadura blanda, es la habilidad de la junta de resistir una fuerza aplicada al paralelo de la junta.

shielded metal arc welding (SMAW) An arc welding process with an arc between a covered electrode and the weld pool. The process is used with shielding from the decomposition of the electrode covering, without the application of pressure, and with filler metal from the electrode.

soldadura de arco metálico protegido Un proceso de soldadura de arco con un arco en medio de un electrodo cubierto y el charco de soldadura. El proceso se usa con protección de descomposición del cubrimiento del electrodo sin la aplicación de presión, y con el metal de relleno del electrodo.

slag A nonmetallic product resulting from the mutual dissolution of flux and nonmetallic impurities in some welding and brazing processes.

escoria Un producto que no es metálico resultando de una disolución mutual del flujo y las impuridades no metálicas en unos procesos de soldadura y soldadura fuerte.

soapstone A metamorphic rock that can be cut or ground into rectangles or rounds and are commonly used for marking metal; also known as steatite or soaprock.

creta hispánica Una roca metamórfica que puede ser cortada o molida en rectángulos o ruedas y que se suele utilizar para marcar metales; también conocida como esteatita o saponita.

soft slag The most porous form of slag which is most easily removed.

escoria suave La forma más porosa de escoria que se quita con la mayor facilidad.

solid state lasers A solid-state laser is a laser that uses a gain medium that is a solid rather than a liquid, such as in dye lasers or a gas as in gas lasers.

láseres en estado sólido Un láser en estado sólido es un láser que utiliza un medio activo que es un sólido y no uno que sea líquido como un láser de teñido o gas como en láseres de gas.

specification A specification is an explicit set of requirements to be satisfied by a material, product, or service.

especificación Una especificación es un conjunto explícito de requisitos que serán satisfechos por un material, producto o servicio.

stack cutting   Thermal cutting of stacked metal plates arranged so that all the plates are severed by a single cut.

corte de metal apilado   Un corte termal de hojas de metal apilados arregladas para que todas las hojas sean cortadas por un solo corte.

standard   A technical standard is an established norm or requirement. It is usually a formal document that establishes uniform engineering or technical criteria, methods, processes and practices.

estándar   Un estándar técnico es una norma o requisito establecido. Generalmente, es un documento formal que establece criterios, métodos, procesos y prácticas sistemáticos con respecto a la tecnología y la ingeniería.

standoff distance   The distance between a nozzle and the workpiece.

distancia de alejamiento   La distancia entre la boquilla y la pieza de trabajo.

synchronized wave form   An alternating current (AC) type where the positive and negative sides of the AC cycle may be manipulated.

forma de onda sincronizada   Un tipo de corriente alterna (CA) en la que los lados positivo y negativo del ciclo de CA pueden ser manipulados.

tack weld   A weld made to hold the parts of a weldment in proper alignment until the final welds are made.

soldadura con punteadora   Una soldadura hecha para sostener las partes de un ensamble de soldadura alineadas adecuadamente hasta que se haga la soldadura final.

tip cleaners   Tools used to clean the orifices of oxyfuel torch tips made of abraded steel wires of specific sizes.

limpiadores de puntas   Herramientas usadas para limpiar los orificios del soplete de gasolina oxigenada; puntas hechas de alambres desgastados de tamaños específicos.

tolerances   The allowable deviation in accuracy or precision between the measurement specified and the part as laid out or produced.

tolerancias   Desviación permitida en la precisión entre la medida especificada y la pieza instalada o producida.

torch brazing (TB)   A brazing process that uses heat from a fuel-gas flame.

soldadura fuerte con antorcha   Un proceso de soldadura fuerte que usa calor de una llama de gas combustible.

tolerances   The allowable deviation in accuracy or precision between the measurement specified and the part as laid out or produced.

tolerancias   Desviación permitida en la precisión entre la medida especificada y la pieza instalada o producida.

type A fire extinguisher   An extinguisher used for combustible solids, such as paper, wood, and cloth. Identifying symbol is a green triangle enclosing the letter A.

extinguidor para incendios tipo A   Un extinguidor que se usa para combustibles sólidos como papel, madera, y tela. El símbolo de identificación es un triángulo verde con la letra A adentro.

type B fire extinguisher   An extinguisher used for combustible liquids, such as oil and gas. Identifying symbol is a red square enclosing the letter B.

extinguidor para incendios tipo B   Un extinguidor que se usa para liquidos combustibles, como aceite y gas. El símbolo de identificación es un cuadro rojo con la letra B adentro.

type C fire extinguisher   An extinguisher used for electrical fires. Identifying symbol is a blue circle enclosing the letter C.

extinguidor para incendios tipo C   Un extinguidor que se usa para incendios eléctricos. El símbolo de identificación es un círculo azul con la letra C adentro.

type D fire extinguisher   An extinguisher used on fires involving combustible metals, such as zinc, magnesium, and titanium. Identifying symbol is a yellow star enclosing the letter D.

extinguidor para incendios tipo D   Un extinguidor que se usa para incendios de metales combustibles, como zinc, magnesio, y titanio. El símbolo de identificación es una estrella amarilla con una letra D adentro.

ultrasonic inspection (UT)   In ultrasonic testing, very short ultrasonic pulse-waves with center frequencies ranging from 0.1-15 MHz and occasionally up to 50 MHz are launched into materials to detect internal flaws or to characterize materials.

inspección ultrasónica (UT)   En la prueba ultrasónica, ondas de pulso ultrasónicas, muy cortas, con frecuencias de centro de 0.1 – 15 MHz y, ocasionalmente, hasta 50MHz que se lanzan a materiales para detectar fallas internas o para caracterizar los materiales.

valve protection cap   A protective cover which fits on a compressed gas cylendar.

tapa de protección de válvulas   Una cubierta protectora que cabe en un cilindro de gas comprimido.

ventilation   The intentional movement of air from outside a building to the inside.

ventilación   El movimiento intencional de aire de afuera de una construcción hacia el interior.

venturi   A device that consists of a gradually decreasing nozzle through which a gas or fluid in a pipe is accelerated, followed by a gradually increasing diffuser section that allows the fluid to nearly regain its original pressure.

venturi   Un dispositivo que consiste de una boquilla que gradualmente decrece a través de la cual el gas o el líquido en una pipa se acelera, seguida por una sección de difusor que gradualmente aumenta. lo que permite al líquido casi recuperar su presión original.

visible light   The visible spectrum (or sometimes called the optical spectrum) is the portion of the electromagnetic spectrum that is visible to (can be detected by) the human eye.

luz visible   El espectro visible (o a veces llamado el espectro óptico) es la porción del espectro electromagnético que es visible al (puede ser detectada por) ojo humano.

welding procedure qualification record (WPQR)   A record of welding variables used to produce an acceptable test weldment and the results of tests conducted on the weldment to qualify a welding procedure specification.

registro de calificación de procedimiento de la soldadura   Un registro de los variables usados para producir una probeta aceptable y los resultados de la prueba conducida en la probeta para calificar el procedimiento de especificación.

warning label   A form of hazard communication that attaches directly to an object.

etiqueta de advertencia   Un formulario de comunicación de riesgo que se adhiere a un objeto.

water jet cutting   A water jet cutter is a tool capable of slicing into metal or other materials using a jet of water at high velocity and pressure, or a mixture of water and an abrasive substance.

corte de chorro de agua   Un cortador de chorro de agua es capaz de rebanar un metal u otros materiales usando un chorro de agua a alta velocidad y presión o una mezcla de agua y una sustancia abrasiva.

water table   A special table designed for Plasma arc or oxyfuel cutting operations, where the torch head is submerged under water in order to reduce smoke and noise.

mesa de agua   Una tabla especial diseñada para el arco de Plasma u operaciones de corte de gasolina oxigenada, donde la cabeza de la antorcha es sumergida en agua para así reducir el humo y el ruido.

weld   A localized coalescence of metals or nonmetals produced either by heating the materials to suitable temperatures, with or without the application of pressure, or by the application of pressure alone and with or without the use of the filler material.

soldar   Una coalescencia localizada de metales o metaloides producida al calentar los materiales a una temperatura adecuada, con o sin la aplicación de presión, o por la aplicación de presión solamente y con o sin el uso del material de relleno.

welding helmet   A piece of safety equipment designed to protect the welders face and head from radiation, sparks, spatter and fumes associated with welding and cutting operations.

casco para soldar   Una pieza de equipo de seguridad diseñada para proteger la cara y cabeza del soldador de radiación, chispas, salpicaduras y vapores asociados con las operaciones de soldadura y corte.

weld joint   The junction of members or the edges of members that are to be joined or have been joined by welding.

unión de soldadura   La unión de miembros, u orillas de miembros que se van a unir o ya se unieron mediante soldadura.

weld test   A welding performance test to a specific code or standard.

prueba de soldadura   Una prueba de ejecución de soldadura según una norma o código específico.

welding position   The relationship between the weld pool, joint, joint members, and the welding heat source during welding.

posición de soldadura   La relación entre la soldadura en fusión, unión, miembros de la unión, y la fuente del calor de la soldadura mientras se suelda.

welding procedure specification (WPS)   A document providing in detail the required variables for specific application to assure repeatability by properly trained welders and welding operators.

calificación de procedimiento de soldadura   Un documento que provee en detalle los variables requeridos para la aplicación específica para asegurar la habilidad de repetir el procedimiento por soldadores y operadores que estén propiamente preparados.

welder certification   Written verification that a welder has produced welds meeting a prescribed standard of welder performance.

certificación de soldador   Certificado escrito que indica que un soldador ha confeccionado soldaduras que cumplen con la norma establecida del desempeño de un soldador.

welding code   A document or specification governing aspects process and procedures for the joining of materials by welding.

código de soldadura   Un documento o especificación que rige los aspectos de procesos y procedimientos para la unión de materiales por medio de la soldadura.

welder performance qualification   The demonstration of a welder's or welding operator's ability to produce welds meeting prescribed standards.

calificación del desempeño del soldador   La demostración de la habilidad de un soldador u operador de soldadura para confeccionar soldaduras que cumplen con las normas establecidas.

welding schedule   A written statement, usually in tabular form, specifying values of parameters and the welding sequence for performing a welding operation.

programa de soldadura   Una declaración por escrito, generalmente en forma tabular, especificando los valores de los parámetros y la secuencia de soldadura para llevar a cabo una operación.

welding symbol   A graphical representation of a weld.

símbolo de soldadura   Una representación gráfica de una soldadura.

workmanship standards   The qualitative requirements for a part to be produced.

normas de mano de obra   Los requisitos de calidad para producir una pieza.

YAG laser   YAG (neodymium-doped yttrium aluminum garnet; Nd:Y3Al5O12) is a crystal that is used as a lasing medium for solid-state lasers.

láser YAG   YAG (neodimio-dopaje Itrio-aluminio-granate Nd:Y3Al5O12) es un cristal que se usa como elemento conductor de láser para láseres en estado sólido.

Index

Italic page references indicate material in figures or tables.